Curatorial Practices for Botanical Gardens

Curatorial Practices for Botanical Gardens

Second Edition

Timothy C. Hohn

ROWMAN & LITTLEFIELD
Lanham • Boulder • New York • London

Published by Rowman & Littlefield
An imprint of The Rowman & Littlefield Publishing Group, Inc.
4501 Forbes Boulevard, Suite 200, Lanham, Maryland 20706
www.rowman.com

86-90 Paul Street, London EC2A 4NE

British Library Cataloguing in Publication Information Available

Library of Congress Cataloging-in-Publication Data
Names: Hohn, Timothy C., 1951- author.
Title: Curatorial practices for botanical gardens / Timothy C. Hohn.
Description: Second edition. | Lanham, Maryland : Rowman & Littlefield
    Publishers, [2021] | Includes bibliographical references and index. | Summary: "This
    important, one-of-a-kind handbook has now been expanded and updated to include
    critical information on national and international guidelines and rules for collecting,
    exchanging, and preserving endangered species and preserving biological diversity"
    —Provided by publisher.
Identifiers: LCCN 2021048909 (print) | LCCN 2021048910 (ebook) | ISBN
    9781538151778 (cloth) | ISBN 9781538151785 (paperback) | ISBN
    9781538151792 (epub)
Subjects: LCSH: Botanical gardens--Curatorship. | Handbooks and manuals.
Classification: LCC QK71 .H65 2021 (print) | LCC QK71 (ebook) | DDC
    580.73—dc23/eng/20211013
LC record available at https://lccn.loc.gov/2021048909
LC ebook record available at https://lccn.loc.gov/2021048910

Contents

~

Acknowledgments

The inspiration for this text came from a deep respect for, and interest in, the role of botanical gardens and their contribution to our understanding of plants. I credit the Clarence E. Lewis Landscape Arboretum and W. J. Beal Botanical Garden of Michigan State University with kindling my interest in botanical gardens. Also Dr. Michael Dirr, who pointed me toward the Longwood Graduate Program of the University of Delaware and Longwood Gardens, both having provided me with invaluable training in public garden management and a wonderful educational experience on the subject. Sincere thanks to Dr. Richard Lighty for his inspired guidance through the Longwood Graduate Program and Dr. James Swasey for his assistance in finishing.

The staffs of many public gardens throughout North America with whom I have become acquainted continue to fuel my interest and nurture my respect for these institutions. The varied gardens for which they work have held me in rapt attention for hundreds of hours examining and appreciating their extensive plant collections and interpretive displays providing an ongoing source of mental and spiritual sustenance.

The first edition manuscript of this text was made possible by the support of the American Public Gardens Association (AGPA) and the Stanley Smith Horticultural Trust. In the early going, Sharon Lee, then publications editor for APGA, was instrumental in helping me complete the first draft along with the many members of the 1994–1996 APGA Collections Committee. My sincere thanks to Nora Smith of the Edmonds Community College library for her help in locating journals and books through interlibrary loan. I would also like to thank the editorial team at the Rowman & Little-field Publishing Group for their interest in this project and their expertise in editing and shaping the text.

Regarding the second edition, I'm indebted to Charles Harmon, Executive Editor, Rowman & Littlefield Publishing Group, who encouraged me to write a second edition of *Curatorial Practices* and helped steer me through the first draft with the technical assistance of Erinn Slanina. Also, many thanks to Megan Murray, Associate Editor, and Copy Editor Nancy Syrett, and the production team at Rowman & Littlefield.

Several public garden professionals were generous with their time and expertise. Kathy Musial, Curator of Living Collections at the Huntington Botanical Gardens, and Paul Meyer, F. Otto Haas Executive Director of the Morris Arboretum (retired), reviewed and endorsed the second edition proposal. Michael Dosmann, Keeper of the Living Collections at the Arnold Arboretum, and Pamela Allenstein, Manager, Plant Collections Network of the American Public Gardens Association, provided an editorial review and endorsement of the second edition manuscript. My sincerest thanks and appreciation to them for their advice, assistance, endorsement, and professional support.

Three professional organizations and their extensive resources were absolutely essential to the writing of this second edition: the American Public Garden Association (APGA), Botanic Gardens Conservation International (BGCI), and the Center for Plant Conservation (CPC). I urge everyone in the botanical garden community to participate in and support these important professional organizations.

Finally, my deepest gratitude to my wife, Lynda, and my children, Sam, Bryce, and Deah, for their patience and sacrifices in support of the time I spent working on this project through both editions. A special thanks to Deah for the images she provided for the cover of the second edition.

~

PART I

THE CURATORIAL PROGRAM

From the standpoint of someone trained in horticulture, botany, or other plant science, I can think of few professions that are more enticing, expansive, diverse, and expressive than being a botanical garden curator. In this work, the curator is able to immerse themselves in a world of plants, both native and exotic, as well as a plethora of information about them. It seems to me that for a plant lover it's an ideal practice and vocation.

What is a curator? What does a curator do? In this section, I try to answer these important questions and provide a detailed description of all the aspects of and expectations for this fascinating profession. The following is a synopsis of the first part of *Curatorial Practices for Botanical Gardens*.

Chapter 1 is an introduction to museums, public gardens, collections, and curatorship with definitions and descriptions.

Chapter 2 details governing and guiding the development and curatorship of collections with policies, plans, and operations manuals.

Chapter 3 explains the methods for building collections through field acquisitions, sharing, purchases, gifts, and propagation.

Chapter 4 describes methods for documenting, characterizing, mapping, and tracking collections.

Chapter 5 illustrates the process for preserving and caring for collections and explains horticultural stewardship.

Chapter 6 deals with collections and public programs, as well as interpretation, education, displays, and exhibits.

~

CHAPTER ONE

Introduction

Museums, Public Gardens, and Collections

I feel certain that in every human habitation, from the dawn of humankind to current time, there is a collection of something—tools, cookware, family heirlooms, artwork—an assortment of items both utilitarian and memorable. Regardless of their purpose, people simply like to collect things. These collections help define who we are as individuals, family groups, regional populations, and societies. If you're a gardener, you will inevitably grow a collection of your favorite plants. I have a penchant for monocots, and my garden is strewn with an eclectic array of these texturally interesting plants from around the world. This apparent human proclivity to collect helped shape the first function of our most comprehensive public gardens: the botanical garden.

Botanical garden collections, however, are more than just an institutional expression of a human trait. Collecting and collections are the heart and soul of a botanical garden's purpose to save, understand, and interpret plants. This purpose is more widely shared by museums in an effort to understand and interpret who we are, how we live, our history, our natural surroundings, and our technological and creative endeavors.[1] "Taken as a whole, museum collections and exhibition materials represent the world's natural and cultural common wealth."[2]

My preoccupation with botanical gardens can be attributed to the countless enjoyable hours I spent on the campus of Michigan State University and, most particularly, at the W. J. Beal Botanical Garden. I marveled over the huge and seemingly ancient specimens of *Cercidiphyllum japonicum* and *Phellodendron amurense* that were surrounded by serpentine beds of taxonomically ordered herbaceous and woody specimens. My first encounter with *Sarcococca hookeriana* var. *humilis* sporting its fragrant flowers on a mild March afternoon left an indelible impression. The campus arboretum and the botanical garden served as a captivating context for discovery and learning about plants; there were so many to see, touch, study, and watch through the seasons. This work is an offshoot of what is an ongoing love affair with botanical gardens.

This text is written from the perspective that botanical gardens are a special type of living museum. Many of the approaches and practices presented here are derived and adapted from the field of museum studies, or museology. From this perspective, I will first try to define and describe a museum and its collections before turning my attention to botanical gardens as a subset of these institutions.

So, what exactly is a museum? Museums are unique institutions that acquire and preserve collections. The collections held in public trust by museums are what set them apart from other cultural or scientific institutions. After spending some time searching through the museology literature and several dictionaries, I prefer the following amalgamated definition of a museum: a permanent institution for the purpose of acquiring, preserving, researching, and interpreting to the public for its instruction and enjoyment objects and specimens of cultural, scientific, historical, technological, and natural history value.[3] Perhaps we could simply agree that a museum exists to make important educational, research, and aesthetic use of permanent collections.[4]

Among the vast array of museums, a large segment includes those institutions that focus exclusively on natural history. This group contains a special niche occupied by natural history museums that collect and preserve living specimens: zoos, aquariums, botanical gardens, and other similar institutions. I turn to a slightly modified version of the earlier definition to define a botanical garden: a permanent institution for the purpose of acquiring, preserving, researching, and interpreting to the public for its instruction and enjoyment plants of cultural, scientific, historical, technological, and natural history value.

With this definition, you can see that botanical gardens, like museums, are set apart by the centrality of their collections to their purpose and programs—they are living museums. Dr. Richard Lighty points out some useful distinctions between botanical gardens and other similar institutions:

> An institution primarily involved in research on plants over which it does not exercise ownership and continuing stewardship should be called a botanical

institute. A display garden with unrecorded or temporary collections is really a park. And an institution that teaches people about plants without using a carefully assembled and recorded group of objects to do this is a school.[5]

As living museums, botanical gardens are subject to many of the same tenets of museum practice, or museology, common to all museums. Very often the challenge is even greater in that living museums must employ a diverse and specialized approach to the task of developing and managing— curating—their collections. With this text, my goal is to familiarize you with a range of both standard and specialized museological concepts and practices useful for the management, or curation, of botanical collections.

Before moving ahead with a discussion of curatorial practices, I would like to define just what a "collection" is and identify both the elements of curatorial practice and the garden professionals responsible for it. First of all, the use of the term *collection* often occurs without regard for the possible difference between the meaning of its singular and plural forms. Museum *collections* may be defined as the collected objects a museum has brought together to fulfill the institution's stated purposes: woody plants in an arboretum. A *collection* is a subset of the collections, consisting of objects having something of importance in common relevant to the museum's programs, such as the oak collection at the arboretum—plants sharing a taxonomic affiliation.

Within their collections, a botanical garden may have a collection of orchids, economic plants, and monocots. The concepts that are used to define a collection may be overlapping, and the plants associated with one collection may also conform to the concepts or themes of another; orchids, for example, are monocots. The conceptual basis for organizing a collection has a significant impact on its overall management, public presentation, and use.

Collections, then, exist for museums and botanical gardens as both physical objects and conceptual entities. A comprehensive curatorial program for botanical gardens will address the planning and sustainable development, preservation, and use of collections as both living assemblages and conceptual entities. In addition, the need for a comprehensive, rational, and sustainable approach to collections management is mandated by the fact that botanical gardens, as they were defined earlier, hold collections in the public trust.

For the purposes of this text, I take the all-inclusive view of curatorial practices to include the acquisition, documentation, preservation, and use of the collections. Living collections, as anyone who gardens can tell you, require constant, diligent attention. Plants are continuously growing and changing in a dynamic relationship with the environment. The relationship between particular types of plants and the environments to which they are adapted must be accurately understood and competently facilitated or modified. Fortunately, plants have the advantage over living collections of animals in that many can be vegetatively propagated. On the other hand, the promiscuous nature of many plants may cause hybridization among closely related taxa in the collections, complicating collection use and preservation. These special collections management considerations, along with the limitations of space and other restrictions covered in the following chapters, dictate that collections of living plants be synoptic rather than comprehensive in order to be useful and sustainable.[6]

Unfortunately, the elements of curatorial practice are not easily articulated or categorized due to their changing interpretations within the field. I have attempted to use the most descriptive and inclusive elements common to museology. These elements are as follows:

- Governing collections: Well-managed collections may be governed by three documents that will guide and limit what a botanical garden collects—a collections management policy, a collections management manual, and a collections plan. These documents help ensure that collections are meaningful and relevant to the mission of the institution.
- Building collections: Botanical gardens must invest in the acquisition of new plants to ensure the viability of their collections for future generations. Gardens acquire plants in many different ways: exchange, field collection, gifts, hybridization, purchase, and selection.
- Documenting collections: Perhaps no single facet of a botanical garden so thoroughly distinguishes it from a park than the documentation it maintains on its collections. To paraphrase Carl Guthe, the significance of collections lies not in themselves alone but also in the information relating to them.[7] Without proper documentation, collections have a limited story to tell and little reference value. Moreover, all museums are ethically and legally obligated to maintain basic information about their collections according to accepted professional standards.[8]
- Preserving collections: Collecting wisely and preparing good documentation would hardly make sense if the collections were then allowed to deteriorate or die.[9] The first obligation of botanical gardens—the adequate management of their collections—is strongly predicated on their preservation and care. To adequately preserve collections, we must expand the scope, meaning, and perspective commonly associated with plant care and horticultural practice.

- Collections and research: Research is a fundamental part of botanical garden work. It is one of the principal justifications for our collections. Research on the collections involves both consideration of the plant for what it alone can reveal, as well as what can be discovered about its context from additional sources of information. It is the endeavor to *discover* new knowledge and *compile* facts, to *interpret* the information, and establish or revise accepted conclusions, theories, and laws.[10]
- Collections and public programs: The relationship between curatorial practice and public programming should be a symbiotic one. The collections are the center and foundation of the botanical garden experience. Therefore, a consideration of public programming needs impacts policies affecting collections development and management. Concomitantly, public programs should be based upon the collections.

The management of the collections is most often the charge of the museum or botanical garden curator. In fact, no other title and position are as firmly attached to the museum as that of the curator. As such, you might presume that the position, responsibilities, and privileges of a curator would be tightly defined—not so![11] According to the *Standard Practices Handbook for Museums* of the Alberta Museums Association, "being a curator is the most impossible profession on earth because, theoretically, the professional demands involve almost every museum area. No one can possibly participate in all those areas and still do the things which curators do uniquely, which is to work with the collection."[12] I use the following definition for the term *curator*: a botanical garden staff member responsible for the acquisition, documentation, and preservation of collections for current and future research, conservation, educational, and exhibition/display needs.

Undoubtedly, some readers will take exception to this definition based upon the wide range of activities that are now commonly assigned to curators. Also, our understanding of what a curator does has been changed or clouded by the proliferation of program specialists and a heightened sensitivity to public scrutiny and politics. All of this creates an environment in which some of a curator's traditional responsibilities are passed on to specialists and the validity of other activities is questioned.[13]

The validity of the above definition may be found in the expectation that qualified curators have been intensively trained and can be expected to know the principles reflected in the collections so that they will best serve the purposes of the institution. Trained in scientific inquiry and active in their own research, they will be equipped to anticipate the needs of researchers and can assemble and preserve collections responsive to those needs.

Further, the curator may be expected to provide good judgment on access to the collections.

Regardless of the various nuances of curatorial responsibility and work obligations among the diverse and varied types of botanical gardens, curatorial work should be guided by the following universal values articulated by the American Alliance of Museums in *A Code of Ethics for Curators*:

- To serve the public good by contributing to and promoting learning, inquiry, and dialogue, and by making the depth and breadth of human knowledge available to the public.
- To serve the institution by responsible stewardship of financial, material, and intellectual resources; by pursuit of the goals and mission of the institution with respect for the diversity of ideas, cultures, and beliefs; and by integrity of scholarly research.
- To serve the museum profession by promoting and practicing excellence, honesty, and transparency in all professional activities.[14]

Professional ethics will be touched on again in several of the upcoming chapters.

In all but the smallest museums and botanical gardens, the curator delegates certain day-to-day responsibilities to other curatorial staff. Documentation is often the charge of the registrar or recorder. The regular care and preservation of the collections is often delegated by the curator to a collections manager or horticulturist. One staffing scenario for the average botanical garden would include a curator as head of the living collections program reporting to the director and responsible for the registrar, horticulturist, propagator, and their supporting staffs. Large institutions with extensive collections subdivided into several large and comprehensive groups (collection) may employ a staff of curators. A general or head curator may handle the overall coordination and administration of such a large collections program. In any case, the curator is ultimately the authority and the individual responsible for the development, preservation, and use of the collections.

The curator will play a major role in education programs and exhibits. The curator must ensure that accessions are properly handled in education programs and exhibits. They will also serve to provide authority for exhibits and act as a valuable resource for education programs. By the same token, curators have a responsibility to support garden educators.

Regarding research, Richard Laub makes a relevant point:

It is through research that curators' knowledge and understanding of their collections keep pace with the rapid developments in their fields. The question of research seems to arise most often in small museums, where budgets may be relatively limited and concern must be given to what sort of activities the public can legitimately be expected to fund. In agreeing to accept charge of specimens, however, a museum takes on a responsibility to place them in the care of competent curators, scientists whom the museum will allow to be scientists. The degree to which a museum can support research financially is one thing. The appropriateness of that support, however, and the validity of research as a curatorial activity should not be in question.[15]

The registrar works on the curator's staff and is generally regarded as the museum's authority on documentation and is responsible for documenting the collections. This includes all registration, catalog, and cartographic records. This work involves extensive record keeping and archiving in an office context using computer hardware, diverse types of software, and hard-copy data. It also entails important fieldwork to develop, maintain, and enhance the various linkages between documentation and collections. The nature of the registrar's work will be more fully explored in the chapter on collections documentation. The horticulturist, or collections manager, also works on the curator's staff and is charged with caring for and preserving the collections, which is, at its root, a horticultural endeavor. The horticulturist may be considered the museum's authority on developing and delivering effective horticultural programs for preserving the plant collections. The curator plays an integral and authoritarian role in determining the cultural requirements for taxa in the collections. However, preserving the plants themselves is not the whole of our obligation; we must also preserve their genetic and programmatic integrity. It is this complex of preservation requirements that adds a uniquely curatorial dimension to an otherwise horticultural activity. The tasks of the horticulturist are more completely described in the chapter on collections preservation.

As I just mentioned, the garden's preservation program must be structured to support the genetic and/or programmatic integrity of the collections. The role of the plant propagator is particularly important. The propagator is considered the garden's authority on methods and techniques for propagating plants. Again, the curator plays a key role in determining which methods or techniques may be most effective with any given taxon in the collections and ensuring overall programmatic integrity. Assuming that genetic diversity exists in the collections and is well documented, the propagator should be prepared to collect and grow vegetative material from the collections and to discard open-pollinated seed or volunteers except for use in selected displays or sales. The propagator will often develop a routine of replenishing stocks of aging or otherwise declining plants.

With this introduction, some readers may not find the characteristics or definitions presented here adequately representative of the institutions where they work. Admittedly, the field of botanical gardens is populated by a diverse set of institutions with unique regional mandates, strong central themes, narrowly defined collections, or specialized governing arrangements. Working from a museum-oriented, generic model of botanical gardens that conform to the definition offered at the beginning of this chapter, my aim is to provide basic, fundamental information on curatorial practices that will be broadly useful to all of these institutions. The implication is that staff members at each individual institution will adapt and customize the details to accommodate their particular mission, mandates, and programmatic specialties. In doing so, I hope that botanical gardens will develop and implement basic curatorial programs and practices with greater comprehension and continuity.

It is also my aim in presenting this broad and fundamental approach to curatorial practices that each garden will be well prepared, no matter their specialties, to meet the curatorial and collections management standards of several noteworthy professional organizations, for example, American Public Gardens Association, Botanic Gardens Conservation International, and the American Alliance of Museums.

The subject of professional organizations is a good segue to the importance of networking. The staffs of botanical gardens—particularly those charged with collections management—should examine their collections, management practices, and programs from a networking standpoint. To be truly effective, botanical gardens will need to band together as networks of institutions that collectively manage and preserve plants throughout the world. Dr. Peter Raven's advice is poignant:

> In our modern era that must mean not only exemplars of plants from throughout the world, but genetically diverse samples maintained by gardens for the perpetuation of those species, linked of course with seed banks, tissue culture facilities, propagation facilities and all the other necessary methods—if really comprehensive collections of plants are to be maintained properly and available in the future.[16]

As you use this text, keep Dr. Peter Raven's recommendation in mind, as it will impact all of your curatorial practices and programs.

Finally, there is another element of curatorial practice, one also important to the administration and management of the entire institution, that deserves elaboration: sustainability. This term was introduced at the beginning of this chapter: "A comprehensive curatorial program for botanical gardens will address the planning and *sustainable* development, preservation, and use of collections as both living assemblages and conceptual entities." Why such a concern for sustainability? A concern for sustainability arose in the early 1970s as growing numbers of people realized that the degradation of the environment would seriously undermine our ability to ensure expanding prosperity and economic justice.[17] Human beings—now a geophysical force on the planet like no other—appropriate large volumes of Earth's resources and are rapidly polluting what remains. For every person in the world to reach present U.S. levels of consumption with existing technology would require the resources of four more planet Earths. Global climate change is one of the many symptoms of the environmental blowback spawned by our unsustainable use of natural resources. Obviously, the capacity of Earth to support us is approaching its limit. We are entering an ecological bottleneck from which we may safely emerge only by changing our way of life. It is time to sort out our priorities and calculate what it will take to provide a satisfying and sustainable life for everyone into the future. A principal, and perhaps most important, question of this century is this: how best can we shift to a culture of permanence, both for ourselves and for the environment that sustains us?[18]

Clearly, fundamental changes are needed in our values, institutions, and ways of living. Botanical gardens have a major role in conservation of natural resources, research, education, and information exchange necessary to make these changes possible. The tenets of sustainability reach into every corner of the mission and operation of botanical gardens. Thus, botanical garden administrators and staff must initiate and support mobilization of internal and external resources so that their institutions respond to this urgent challenge. I consider this an important mandate for botanical gardens, and I ask that you keep it in mind as you consider the recommendations in this text, develop your collections management and curatorial programs, and conduct your work.

Notes

1. American Association of Museums, *Museums for a New Century* (Washington, DC: AAM, 1984), 35.

2. American Association of Museums, *Code of Ethics for Museums* (Washington, DC: AAM, 2000).

3. Edited version of definitions used at one time by the American Association of Museums and the Canadian Association of Museums.

4. G. E. Burcaw, *Introduction to Museum Work* (Nashville, TN: American Association for State and Local History, 1975), 4.

5. Richard W. Lighty, "Toward a More Rational Approach to Plant Collections," *Longwood Graduate Program Seminars* 16 (1984): 8.

6. Peter Ashton, "Museums and Botanical Gardens: Common Goals?," in *Museum Collections: Their Roles and Futures in Biological Research*, ed. E. H. Miller, Occasional Papers 25 (Victoria, BC: British Columbia Provincial Museum, 1985), 209.

7. Carl Guthe, *Documenting Collections: Museum Registration & Records*," American Association for State and Local History Technical Leaflet 11 (Nashville, TN: American Association for State and Local History, 1970).

8. American Association of Museums, "Stewards of a Common Wealth," in *Museums for a New Century* (Washington, DC: AAM, 1984), 46.

9. American Association of Museums, "Stewards," 40.

10. Alberta Museums Association, *Standard Practices Handbook for Museums* (Edmonton, AB: AMA, 1990), 186.

11. Alberta Museums Association, *Standard Practices*, 186.

12. "Being a Curator Is the Most Impossible Profession on Earth!" *Muse*, Winter 1986, 15–18.

13. Richard S. Laub, "The Natural History Curator: A Personal View," *Curator* 28, no. 1 (1985): 48.

14. American Alliance of Museums, *A Code of Ethics for Curators* (Washington, DC: AAM, 2009).

15. Laub, "Natural History Curator," 47.

16. P. Raven, "A Look at the Big Picture," *Public Garden* 12, no. 2 (1997): 9.

17. Richard M. Clugston and Wynn Calder, "Critical Dimensions of Sustainability in Higher Education," in *Sustainability and University Life*, ed. W. L. Filho (New York: Peter Lang, 1999).

18. E. O. Wilson, *The Future of Life* (New York: Knopf, 2002), 22.

CHAPTER TWO

~

Governing Collections

Policies, Plans, and Manuals

"We can never have enough of nature. We must be refreshed by the sight of inexhaustible vigor."

—Thoreau

One of the greatest challenges anyone faces in curating collections is maintaining a disciplined collecting program. In the face of our own preferences and prejudices, the temptation to collect plants with no discernible use for our gardens, or to carelessly reject those of potential value, can be overwhelming. This temptation may be exaggerated if it is spread among many garden staff members having tacit approval to acquire plants for the collections. Botanical garden collections must be governed by policies that ensure that the plants acquired by the garden are those needed to fulfill its purposes. Collections developed without regard to the mission and purposes of the garden are wasteful and subvert its useful function. Every public garden has its "attic" collection of plants that seemingly appeared out of nowhere and is not particularly useful, yet everyone is loath to discard it.

The well-managed collection is principally governed by three documents that will guide and limit what a garden collects—a collections management policy, a collections management plan, and a collections management procedures manual. There may be a fourth document, an ethics policy, that specifies the professional ethics for the conduct of curatorial practice at the institution. The alternative to an ethics policy document is to embed those guidelines in the appropriate sections of the other three collections documents. I have taken the approach in this text that the subject of ethics is discussed in each chapter where it applies to the subject at hand. The collections policy, plan, and manual provide guidelines that help ensure that collections are meaningful and relevant to the mission of the institution. They assist in interpreting the purposes of the garden in terms that are useful for the daily management of the collections. "A museum cannot contribute to a body of knowledge or responsibly enlighten the public if its trustees and staff do not devote adequate attention to shaping the collection and improving it."[1]

These governing documents must be vested with authority by the garden's governing board—or legal authority—to carry the necessary weight of accountability. Before we get ahead of ourselves, the question arises: by what process are these documents composed and their goals and objectives conceived? I would hope that work will be done by a group composed of individuals representing the legal authority of the institution and the professionals responsible for the institution's operation to produce a strategic plan. It's possible that such a process might involve a subgroup of curatorial staff to work out the collections component of the strategic plan. A strategic plan outlines the future for the institution and what is required in its operations and programming to achieve that future. A strategic plan articulates the botanical garden's vision, mission, operational and programmatic goals and policies, and measures of success.[2] If no such work has been done, it is incumbent upon the curator and their staff to undertake a smaller scale strategic planning process to chart the development and program direction for the collections program and articulate that in the three governing documents that will be outlined in this chapter.

Once these governing documents are approved and adopted, the garden's staff are accountable for conforming to the policy direction provided by them. Approval of such governing documents may be the most important decision that any board or legal authority will make.[3] Like the collections they govern, these are living documents that must be consulted regularly to ensure that they maintain an institutional relevance based on the current programmatic directions of the garden—and if for no other reason than for the board and staff to remind themselves of the collecting goals of the institution.

Collections Management Policy

"If we would know where we are and whither we are tending, we could then better judge what to do and know how to do it."

—Abraham Lincoln

The collections management policy is the most important and fundamental document for governing the development and management of collections. It specifies the delegation of authority and responsibility for the growth and protection of the collections. The collections management policy also certifies that the garden rectifies its collections with the purposes of the institution and uses proper procedures and records to preserve and interpret them. In short, the collections management policy should relate the garden's objectives to its processes to produce coherent, meaningful collections.

The American Alliance of Museums (AAM) publication *Museum Ethics* makes clear the necessity for a collections management policy. To emphasize the importance of this document, the AAM's Accreditation Commission requires that museums have such a policy in order to be accredited. Further, the collections management policy helps safeguard employees and the board to which they report from the unethical, unauthorized, or illegal collection activities of any individual associated with the garden.[4] The public has a vested interest in the collections and the real estate of the garden because of its tax-exempt status. A well-crafted and well-implemented collections management policy helps satisfy the fiduciary responsibilities of the botanical garden board.

Other purposes and benefits of a collections management policy are the continuity of purpose that it establishes as the staff and personnel of a botanical garden change. Toward this end, it can also be used as an effective tool in building teamwork and commitment by helping the staff gain a clear understanding of the purpose of the collections and their role in managing them. The introduction to the Royal Botanic Garden Edinburgh's *Collection Policy for the Living Collection* articulates this point:

> The Policy lays down some challenging targets that will be hard to reach and hard to sustain. However, the authors believed that it was important to present these standards and aspirations even if they cannot all be met all of the time. Were they to have presented the status quo there would have been no chance of raising standards. At least by presenting a vision of how the Collection should be shaped and managed over the coming years it will present staff and external parties with a view of current thinking. The Policy is intended as a guide and each Curator has the responsibility to develop and progress the

elements of it which are of most importance to their parts of the Collection. However, in trying to meet as many of the standards and challenges as possible it will be necessary to make the very best use of available staff and resources.[5]

A garden's governing responsibilities lie within a larger context of governance that impacts and also describes what the garden is expected to do. This context should be acknowledged within the collections management policy as a reminder of the overarching governing authorities to which the garden responds. This will include reference to articles of incorporation, founding documents, and other legally binding garden authorizations that legitimize its existence and purpose in developing collections. In addition, there are several global multilateral environmental agreements and network strategies of extreme importance to the conservation and preservation of plant life that should be adopted by botanical gardens and acknowledged as part of the larger governing context. These include the following:

- The Convention on International Trade in Endangered Species of Wild Fauna and Flora (CITES): CITES provides a legal framework for the regulation of trade in those endangered plant and animal species that are commercially exploited. Any country that is a signatory to the Convention has an obligation to maintain an authority with monitors and regulates the trade in endangered species between it and other nations.
- The Convention on Biological Diversity (CBD): CBD is an important international instrument for the conservation and sustainable use of biodiversity. Although the United States is not a signatory to the Convention, many U.S. botanic gardens are actively implementing the CBD, and they are affected by other countries' CBD-related laws when they work internationally. The Global Strategy for Plant Conservation (GSPC) and the Nagoya Protocol are particularly important treaty elements of the CBD.
- The International Treaty on Plant Genetic Resources for Food and Agriculture (ITPGRFA): a treaty that works in tandem with the CBD, its objectives are the conservation and sustainable use of plant genetic resources for food and agriculture as well as the fair and equitable sharing of benefits arising from their use for sustainable agriculture and food security.
- The 2030 Sustainable Development Agenda: This agenda and the associated Sustainable Development Goals (SDGs) were adopted in 2015 by the international community through the United Nations. With 17 goals and 169 targets, the goals of the Agenda are comprehensive in

recognition of the interrelationships between human development and the environmental, economic, social, and political context in which it occurs.

These examples represent the range of conventions, treaties, strategies, and agendas that may impact botanical garden missions, collections, and collections programming, thereby needing recognition and accommodation in the collections governing documents of participating gardens—it is not an all-inclusive list. An emerging and overdue concern expressed in some of the abovementioned international agreements touches on the decolonization of museums and collections, particularly in regard to indigenous rights, knowledge, and benefits sharing. I urge all gardens to give serious consideration to participating in these and other critically important and empowering networking opportunities—particularly those that focus on conservation. Several of these will be discussed in greater detail in other sections of this book pertinent to their role in curatorial practice.

At the very least, a practical collections management policy should include a statement of policy and collections purpose as well as guidelines and standards for acquisitions, accessioning/deaccessioning and other documentation, collections preservation, and disposal of plants in the collections. These components define the difference between a collections *management* policy and a simple acquisitions or accessions policy. Standards are particularly important when it comes to establishing a rigorous level of comprehension in the policy. Consider acquisitions, for example: I mentioned earlier that it might be specified in the collections management policy that new acquisitions conform to specified criteria relating to cataloging and organizational themes such as taxonomy, geography, or ecology. Those accessions meeting the largest number of criteria (multifunctional accessions) may be considered of highest priority for the garden. Before moving on to a specific example of accessions criteria, I want to pause on the subject of collections standards. The American Public Gardens Association (APGA) publication, *Compass for Progress: Standards of Excellence in Plant Collections Management*,[6] is an excellent resource and guide on the subject. From Pamela Allenstein, Plant Collections Network Manager at APGA:

> These are intended to articulate our field's best practices and point gardens towards targets. We also have a companion self-assessment tool that's a downloadable spreadsheet gardens can use as a workbook and planner. Gardens can use the self-assessment for auditing their overall collections program or to focus on a core collection prior to applying for accreditation. [personal communication]

Returning to accessions criteria as types of standards, the National Tropical Botanical Garden in Hawaii uses the following collections themes and criteria that best serve their mission and programmatic needs for developing and governing their collections:

- Economic (Ethnobotanical) Plant Collections
  - Tropical Fruit including bananas, citrus, mangos, avocados (McBryde, Bamboo Bridge, Fruit Orchard, Allerton Orchard, Kampong)
  - Spices and Perfume Plants – McBryde Garden, Spice of Life Trail
  - Medicinal Plants- (used to be the whole Mid-Valley area of McBryde, DL thinks not enough focus has been placed on it to give it priority)
  - Polynesian Canoe Plants; Traditional Pacific Ethnobotanical (Kahanu, Limahuli, McBryde Garden, old and new Canoe areas in Big Valley and Middle Valley) a. Hala, *Pandanus* spp., (Allerton) b. Awa, *Piper methysticum*, (most died, some at Pump Six area, Kahanu)
  - Bananas (Kahanu) and Breadfruit (Kahanu and McBryde) d. Taro, *Colocasia esculenta*, (Limahuli and Kahanu)
  - Kitchen Garden (Visitor's Center)
- Synoptic Collections
  - Geographic Groups
    - Pacific Island (Big Valley to east of Canoe Garden, and Waterfall areas)
    - Polynesia (Marquesas, Tonga, Samoa) ii. Micronesia (Caroline Islands, Marshall Islands, Guam, Marianas) iii. Melanesia (Fiji, New Caledonia, New Guinea)
    - Other island groups
      - Mascarene, including Mauritius (Limited area within the Middle Valley area) c. Indonesian (Kampong)
- Taxonomic Groups
  - Zingiberaceae
  - Erythrina
  - Aroids
  - Orchids
- Conservation Collections
  - Breadfruit (*Artocarpus altilis*)
  - *Pelagodoxa* palms, from Marquesas Islands
  - *Pritchardia* palms
  - Native Hawaiian Conservation Collections, Kauai endemics, Other Hawaiian island endemics
  - Species Listed as Threatened and Endangered for Conservation and recovery under the U.S. Endangered Species Act (Federal T &E)

- Species listed as priority for collection, propagation and restoration by the HRPRG Plant Extinction Prevention Program (PEP) (Perlman)
- Center for Plant Conservation designated species
- Representative of ecosystem/plant community types, including common plants (see restoration locations on maps)
- Historical Landscape Collections

(National Tropical Botanical Garden, "Collections Policy, June 20, 2012," https://ntbg.org/wpcontent/uploads/2020/02/ntbgcollectionspolicy.pdf)[7]

What makes these organizational themes and criteria powerful tools for developing useful collections is their high degree of integration and their emphasis on ecologic, economic, and cultural significance to the service region of the institution in accordance with its mission.

Creating and implementing such a policy is hard work. First of all, garden staff, trustees, and other key stakeholders often forget that collections are central to the institution's purpose. The creation of a collections policy necessitates a return to this perspective, which can be a difficult and confrontational effort. Furthermore, when there is no consensus on the value and importance of the collections to the institution, policy discussions will be digressive and unproductive. Care should be taken to prepare and orient the participants of any committee or working group charged with drafting a collections management policy. Finally, divide the task in two: (1) during the first phase, concentrate just on the policy (that is, strategy and operational principles); (2) the second phase may then concentrate on procedures. Tackling these together will make the job needlessly daunting. In some cases where large and comprehensive collections are involved, the collection management policy may be one of strategy and principles only, while a separate collections management manual will outline procedures. Manuals are discussed later in this chapter.

Collections management policies should be carefully worded, comprehensive, and oriented with a perspective on the future. Gardens should not find themselves in the position of having to revise their policies on a continual basis simply because they failed to cover all relevant matters or because a handful of people have changed their minds about scope or authority. Comprehensive collections management policies do not require changes except for good reasons.

As I alluded to above, botanical gardens with diverse and extensive collections will have lengthy collections management policies. These gardens will require such policies to cover their broad range and variety of collections management activities while smaller gardens will find utility in a shorter, less formal policy. A short collections management policy for a small garden may simply outline how plants are acquired, documented, preserved, and deaccessioned. The policies of larger gardens will contain these provisions in greater detail and may contain other sections outlining collections access, design considerations for the exhibition of collections, conservation measures, and so forth. For example, with one of the world's largest plant collections, spread among four different sites, the Royal Botanic Garden Edinburgh has a long and explicit collections policy (*Collection Policy for the Living Collection*) divided into five detailed sections with six appendixes. On the other hand, the Bellevue Botanical Garden (Bellevue, Washington) is a small urban garden with a focus on collections primarily for horticultural display and education. The BBG Collections Standards policy document is short and concise with the basic information to help govern a small and less complex array of collections.

The participants in a process to write a collections management policy should be carefully chosen. The number of group members should be representative but kept to a manageable and productive limit. Gardens may use a small working group and a larger group for review and revision. The process of writing a collections management policy should provide junctures for reference to the garden's mission statement. Also, policy writers should be directed to identify their guiding principles for collections management.

Best Practice 2.1: Writing a Collections Policy
In their article "How to Write a Collections Policy," in *Public Garden*, Donnelly and Feldman suggest the following set of processing questions to help you get started on writing a collections management policy:

Purpose

- What is the purpose of your collections management policy?
- Who is responsible for the implementation, interpretation, and periodic review and revision of the policy?
- What is the overall purpose of the collections as related to the institution's mission?

Building Collections

- What are the criteria for selecting additions to the collections?
- Who is responsible for initiating acquisitions to the collections? Who is responsible for approval?

- What are the legal and ethical considerations to be adhered to in collecting?
- Under what conditions, if any, will the institution accept gifts and loans?

Documentation

- What is the institution's commitment to collections documentation?
- Who is responsible for the administration and accuracy of the documentation system?
- What items of the collections are to be accessioned?
- What types of materials, if any, will not be accessioned?
- What essential records, files, or fields of information will be collected and preserved for each item in the collections?
- Who has responsibility for retrospective inventories of the collections?
- What criteria are necessary to warrant deaccessioning an item from the collections?
- What restrictions apply, if any, to the deaccessioning of a collections item?
- Who must review and approve deaccessioning recommendations?

Preservation

- What minimum standards of care are to be applied to the collections?
- What is the institution's commitment to safeguarding the collections and their documentation?

Insurance

- Will risk management techniques be applied to the collections and by whom?
- Are collections to be insured at full value or at a fraction of their value?
- What records must be kept regarding insurance and by whom?
- Who has authority to approve deviations from established insurance procedures?

Research

- What provisions will be made for the ongoing evaluation of the collections?

Disposal

- What principles govern disposal of deaccessioned material?

Access

- Who may use and have access to the various collections and documen-
  tation, and for what purposes?
- Will loans of accessions or other plants be made?
- Are propagules from plant collections to be taken and by whose
  authority?
- Who must approve special access requests?[8]

In addition to the abovementioned issues, gardens should also specify how
their documentation will be preserved. This will include a consideration of
duplicate computer and hard-copy files as well as archival guidelines. Also,
some thought should be given to documenting the record-keeping process
and the collections management program itself. These concepts will be taken
up in greater detail in the next chapter. Also, under research, I would add
what research will be undertaken in the area of plant conservation, a mandate now for all botanical gardens. Finally, curators and garden administrators should be familiar with risk management techniques because they put
insurance in the proper perspective. Risk management is a comprehensive
and systematic way of dealing with the identification, analysis, and evaluation of risk along with the best ways of dealing with it. It is an aggressive
approach that first recommends avoiding, controlling, reducing, or accepting
risk before insuring against it. Once certain risks are identified, the garden
can intelligently bargain with insurance providers over the terms of insurance contracts. Chapter 5 offers more on this topic.

This is a good point at which to review the concept of standards and
criteria for acquisitions. The acquisitions section of any collections management policy should contain criteria to provide adequate guidance to the
institution. A consideration of global multilateral environmental agreements
(or MEAs) that shape many national laws and conservation initiatives must
be included here. Perhaps the two most important of these are the Convention on International Trade in Endangered Species of Wild Fauna and Flora
(CITES) and the Convention on Biological Diversity (CBD). All botanical
gardens should adopt standards for acquisition that respect the international
laws imposed by these two important conservation agreements.

The convention [on biological diversity] affirms the sovereign rights of nations over their genetic resources, a right that has not previously been recognized legally or subject to such international legislation. Hereby rests one of the Convention's major impacts on botanical gardens. Access to genetic resources and benefit sharing are relatively new issues for most botanical gardens and ones for which new policies are needed. Botanical gardens have traditionally enjoyed virtually free and open access to plant materials from any parts of the world. Often this material has been received and continues to be obtained via the international Index Seminum scheme whereby botanical gardens have offered seeds of plants from their collections or from the wild to other botanical gardens on an exchange basis. It is clear that a fundamental and radical reorganization of the international botanical garden seed exchange scheme is required to bring it into line with the Convention and vigorous debate on how this may be achieved is already underway.[9]

Clearly, botanical garden staff must obtain and become familiar with CITES and the CBD and agree on how they are to become part of their collections management policy and implemented within their garden's programs. A focal point of the CBD for botanical gardens is the Global Strategy for Plant Conservation (GSPC). A helpful document for guiding a botanical garden's participation in these two conventions is the *International Agenda for Botanic Gardens in Conservation* (second edition) by Botanic Gardens Conservation International. Above all, botanical gardens should not shrink from the challenges articulated by these two conventions but should embrace them as opportunities to showcase their tremendous potential as conservation institutions.

Here are some items of particular relevance to governing collections taken from a checklist for botanical gardens to follow relevant to the CBD developed by Peter Wyse Jackson, secretary general of Botanic Gardens Conservation International:

- Obtain and read a copy of the text of the CBD and make it available to others in your botanical garden.
- Ensure that staff of your garden know about the CBD and understand its provisions and implications.
- Initiate a debate in your garden toward the formulation and agreement of an official policy on the CBD and a strategy for its implementation.
- Prepare and follow an institutional code of conduct on collecting and the acquisition of plant material.
- Develop Material Transfer Agreements to ensure that benefits arising from plant material distributed [are] fairly and equitably shared.

- Review your garden's current activities that are relevant or contribute to the implementation of the CBD; undertake a "CBD-audit" or strategic review for your garden and its collections.
- Consider how the mission of your garden is relevant to the CBD and to biodiversity conservation in general or/and consider reviewing your mission to become more involved in biodiversity conservation. Remember that the CBD is relevant [to] the national situation, that it is not just for gardens with international programs.
- Make sure that all staff are aware of and follow the garden's policies, procedures, and practices relating to implementing the CBD.
- Ensure that all the actions of your botanical garden are in line with the spirit and letter of the convention.[10]

There are several objectives and targets in the Global Strategy for Plant Conservation in the CBD that are of particular relevance to botanical gardens and, more specifically, how they govern the development and management of their collections. Some of the more important are as follows:

- Objective I: Plant diversity is well understood, documented and recognized: This constitutes an important role of botanical gardens and is, in fact, a large part of the work many of them currently do.
- Objective II: Plant diversity is urgently and effectively conserved. Target 7: At least 75 percent of known threatened plant species conserved *in situ*: *in situ* conservation: A practice of widely accepted efficacy and importance to botanical gardens.
- Objective II: Plant diversity is urgently and effectively conserved. Target 8: At least 75 percent of threatened plant species in *ex situ* collections, preferably in the country of origin, and at least 20 percent available for recovery and restoration programs: *ex situ* conservation. This has been the raison d'etre of many botanical garden collections programs.
- Objective III: Plant diversity is used in a sustainable and equitable manner: Which members of our collections have unknown or underutilized economic value and how do we make sustainable use of these and distribute the benefits equitably?[11]

A primary CBD obligation for botanical gardens is to share the benefits accrued to them of biological diversity with the country of origin; commonly referred to as access and benefit sharing (ABS). This obligation is implicit in the mission of the garden as it applies to the country where the garden is located. It is equally important that gardens observe this obligation toward foreign countries of origin for plants in their collections. The shared benefits may range from economic to informational in nature. Although the framework provided by the CBD is not retroactive past 1993, many gardens are proceeding to implement the spirit of the CBD without regard for the date of acquisition of subject material. Access and benefit sharing will be addressed in greater detail in chapter 3, "Building Collections."

In keeping with the code of the CBD, your collections management policy should specify the use of agreements or contracts for the acquisition or disbursement of plants. These documents should specify a means for sharing the potential benefits that may arise from the research, development, and use of these genetic resources. A common type of document used between gardens is a material transfer agreement, material acquisition agreement, or material supply agreement. These are now commonly used as part of the *Index Seminum* seed exchange program. The International Plant Exchange Network provides a system for the exchange of plants between institutions. The next chapter offers more information on this network. The provisions for access and the sharing of benefits in the CBD present some vexing issues and serious challenges for botanical gardens. Botanic Gardens Conservation International is an indispensable resource for botanical gardens seeking assistance to interpret and conform to the code of the CBD.

At this point, it is pertinent to focus on setting policy for the introduction and popularization of exotic species. Even though this has, and continues to be, the raison d'être of many botanical garden collections and collections-building programs, growing concerns about the conservation of native plants are an important priority. Many invasive plants have been introduced as ornamental plants and therefore owe their introduction to botanic gardens and nurseries. Therefore, curators, collection managers, and the entire botanical garden staff must be cognizant of the risks to biological diversity of exotic species introductions. According to Dr. Peter S. White, director of the North Carolina Botanical Garden, the risks and impacts of exotic plants on biological diversity may be divided into two broad categories:

[1] the spread of exotic genes into a native species through hybridization and [2] impacts to natural habitats and native populations. The first of these . . . is a function of close relationship between the introduced and native species and also usually involves congenerics. Outbreeding depression, sometimes called genetic pollution, occurs when the hybrid genotypes are less fit than the parents.

[Regarding the second of these,] some exotics reduce or eliminate populations of native plants and some even change the way ecosystems function. For example, some exotic tree invaders in south Florida transpire more water than native vegetation, thereby lowering the water table, promoting decomposition of formerly saturated organic matter (thus lowering elevation profiles) and promoting intense wildfires.[12]

Dr. White also suggests two sets of policy standards on exotic species that botanical gardens may use in crafting collections management policies and other important governing documents. One set of standards is limited and more general and may be appropriate for gardens initiating conservation programs and measures, while the other set is more comprehensive and exclusive. These have been combined below with a list of five actions gardens may take in response to potentially invasive species suggested by Dr. Sarah Hayden Reichard of the University of Washington.

Best Practice 2.2: Standards and Actions on Exotic, Invasive Species
A general/limited set of standards is as follows:

- Follow all applicable laws on the prohibitions on the introduction of soils and plants, and follow quarantine procedures; establish stricter policies, if legislation is deemed inadequate to prevent new exotic species problems.
- Avoid introducing close relatives that will hybridize with native species and create substantial gene flow to those populations.
- Do impact and risk analysis; predict the danger of exotic species impacts from current plantings and introductions; use exotics only if risk is low and remove known invaders from collections; if prediction is uncertain, develop sound and peer-reviewed monitoring protocols.
- Do impact and risk analysis for the distribution of gene pools beyond the region in which they were collected; export plants only to institutions with a comparable exotic species policy.
- Develop sterile exotic plant material.
- Assume responsibility for impacts in natural areas; form management partnerships with natural areas.[13]

The comprehensive/exclusive set of standards is as follows:

- Do not transport species and genes across natural barriers to dispersal; do not transport species and genes beyond natural range (unless at some time in the future, climate change causes a resetting of the geographic range of species, and then transport species only within one continent

to sites of appropriate climate); hence, perform no exotic species introductions and do not distribute plants or seeds outside native range.
- Grow and promote native plants of a region or physiographic province.
- Select for native species and genotypes for specific landscape situations and promote these in horticulture.
- Remove exotic plantings.
- Assume responsibility for impacts in natural areas; form management partnerships with natural areas.[14]

Five responsible actions for botanical gardens to take regarding the introduction of potentially invasive exotic species that may be specified in collections management policies are the following:

- Screening introductions for invasive potential using traits of the species, their geographic distributions including knowledge that they invade elsewhere, and their taxonomic relationship to other invaders. There must also be the resolve to discard or delay introduction of species that show invasive potential.
- Holding species judged to have moderate invasive potential . . . for at least five years before making a decision to release them for public use and removing those that do show an invasive potential from the garden.
- Encouraging the development of sterile hybrids and superior cultivars of native plants.
- Removing from display species that are already invasive in the area or including in interpretive material the suggestion that the species not be used in local landscape.
- Including in collections policies requirements to screen new introductions and work to prevent the spread of invasive species.[15]

The staff, board, and stakeholders of each botanical garden must decide for themselves what path they will follow in regard to the conservation of biological diversity. What seems clear is that each garden must adopt a proactive approach to this problem, one that places it in a position of leadership with regard to its constituents and publics.

Turning our attention to the development of other standards and criteria for acquisition, the process used at the Royal Botanic Garden Edinburgh (RBGE) serves as a useful and comprehensive example and is well documented in the article "Use of Records Systems in the Planning of Botanic Garden Collections" by James Cullen.[16]

Two factors were initially considered: (1) environmental and staffing constraints and (2) the purposes of the garden. The purposes of the garden were broken down into several functional categories that acquisitions must serve:

- Amenity
- Direct education (the provision of material for classes)
- Indirect education (material used for interpretive purposes in displays, etc.)
- Direct research (provision of material for ongoing research)
- Indirect research (reservoir of material for potential research)
- Conservation

After a review of the garden's purposes, the RGBE staff decided that taxonomic collections for direct research and education programs would be given priority. At this level, the garden obtains as much material of wild origin within several taxa at different ranks with which they conduct research. Within this approach, they have also placed a priority upon replacing stocks of unknown origin with fully documented plants. To satisfy their indirect research and education functions, the garden staff decided that they needed what is often referred to as "a general representation of the world's flora." Of course, the problem with this requirement is deciding what a "general representation" is and how it can be achieved within the capabilities of any one institution. The Edinburgh staff considered two vexing questions in coming to grips with the matter of representation:

- What taxonomic rank should be the basic unit?
- How can the level considered adequate be related to the size of the group in the wild?

Ultimately, the Edinburgh staff decided to use the family as the working taxonomic unit and the number of genera in a family as a working criterion. Within this parameter, they have as a goal 10 percent of the genera in large temperate families, 4 percent in large tropical families, and increasing percentages in smaller families as a goal. Family priorities are based on current research priorities: families already well represented in the collections, families of potential research value, and so on. The lowest level of priority is given to those families that are impossible or extremely difficult to represent at all: parasites, saprophytes, and other difficult groups. Needless to say, this system is not without problems, and the staff, I am sure, would admit that these guidelines are meant to be implemented with flexibility.

The following is an excerpt of the acquisition priorities for the Royal Botanic Garden Edinburgh that were established in 1994. This hierarchy of priority is structured around hardy (H) versus tender (T) families of plants. These two primary groups are then broken down into a secondary, numbered hierarchy by programmatic priority, for example, H1, H2, H3, and so on:

H1 Families with a substantial hardy content in which RBGE should be prepared to grow *multiple wild-origin collections of all genera and species*. Examples: *Arisaema, Berberidaceae, Gentianaceae, Iris, Nothofagus, Illiciaceae, Magnoliaceae, Meconopsis, Pinaceae, Primulaceae, Rhododendron, Rosaceae* (European and Asian), *Saxifragaceae* (woody genera), *Umbelliferae, Winteraceae*.

H2 Families with a substantial hardy content which are related to H1 families and families in which RBGE has had a long-standing interest but which are not actively worked on at present. *Minimum 50 percent of genera and 25 percent of species*. Multiple wild-origin accessions will not normally be grown apart from species with a very wide geographic distribution. Examples: *Acer* (N. American), *Alnus, Papaveraceae* (except *Meconopsis*), *Rosaceae* (other than European and Asian).

H3 Mainly hardy families for which RBGE requires minimal representation. *A few genera* of each with *one or two species* of each, will be sufficient. Examples: *Boraginaceae, Campanulaceae, Caryophyllaceae, Juglandaceae, Onagraceae, Ranunculaceae*.

T1 Tender families and also tender genera of H1 families in which RBGE has decided to specialize. Depending on culture requirements, space availability and size of the family or genus, RBGE should be prepared to grow *multiple wild-origin collections of all genera and species*. Examples: *Aeschynanthus, Agapetes, Dendrobium, Musaceae, Rhododendron* (section Vireya).

T2 Families which are related to T1 families, and families in which RBGE has had a long-standing interest but which are not actively worked on at present. Minimum 10 percent genera and 5 percent of species, depending on family. Multiple wild origin accessions will not normally be grown apart from species with a very wide geographic distribution. Examples: *Acanthaceae, Cannaceae, Droseraceae*, tender ferns, *Gesneriaceae* (other than *Aeschynanthus*), *Zingiberaceae*.

T3 Tender families for which RBGE requires minimal representation. A few genera of each with one or two species of each will be sufficient. Examples: *Bromeliaceae* (other than *Vriesea*), *Orchidaceae* (other than selected genera).[17]

A revised policy was developed in 2006 that retains this hierarchical scheme of criteria with an additional set of codes within each of the (H) and (T) categories to further prioritize acquisitions with greater comprehension and detail. This third hierarchy of codes will take the following into account (from the 2006 revised policy):

- Specific research projects (R)
- Conservation projects or of conservation interest (C)
- Education and interpretation in the widest sense (E)
- Teaching (e.g., MSc and HND [T])
- Historic collection or significance (H)

These letters will be appended onto the existing H1, H2, H3 and T1, T2, T3 system (shown below) which has been revised (the plants within each group, not the groups themselves) since the 1994 policy in line with new priorities.[18]

As an example of how this specific coding works, ferns and fern allies are extremely important to the mission of the RBGE, and therefore, a large number of families are prioritized in the H1 and T1 acquisition categories. In addition, native ferns in the H1 category are coded R (research priority) and C (conservation priority), with some representative families coded T (education [E] and teaching priority).

What about public access to the collections? Botanical garden staffs should not forget that public disclosure is an important part of museum and collection access. The collections management policy should be published and distributed to donors and other responsible parties with interests in the botanical garden. There are many other ethical considerations in addition to disclosure that should be articulated in the collections management policy. These would include setting forth general rules for those who control or otherwise work with the collections: trustees, staff, and their immediate families. Violations of these ethics rules carry no force of law but may be cause for censure, reprimand, or dismissal. Consider the following questions of ethical significance when drafting your collections management policy:

- May staff members acquire, collect, and own plants or other objects of the same or similar nature as those in the garden's collections?
- Should staff members disclose their collecting activities?
- May staff and trustees commingle garden and personal collections either on or off the garden's premises?
- May official connections be used to develop personal collections?

- May staff who create collections documentation have ownership of the documentation?
- Do the staff and trustees have ethical obligations to improve the collections and their documentation during their tenure with the garden?
- Do we maintain high standards for collections preservation and care?
- Should the garden avoid illicit trade in plants or plant propagules?
- May trustees overrule experienced staff in the acquisition and management of the collection?[19]

Grappling with ethical questions can be very difficult. If this creates serious impediments to the policy-making process, skip it and go on. Come back to this section after the rest of the work has been completed.

Some additional thoughts on the deaccession, disposal, and review of collections: many gardens often sell deaccessioned plants. You may wish to include guidelines for the use of funds obtained through the sale of deaccessioned plants (subject to the CBD and any agreements made with those from whom they were acquired). Typically, these proceeds are funneled back into the collections management program. It is advisable to include designated policy review intervals to incorporate day-to-day policy implementation decisions, written records required for ad hoc policy decisions, and a general analysis of the garden's conformance to its collections management policy standards. This falls directly within the purview of a collections committee, an important advisory group for external validation of curatorial practice and collections management, as well as an important body for public access and transparency for the botanical garden.

Larger institutions may consider the addition of a separate policy section outlining accountabilities and authority as they relate to the board and staff. "A good policy retains important approval and oversight responsibilities for the full board and specifies the functions that the board can delegate to others. Customarily, the board appoints a collections committee and assigns its duties."[20]

Making board assignments and accountabilities a matter of policy will help mitigate any buck-passing during times of indecision or, more important, crisis.

Many gardens have board-appointed collections committees to oversee the staff's collections management activities as well as advise the board and director on collections issues. I recommend that all botanical gardens appoint a collections committee for external assessment, guidance, and validation of collections development and management. The committee is composed of laypeople best suited to function in a deliberative capacity. No member of the committee has authority to acquire, deaccession, or initiate any other activity regarding the collections. The committee's connection to the garden is through the director, who may act as a gatekeeper between the staff and the collections committee.

Collections Management Plan

The collections management plan is an assessment of the botanical garden's collections management program in light of the standards and guidelines established in its collections management policy. A plan leads the botanical garden staff in a coordinated and uniform direction over a period of years to increase collections and refine and expand their value in a predetermined way. A collections plan helps the garden gain intellectual control over collections.[21] The Denver Botanic Gardens articulates their purpose well:

> Adhering to the standards and guidelines articulated in the Living Collections Management Policy, this Plan provides staff taking care of these collections coordinated and uniform direction to assist in the maintenance, expansion, refinement and development of the collections over the specified period of time.[22]

The collections plan should address all programmatic areas of concern within the collections management policy. The plan may include the following:

- Profiling inventories of collections containing descriptions of their size, character, breadth of scope, and conformance to standards and criteria in the collections management policy
- A method for establishing propagation priorities and plans for a cyclical program to repropagate existing plants in the collections
- An assessment of the overall collections management program based on the standards and requirements established in the collections management policy
- Specific recommendations for correcting deficiencies identified in the assessment including staffing, planning, and programming requirements[23]

The collections management plan should be prepared by a staff team with assistance from outside consultants or professionals having expertise appropriate to particular collections or programmatic activities, such as select members of the Collections Committee. The team may be composed of a cross section of garden staff with direct responsibility for the collections or specific interests in its use and headed by a team coordinator. Gardens with extensive collections management programs may choose to create several teams that will focus on specific areas of the operation. In this case, each team should designate a coordinator to organize and lead the team's activities.

Consult with the Institute of Museum and Library Services (IMLS) (https://www.imls.gov) before initiating any planning activities. The IMLS Museum Assessment Program is a specialized consultancy that may combine well with your institution's needs for a collections management plan.

Garden planning team members should provide adequate orientation for outside planning team members regarding the garden's history, policies, and so forth. The team should then prepare an inventory protocol and program assessment checklist before conducting any on-site work. In making these preparations, the team members should solicit and submit specific collection management programming issues for review to the team. They should also refer to the collections management policy to help identify checklist items of particular importance; in fact, there should be an ongoing correspondence between the work of the review team and the collections management policy. Team members should prepare to conduct their work at a generic level, earmarking needs for specific, detailed inventories or reviews for collection plan recommendations, for example, conservation assessments of specific collections.

On-site work should be scheduled in sufficiently large blocks of time to complete whole sections of the plan assessment. The planning team may need to prepare a schedule of several site visits, with each site interval lasting one to several days and focusing on some specific aspect of the collections management program.

The team coordinator plans and schedules on-site activities ensuring that all sections of the collections management plan are adequately researched. The team coordinator should also document basic findings and recommendations as the process moves forward. The coordinator should also arrange for the implementation of provisional measures needed to expedite the planning process. Finally, the coordinator will review, edit, and compile draft sections of the final plan with the assistance of team members.

The team members will review all relevant documents, assist in the organization of all procedures, participate in on-site work, write assigned sections of the plan, and submit drafts to the team coordinator.

The recommendations of the plan may be implemented in-house or require the services of a consultant—or a combination of both. For instance, a certain collection may require a detailed condition survey or verification by an authority on that group. Another example would be the need for a consulting archivist to assess the preservation needs of a portion or all of the hard-copy collection documents.

Implementation of the collections management plan will likely occur over several years. Plans should be reviewed every five years, whether or not they are fully implemented, to ensure that they have not strayed from their goals and that those goals are still relevant. The Denver Botanic Gardens' *Living Collections Management Plan* serves as a good example. The plan begins with a specification for regular reviews and revisions of the plan at a minimum of every five years. The majority of the 53-page document is composed of a set of individual analyses of the gardens' seven major plant collections. Below is the rationale for the collections analysis, the categories of information that were collected, and the actions necessary suggested by that information:

> Since each of the seven collections have unique characteristics, it would be impossible to combine them and perform a single analysis. Hence, each collection has been treated separately and the following addressed for each collection:
>   a) Description
>   b) Collections Content
>   c) History of Collection
>   d) Justification
>   e) Strengths
>   f) Weaknesses
>   g) Improvements to Collection since 2006
>   h) Opportunities and Collection Priorities
>   i) Implementation Strategy (with timeline)
>   j) Evaluation[24]

Taking a closer look at the Denver plan, the Alpine Collection is described from an ecological and taxonomic perspective with the contents organized in a taxonomic table showing the overall number of taxa and then a breakdown of that category into specific ranks and total number of accessions. There is a useful explanation of the collection overlap that exists at the Denver Botanic Gardens regarding alpine plants, a status common to many living collections:

> Due to the nature of collections, several of the other collections overlap with the Alpine Collection and vice versa; Native, Steppe and Amenity all have some overlap with this collection. Alpines native to western North America overlap with those in the Native Collection, and chasmophyte and facultative

alpines from steppe regions cross over with the Steppe Collection; however, these species need a rocky environment to survive in the garden and they are displayed in the Rock Alpine Garden or a traditional setting for alpines, such as troughs. In 2016 a new Steppe Garden was added with these microclimates in mind. Alpines of cultivated origin, the various Primula and Saxifraga cultivars, overlap with the Amenity Collection, but are best served in the Alpine Collection.[25]

The common relationship described above among institutional collections should be included in the definition of collections in the collections management policy.

The history of the Alpine Collection provides a very useful context for interpreting and understanding its current status and exhibition. There is useful documentation presented regarding the inability to fulfill the original purpose to acquire and display a large proportion of true alpine plants. Denver's climate has proven too rigorous for many of these plants, and the collection has evolved over time to contain a larger percentage of plants native to Central Asia. This becomes a part of useful catalog information on this collection overall—and for some accessions in particular.

The section on strengths lists impressive numbers of accessions and rarities, including some benchmarking information on the status of the collection on a national and international scale. Weaknesses are accounted for without reservation including some that are common to the entire field, such as insufficient staffing. Even a mundane but what may be crucial weakness is accounted for, such as a properly functioning irrigation system. This worthwhile accounting serves to identify the action items for the current edition of the collections management plan.

Improvements to the collection are accounted for since the last plan was developed in order to help establish a track record of progress. The core of the plan is articulated in the next two sections: "Opportunities and Collections Priorities" and "Implementation Strategy." In "Opportunities," the findings of the previous parts of the analysis are translated into action items. This section also contains a desiderata of taxa to acquire for the collection to increase its relevance to the mission and programs of the institution. Consistent with the mounting conservation obligations of botanical gardens worldwide, there is an item noting the need to focus on the acquisition of plants of documented wild origin.

Collection plans seem to be a rare commodity among botanical gardens, and the Denver Botanic Gardens maintains a fine example.

Collections Management Procedures Manual

Once gardens have their policies and plans in place, it may be useful to produce a procedures manual detailing all the necessary steps to implement these guidelines. Moreover, a collections management procedures manual is a thoughtfully prepared set of steps and guidelines for establishing and conducting the various facets of the collections management program. It also provides benchmarks for the establishment of consistent standards for collections management as well as guiding the day-to-day management of the collections.[26] Finally, it will provide an important level of continuity over time as inevitable personnel changes occur.

It is not uncommon for the professional staff of nonprofit organizations to spend 25 percent or more of their work time managing and resolving daily crises.[27] Perhaps one reason for this is insufficient program planning and the accompanying uncertainty about how program activities should proceed under the varying day-to-day operating requirements of organizations such as botanical gardens. Developing a collections management manual requires a careful evaluation of how collections management processes and operational tasks are to be accomplished. The manual should then outline these details in a way that will significantly reduce staff uncertainty and disorientation regarding the fulfillment of their responsibilities and programmatic objectives.

A collections management procedures manual will generally address two areas of concern: organizational structure and function, as well as operational procedures. The first area is relevant to the manual in that it contains information that establishes an operational context and provides basic orientation to all staff members regarding the purpose and mission of the institution. There should also be a short section on the purpose of the manual, how it is to be used, and what it contains.

The bulk of the manual describes, in a step-by-step, playscript-text, or item-by-item format, the basic operating procedures for each aspect of the collections management program. These should relate closely to the programmatic areas of concern in the collections management policy, such as the following:

Acquisitions

- Level of authority
- Accepting gifts
- Making purchases
- Field collecting

Documentation

- Accessioning
- Definition of record fields
- Registration of accessions
- Catalog records
- Geographic and disposition data
- Archiving

Collection Preservation

- Maintenance management guidelines
- Propagation procedures
- Evaluation

The body of the manual should be supplemented with a glossary of important and key terms.

Manuals may be written by a staff specialist, an outside consultant, or whomever will produce the best and most timely results. Those procedures that require specialized knowledge may best be written by an authority in that field. Regardless of who does the writing, staff contributions through group process are essential to ensure support and thoroughness.

In the step-by-step format, sections are broken down by headings and subheadings with each explained in a sequential manner. The accessioning process, for example, may be well explained in this manner. The playscript-text format is sequential but is applicable to procedures with more than one player, each with a specific role in the process. For example, this format may be useful in detailing the interaction between the registrar and the propagator regarding accessioning of propagules. Finally, the item-by-item format consists of presenting blocks of information that are related by unit or position. This format works well for delineating the specific responsibilities of any given staff member or unit of the operation.[28]

A collections management manual must be written and produced so that the staff will actually use it. Their involvement in the process will help, but it must also be understood that the manual will continually evolve with the collections management program. That evolution will be most relevant and effective to the extent that the staff are involved in driving it. Needless to say, the manual should be subjected to thorough staff review and updates and must be easily accessed by all employees.

The Arnold Arboretum's *Plant Inventory Operations Manual* is a good and highly detailed example of what the organization of a larger collections management procedures manual might look like. The opening overview provides a thorough orientation to the importance of an inventory to the purpose of the Arboretum and the integrity of its plant collections. It also specifies the purpose of the manual:

> This Manual serves several purposes. The initial and basic intention is as a means of documentation: to put into print for the first time the Arboretum's meticulous procedures related to inventorying living plants. Thus, it becomes a resource for staff to use as they carry out their day-to-day work. Also, because it is a document unique in the public garden community, it also can be used as a resource by peer institutions. The creation of the document also proved meaningful. During the process of writing and editing, curatorial and other staff members had the opportunity to scrupulously assess our current practices in light of a bolder and more expansive curatorial paradigm. As a result, current practices were evaluated to see if they met curatorial needs and requirements; any necessary changes for improvement were then crafted. The end product, this Manual, is a composite of tried-and-true methods and many new approaches.[29]

I particularly appreciate the segments about the manual being a resource for peer institutions and the value of the process of creating it as a means to assess their current practice.

The majority of this document is composed of a step-by-step articulation of the workflow from beginning to end. There is a helpful workflow diagram illustrating the order of the principal operations that begins this section of the manual. This workflow diagram is captioned with a bold statement of the administrative responsibilities for this operation:

> The Manager of Plant Records, in consultation with the Curator of Living Collections, manages inventory activities in the permanent collections. Other curatorial staff are critical contributors in the inventory process and are necessary members of the field check team.[30]

Each of the workflow steps is explained in specific detail that often includes caveats and recommendations for dealing with exceptions—items that are often overlooked or inadvertently excluded from procedures manuals. The following is an example:

> Using a laptop computer, input evaluative data in the appropriate BGBASE PLANTS table fields *or write on a hardcopy inventory (see pg. 6, Section 2.2.2.*

Step 2—Gather Essential Materials and Tools) if broadband connection is down or weather hinders use of electronic devices. [Emphasis added.][31]

And this:

[T]he plant may have recently been coppiced and the DBH entry may contradict in-field observations.[32]

This thorough, 42-page document ends with very important information for the staff and volunteers conducting this work on what to do in a medical emergency. This possibility may involve individuals conducting the inventory or they may come upon arboretum visitors needing help. This is vital information that should accompany anyone working in a public space.

Special Collections and Concepts

Collections policies and plans may contain separate sections or addenda pertaining to the control and management of specialized collections. Conservation, historical, cultural, and artistic collections and displays may require an esoteric set of standards, procedures, and other controls imposed upon them that are best established and articulated separately from the standard collections documents. What follows are some typical examples of special policies.

Collections Preservation Policies

Every museum has collections, and every museum has the capability of making some progress, however small, toward better collections care.[33]

Gardens may find it useful to articulate a comprehensive commitment to collections care and preservation in a preservation policy that augments the collections management policy. Such a policy should contain standards for collections preservation that the garden is committed to implement and uphold. Here are some very general and flexible examples of standards suitable for a collections preservation policy:

- The garden will delegate a specially trained professional to be responsible for collections care and preservation.
- The garden will establish, monitor, and maintain environmentally and genetically sustainable horticultural standards for the care and preservation of the collections.
- The garden is committed to protect the collections from vandalism, theft, fire, and poor handling and other abuses.

- The garden will establish, monitor, and maintain propagation and production standards and protocols for the multiplication, aftercare, and preservation of the collections.
- The garden will consult with and be guided by the advice of qualified experts in areas of special preservation needs, for example, soil chemistry, plant pathology and entomology, and reproductive biology.
- Garden personnel will attend professional development and training programs relevant to collections preservation and care.

Design Policies

There are many public gardens and portions of botanical gardens that are deemed so exceptional in their historical significance, artistic success, or other qualities as to warrant special policy considerations to preserve their integrity and continuity. Instituting a garden design policy is an important step in governing these elements of concern for the garden. Also, and as someone with a horticultural background, it seems to me that a concern for design in a botanical garden, whether it be for the exhibition of collections or simply a pleasing amenity, is intuitive.

> While it is the scientific, conservation and education value of the individual plants and their assemblage into meaningful collections that are of the utmost importance it is also important to recognise that the design and landscape of each Garden is of value too.[34]

A design policy may have its greatest importance for botanical gardens that began life as elaborate private and/or institutional landscapes or display gardens of great amenity value. Those that came into being with an original purpose as a botanical garden may have developed a greater dependence upon botanical exhibits that require the guidance of an exhibit policy involving garden educators as well as curators. This subject is covered further in chapter 6, "Collections and Public Programs."

Design policies will dictate design decisions and/or establish the process by which design decisions are made.[35] They should also identify those people or positions with authority for design decisions and responsibility to implement the design policy. This will certainly include elements of the board of trustees or other governing authority, director, select staff, and other consultants or designers. "Enlightened administration and skilled maintenance depend on interpretation of the garden design and, for this reason, it is useful to consult with an advisor. Such an advisor is needed who, through training, experience and through research, has gained broad, historical perspective upon the garden's design, materials and function, and the designer's intent."[36]

The design policy should ensure responsive and responsible design under changing public garden administrations. One of the best ways to ensure this is, to the extent possible, to document the original design intent of any garden, display, or other important designed space. This design policy should then hold garden boards and administrations accountable for conformance to this documentation to whatever level is deemed adequate in the policy.

The Fairchild Tropical Garden in Coral Gables, Florida, was fortunate to obtain written interpretation of the design intent of the garden from the original designer. This document serves as a useful part of a site master plan to help guide the development and alteration of the garden as it evolves in collections and services. To solidify administrative accountability to the design intents as explained in this document, the bylaws of the garden were amended with Article VI, the Landscape Design Article:

> The landscaping design and the arrangement of planting as described in the Memoirs by William Lyman Phillips, under date November 1, 1959, which are incorporated herein by reference, shall be strictly adhered to continuously unless changes are deemed advisable by a three-fourths vote of the entire Board of Trustees and then such plans shall be drawn and certified by a competent authority and approved by a three-fourths vote of all voting members of the corporation in good standing and voting.[37]

In the case of historical gardens, it may be valid to interpret the design requirements of the garden from a context fixed in time. On the other hand, for a garden of special artistic value, it may be a serious error to attempt to manipulate the design and plantings in reference to a fixed point in time. The most appropriate intent might be to understand the essence of the art and how it may be skillfully and sensitively managed for continuity and integrity.[38]

There are many gardens that choose to embed design and representation standards and criteria in their collections management policy as a subsection closely allied to access, acquisitions, and public programs. The Royal Botanic Garden Edinburgh's *Collection Policy for the Living Collection* contains a section titled, "Landscape, Design, and Representation Policy." At the beginning of this section, an important point is made clear: "Important scientific plants set in an uninspiring and poorly designed landscape make little sense and would not maximise the value of the plants or collection."[39] Cultivars are carefully considered for their ornamental value:

> It is an explicit policy of RBGE and of this Collections Policy that its living plant collections will be overwhelmingly dominated by well-documented, wild origin plants. Cultivars . . . do have a role to play in our Gardens and should be selected as carefully as wild origin species. They are important in creating attractive displays as noted above and are also useful for teaching.[40]

See the next section for a policy consideration of cultivars.

Design also touches on the concept of collections representation as I mentioned at the beginning of this section. As stated in the introduction to this text, both the limitations of space and the requirements of collections management often dictate that institutional collections be synoptic rather than complete. Within any particular synoptic framework, representational themes provide guidelines for how the selected plants will be exhibited or displayed on the grounds. The traditional representational display theme for botanical gardens, one that inspired me at the W. J. Beal Botanic Garden while attending Michigan State University, is taxonomy. A less Cartesian and more ecological influence has given rise to geographic and ecologic display and exhibition themes.

The Royal Botanic Garden Edinburgh's *Collection Policy for the Living Collection* contains specific guidelines for the use of representational display themes at its four satellite gardens. Some of the issues addressed are the following: a specification that each of their four gardens utilize both taxonomic and geographic representations; how RBGE differentiates between geographic and ecologic ("niches" and "habitat types") representations; representations should be created with interpretation in mind; each garden should have a representation of native plants; representations with strong research, conservation, and education designs must be reviewed for relevance every five years; high priority should be given to including species from biodiversity hotspots and the surrounding areas.

Cultivar Collections

An area of specialization for many botanical gardens is the acquisition and use of cultivars. Many find it difficult to govern cultivars in their policies primarily because our collections management policies are often most strongly oriented to botanical taxa—those plants that find a place within the framework of botanical classification. The inclusion of cultivars in the collections of botanical gardens may be justified by their agricultural and horticultural importance as a source of food, the botanical and horticultural significance of hybrid cultivars, the need to document and study their validity, and the representative range of variation within species represented by a series of cultivars.[41]

There is much work for botanical gardens to do regarding cultivars and, in this regard, I'm speaking of both economic crops and ornamental plants. Progenitors of important food, fiber, and other economic crops should be documented and preserved. The same may be said for lineages of important and historical ornamental clones. More will be presented on cultivar collections later in this text. The following excerpt from section B, "Historic and Priority Cultivar Collections," article 2, "Cultivar Collections" of the 2016 Arnold Arboretum collections policy may serve as a useful guideline for the governance of cultivar collections:

1. Distinctive Cultivar Collections
   Early and throughout its development, the Arboretum has established diverse collections of garden selections intentionally or now regarded as cultivars within various plant groups (e.g., *Malus*, *Rhododendron*, *Syringa*, *dwarf conifers*). These collections are maintained, and development is limited to acquisitions that notably expand the group's breadth; they are not to be comprehensive.
2. Cultivars with names proposed prior to 1953
   The Living Collections contain a number of historic cultivars with Latinized names that were proposed in a botanical context (typically formae) prior to 1953. As a general rule, these are maintained, particularly when they represent material unique in cultivation. However, these collections are not actively developed.
3. Arnold Arboretum Cultivar Introductions
   Throughout its history, the Arboretum has selected and introduced a number of named clones to horticulture, many of which were initially regarded as botanical formae but are now recognized as cultivars. Because they arose at the Arboretum, they are maintained and development occurs to repatriate genotypes lost by the Arboretum, as well as to select new cultivars worthy of introduction.[42]

Textbox 2.1.   Recommendations for Governing Collections

Basic
- The garden has a collections management policy governing what the garden will acquire, specifying that the collections will be documented, and outlining when collections are deaccessioned and how they may be disposed of.
- The garden has a collections management procedures manual describing how the above activities are implemented.

Intermediate
- The garden has a comprehensive collections management policy articulating the basic governing parameters for all aspects of the garden's collections program.
- The garden has a collections management plan that profiles the future development of the collections, including propagation priorities.

Advanced
- The garden has governing policies for specific management areas within the collections program such as documentation, preservation, design, and research.

Notes

1. American Association of Museums, "Stewards of a Common Wealth," in *Museums for a New Century* (New York: AAM, 1984), 36.

2. Gail Dexter Lord and Kate Markert. *The Manual of Strategic Planning for Museums* (Lanham, MD: AltaMira Press, 2007), 4.

3. Carl E. Guthe, *The Management of Small History Museums*, 2nd ed. (Nashville, TN: American Association for State and Local History, 1964), 24.

4. D. R. Porter, *Current Thoughts on Collections Policy*, Technical Report 1 (Nashville, TN: American Association for State and Local History, 1985), 2.

5. Royal Botanic Garden Edinburgh, *Collection Policy for the Living Collection* (Edinburgh: RBGE, 2006), 3.

6. American Public Gardens Association, *Compass for Progress: Standards of Excellence in Plant Collections Management*, https://www.publicgardens.org/programs/plant-collections-network/compass-progress-standards-excellence-plant-collections.

7. National Tropical Botanical Garden, "Living Collections," https://ntbg.org/science/collections/living-collections/

8. Gerard T. Donnelly and William R. Feldman, "How to Write a Plant Collections Policy," *Public Garden* 5, no. 1 (1990): 33.

9. Peter Wyse Jackson, "Convention on Biological Diversity," *Public Garden* 12, no. 2 (1997): 15.

10. Jackson, "Convention," 16.

11. Convention on Biological Diversity, "Global Strategy for Plant Conservation," https://www.cbd.int.gspc.

12. P. S. White, "A Bill Falls Due: Botanical Gardens and the Exotic Species Problem," *Public Garden* 12, no. 2 (1997), 23.

13. White, "A Bill Falls Due," 24.

14. White, "A Bill Falls Due," 24.

15. S. H. Reichard, "Learning from the Past," *Public Garden* 12, no. 2 (1997), 25–26.

16. J. Cullen, "Use of Records Systems in the Planning of Botanic Garden Collections," in *Conservation of Threatened Plants*, ed. John Simonds (New York: Plenum Press, 1976), 95.

17. E. Leadlay, ed., *Darwin Manual* (Richmond, Surrey, UK: Botanic Gardens Conservation International, 1998), 9.

18. Royal Botanic Garden Edinburgh, *Collection Policy for the Living Collection*, 31.

19. Leadlay, *Darwin Manual*, 10.

20. Leadlay, *Darwin Manual*, 7.

21. Leadlay, *Darwin Manual*, 2.

22. Denver Botanic Gardens, *Living Collections Management Plan* (Denver, CO: DBG, 2017), 2.

23. National Park Service, *Museum Handbook* (Washington, DC: NPS, 1994), 3–7.

24. Denver Botanic Gardens, *Living Collections Management Plan*, 3.

25. Denver Botanic Gardens, 4.

26. Alberta Museums Association, *Standard Practices Handbook for Museums* (Edmonton, AB: AMA, 1990), 89.

27. T. S. Brady, "Six Step Method to Long Range Planning for Nonprofit Organizations," *Managerial Planning* (1984): 49.

28. Meipu Yang, "Manuals for Museum Policy and Procedures," *Curator* 32, no. 4 (1989): 271.

29. Arnold Arboretum, *Plant Inventory Operations Manual*, 2nd ed. (Jamaica Plain, MA: Arnold Arboretum, 2011), 2.

30. Arnold Arboretum, *Plant Inventory Operations Manual*, 5.

31. Arnold Arboretum, *Plant Inventory Operations Manual*, 10.

32. Arnold Arboretum, *Plant Inventory Operations Manual*, 10.

33. David McInnes, "Commitment to Care: A Basic Conservation Policy for Community Museums," *Dawson & Hind* 14, no. 1 (1987): 18.

34. Royal Botanic Garden Edinburgh, *Collection Policy for the Living Collection*, 19.

35. C. Moss-Warner, "Current Design Policies of Botanical Gardens and Arboreta in the United States," *Longwood Graduate Program Seminars* 10 (1976).

36. D. K. McGuire, "Garden Planning for Continuity at Dumbarton Oaks," *Landscape Architecture*, January 1981, 82.

37. Fairchild Tropical Garden, *Fairchild Tropical Garden Bulletin*, July 1961.

38. McGuire, "Garden Planning," 85.

39. Royal Botanic Garden Edinburgh, *Collection Policy for the Living Collection*, 19.

40. Royal Botanic Garden Edinburgh, *Collection Policy for the Living Collection*, 19.

41. S. A. Spongberg, "The Collections Policy of the Arnold Arboretum: Taxa of Infraspecific Rank, and Cultivars," *Arnoldia* 39 (1979): 370.

42. Arnold Arboretum, *Living Collections Policy* (Jamaica Plain, MA: Arnold Arboretum, 2016).

CHAPTER THREE

~

Building Collections

"Museums collect to preserve objects of apparent or possible value that otherwise might be lost, and to bring objects together for use. [However,] museums are sometimes so busy preparing exhibits, expanding into new . . . fields, getting publicity, and raising funds that their most fundamental job or obligation gets pushed into the background . . . collecting."[1]

Botanical garden acquisitions are driven by the curiosity and acquisitive instincts of individual staff members and governed by specific institutional policies. Good collections are the result of thoughtful collecting based upon logical, intelligent planning. A good measure of that planning goes into the development of collections management policies, plans, and manuals as described in the previous chapter. Thoughtful, ethical collecting and the building of good collections proceed according to comprehensive policies built upon an understanding of some basic tenets of collecting: "1) Museums cannot collect all objects that exist; 2) collecting has to be selective; 3) it is an abstraction from the real world."[2] To this I would add two principal goals of any collecting program: accuracy and completeness.

Before going any further on the subject of building collections, I think it pertinent and of critical importance to point out that the above tenets and goals for a collecting program have particular poignancy and application directed to plant conservation. The fact that most plant collections are a carefully chosen synopsis of what is naturally available due to the limitations of space and resources, and the growing importance of preserving biodiversity, the goals of accuracy and completeness become weightier and of even greater challenge to achieve.

> The conservation of the Earth's biological diversity is the responsibility of all humankind. Throughout their history, botanic gardens have made essential and indispensable contributions to teach, research, understand and preserve diversity of plant life.[3]

In light of this quotation as it applies to building collections, the first priority of all botanical gardens should be the conservation of endangered plants. In so doing, botanic gardens must act within the framework of the Convention on Biological Diversity (CBD) and the Nagoya Protocol, a supplementary agreement to the CBD. Furthermore, gardens should endeavor to comply with other relevant national and international laws as well as the Convention on International Trade in Endangered Species of Wild Fauna and Flora (CITES).

If botanical gardens are to actively build collections, they need to determine what the collections ought to contain and make a strong, ethical, legal, and continued effort to locate and acquire the necessary plants. A passive approach leads to collections that are largely a reflection of what peer institutions, consultants, planners, and the general public think ought to be in the collections. The notions and presumptions of potential donors, for example, are likely to be disconnected in regard to the scope and purposes of the garden's collections. A tightly governed and forcefully implemented collections development program is a critical part of curatorial practice during this emerging period of emphasis on plant rescue and conservation. The American Public Gardens Association provides two very pertinent and helpful guides to assist in collections development: *Collections Development Planning Guide*[4] and *Core Collections Primer and Collections Prioritization Worksheet*.[5] The planning guide is, "a series of questions posed as a starting point for assessing, planning, and goal-setting for each plant collection." The primer is a, "prioritization tool that can be customized by assigning different numerical weights to metrics depending on their relevance to each garden/collection."[6]

In general, the acquisitions section of the collections management policy should specify the qualitative, quantitative, and conceptual limits of the collections as well as the ethics of acquisition to which the botanical garden will adhere. As a standard practice, the garden should only acquire those plants it can properly document, preserve, and provide access to. As I've underscored previously, the garden should not acquire plants that have been obtained in contravention to any laws or regulations, are poorly documented as to origin, or are the cause of habitat destruction. This point cannot be overstated—botanical gardens must not only adhere to laws, regulations, and collecting ethics but also must model and promulgate a respect for them. Finally, the decision-making process for obtaining plants should be a matter of written record.

> The Arboretum will meet these collections goals through the acquisition of nearly 400 target taxa, or desiderata, with each fulfilling at least one (and typically several) goals. For many of the taxa on the list, the Arboretum needs several unique acquisitions (e.g., from multiple locations), so what is initially a list of 395 blossoms into a vibrant garden of 720 actual targets. Each of these targets will require its own acquisition plan and approach. A few might be purchased from nurseries, some may be acquired from cooperative institutions and repositories, while others will be sought out and obtained through the Arboretum's network of colleagues. However, the majority will be obtained on specific plant expeditions in which an Arboretum staff member leads or participates.[7]

This quote from Dosmann and Port points out that a target list of taxa, or "desiderata," is most typically the starting point of acquisitions toward building effective and relevant collections. Desiderata will of course be guided by the collections policy, and within that framework, it is helpful to conduct a comprehensive gap analysis before moving forward with acquisitions. A gap analysis will help curators, collections committees, and other collections stakeholders make decisions about what to acquire. Such work may be conducted at different scales ranging from an analysis of collections among an array of regional, national, or international gardens to collections of specific taxa held at a limited number of regional institutions. Gap analyses done at any scale may also include comparative statistics on the taxa of concern in the wild, such as their diversity and/or scarcity. A gap analysis is often used to ascertain what taxa are needed to make an individual collection more complete based on the collecting goals of the institution: the previous quote states, "The [Arnold] Arboretum will meet these collections goals through the acquisition of nearly 400 target taxa, or desiderata." But it may also be useful in determining a new or young institution's overall collections focus and conservation priorities.

For example, a developing botanical garden in the southeast United States may decide that a representative taxonomic collection of *Quercus* (Oak) is relevant to their programs and audience given the prominence of that genus in the regional flora. A gap analysis of this type may begin with a review of floras and distributional data for the southeast United States and the state of Alabama, where the garden is located, using the U.S. Department of Agriculture PLANTS, Biota of North America Program (BONAP), and the Alabama Plant Atlas databases. At the beginning of such a search, it may become obvious that a plant name(s) is inconsistent and needs verification before a comprehensive analysis can continue. A good general reference for plant names is the World Flora Online, a resource developed by an international consortium of botanical institutions (more on verification in the next chapter). Then, an analysis of *Quercus* collections among existing regional gardens may indicate that a generic collection of significant breadth is warranted and, further, that a comprehensive collection of rare species containing significant genetic diversity is needed for conservation purposes. In the past, such an analysis would ordinarily be done using literature searches and surveys of institutions. Today, there are a number of useful databases, such as those just mentioned, at our disposal to help quantify and qualify our knowledge of existing plant collections. Nevertheless, literature searches and surveys may still be needed depending upon the information required. For instance, where inventory data shows that a particular taxon is rare in collections, a survey and/or query of the few institutions having that taxon may be necessary to obtain any special preservation information for those accessions.

A gap analysis conducted at any scale may begin with the comprehensive databases of Botanic Gardens Conservation International (BGCI): Garden-Search, PlantSearch, and ThreatSearch. To begin a general gap analysis among a group of regional, national, or international botanical gardens, the BGCI GardenSearch database is useful. To build on my earlier example of a developing botanical garden in the state of Alabama, I can specify both the United States and Alabama in GardenSearch and find that, as of this writing, there are nine gardens listed in that state. Of those, I choose the Birmingham Botanical Garden and find that they have no "special" collections of *Quercus* and make a note for my gap analysis. It is likely that they do have a general collection of *Quercus* as part of the institutional landscape, and a query of garden personnel may be necessary to confirm that. A thorough analysis dictates that I look at all the other Alabama institutional entries—and, perhaps, those of gardens in adjacent states—and take note of *Quercus* listings: this is accomplished quickly and easily. GardenSearch listings also identify conservation collections, herbaria, and other facilities and programs of interest.

After browsing collections information in GardenSearch, I decide to use the BGCI PlantSearch database to obtain more specific information on species of *Quercus* held in institutional collections. Not surprisingly, a query of PlantSearch for *Quercus* shows 2065 taxa. My refined search for the endangered species, *Q. oglethorpensis*, shows only three institutions worldwide with an accession(s) of this tree. As a means of verifying its conservation status, I check the BGCI ThreatSearch database that shows this oak having been listed as *vulnerable* by the International Union for the Conservation of Nature and Natural Resources (IUCN) Red List in 1997, but by 2017, it was listed as endangered indicating a declining population and increasingly dire threat assessment. All of this information is helpful in planning a collection of *Quercus* and developing a desiderata of taxa. Next, I may pursue any of the several means for acquiring the necessary taxa described in the remaining sections of this chapter.

Before getting too much deeper into the subject of building collections and what is acquired toward that end, we should be more specific in our references to obtaining "plants." Botanical garden acquisitions, in fact, take several forms: whole plants, vegetative parts of plants (including tissues), and seeds or spores. There are some broad considerations that apply to these choices when it comes to acquisitions. Provenance, or point of origin, is an important piece of information for acquisitions, particularly conservation collections. Under the heading of provenance are questions of source: wild or cultivated. If wild, were they obtained legally and ethically? Are the collection locations well documented? What are the ecological conditions of the source area as they apply to adaptability, preservation, and propagation? If cultivated, are they from controlled or open-pollinated sources, and original sources if selected and/or hybridized?

There are a number of advantages and disadvantages to acquiring seeds versus plants and vegetative parts of plants. Some of the advantages of seeds include the following: they are relatively easy to collect, transport, and store (orthodox seed), and their genetic diversity (if well targeted) is advantageous for conservation collections. Some disadvantages may also be genetic diversity—or true-to-typeness—of clones and cultivated lines, overcoming certain dormancies, viability, and time to maturity. Regarding whole plants and plant parts, some advantages include clonal development, time of availability, more rapid propagation and time to maturity, and rescue collecting when there is an immediate threat. Some disadvantages include lack of genetic diversity, biosecurity issues, and transport and mortality problems. Which of these factors is most important will depend on the institution's acquisition goals and purposes as spelled out in the collections management policy.

Before turning to the various means through which gardens acquired new plants, I will offer a word about biosecurity. All of the ways in which botanical gardens acquire new plants pose biosecurity hazards. By "biosecurity hazards," I mean plant pests and diseases, including plant invasive species, which pose a significant environmental, social, and economic threat to biodiversity as well as the well-being of the plant collection and local ecosystem. Botanical gardens must put in place, before taking receipt of any plant materials that originate outside of the institution, the means to quarantine and evaluate those materials as a biosecurity threat. There is more on that subject at the end of this chapter.

Textbox 3.1. Recommendations for Building Collections

Basic
- The garden will only acquire those plants it can properly document.
- The garden will only acquire those plants it can properly preserve.
- The garden will only acquire those plants for the collections to which it can provide access.
- The garden will only acquire plants that have been obtained in accordance with the provisions of the Convention on Biological Diversity (CBD), the Nagoya Protocol (NP), and national and international laws related to the protection and sustainable use of biological diversity, access to genetic resources, associated knowledge, and benefit sharing.
- The garden will not engage in habitat destruction through its acquisitions or any other efforts.
- The garden will establish and follow quarantine standards for exotic acquisitions.

Collections develop through bequests, purchases, field collections, transfers, exchanges, gifts, and loans. The collections management policy and plan should guide all of these modes of acquisition. Botanical gardens should use a standard form, such as a deed of gift or a material transfer agreement, that will document the conditions of acceptance for the donation, transfer, exchange, gift, or loan of a plant for the collections. Gardens issuing tax receipts for donations must ensure that the dollar figure shown on the receipt is the appraised fair market value provided by an independent appraiser.[8]

Gifts

Only in rare circumstances should gardens accept conditional gifts of plants. Gifts may be accompanied by a deed of gift that becomes part of the garden's legal record. The gift form should include the following:

- The garden's name and address
- The name, address, and telephone number of the donor
- A statement that the owner is giving up all rights to the gift

- A description of the plant
- The date of transaction
- Original signatures of the owner and garden representative
- Signature of museum representative authorized to accept conditions if any
- The conditions under which the garden is accepting the gift[9]

Finally, send a letter of thanks to the donor acknowledging the gift.

The Denver Botanic Gardens' *Living Collections Management Plan* contains instructive specifications regarding gifts:

Plants and/or collections will be accepted as gifts only if they meet the purpose of the collection and are in compliance with the Convention on Biological Diversity. Donated plants should have provenance and a properly identified name including cultivar name if appropriate. Any illegally collected or obtained plants will not knowingly be accepted. The donor may place no restrictions on gifts of plant material. An in-kind donation form will be given to the donor, and a copy will be submitted to the Development Department. Denver Botanic Gardens encourages gifts of plants to be accompanied by sufficient endowment to provide long-term maintenance. As much as possible, donated plants will be used in displays, however, Denver Botanic Gardens has the right to sell or de-accession any gifts. Under IRS regulations Denver Botanic Gardens will not make monetary appraisal of gifts.[10]

A bequest—a gift given under the terms of a will—is accepted by the same process as other gifts. A deed of gift is not required for a bequest of plants.[11]

Textbox 3.2.   Recommendations on Gifts

Basic
- Botanical gardens should use a standard form, such as a deed of gift, that will document the conditions of acceptance of gifts for the collection.
- Gardens issuing tax receipts for donations should ensure that the dollar figure shown on the receipt is the appraised fair market value provided by an independent appraiser.
- Plants and/or collections will be accepted as gifts only if they meet the purpose of the collection and are in compliance with the Convention on Biological Diversity (CBD).
- Gifts will not be accepted with restrictions on their use or status.

Loans

Loans play a negligible role in the collections and exhibits of botanical gardens, unlike those of cultural and art museums. Still, I would urge you not to accept loans of plants for the collections for an indefinite period of time. Build collections with plants that become the property of the botanical garden and relegate plants on loan to temporary exhibits or other temporary programmatic functions. If your garden has plants in its collection that were loaned with unclear requirements, seek to have those plants unconditionally donated to the institution or returned to the lender.

Textbox 3.3. Recommendation on Loans

Basic
- Do not accept loans of plants for the core collections or for an indefinite period of time.

Purchases

Purchase plants with a sensitivity not only to budget but also to institutional needs as well as collections accuracy and completeness, just as you would for all other modes of acquisition. Make sure that information about cultivars regarding botanical identity, origin, and lineage exists and is accurate, or be prepared to research this data. If possible, obtain cultivars and hybrids from the original source or commercial sources recommended by them. Be on the alert to biosecurity problems and expect nurseries to supply documentation regarding pest inspections. Even so, be prepared to closely examine nursery acquisitions and quarantine them if necessary.

If the garden is purchasing shares in or subscribing to a field-collecting expedition, the credentials of the collector(s) and their collecting methods should be thoroughly scrutinized. Gardens should be careful not to contribute to field-collecting efforts that may employ ethically and legally questionable collecting methods. Furthermore, and to reiterate a critical protocol, botanic gardens that purchase plants or field-collecting subscriptions must check that originators act within the framework of the Convention on Biological Diversity (CBD) and the Nagoya Protocol (NP), a supplementary agreement to the CBD as well as complying with other relevant national and international laws as well as the Convention on International Trade in Endangered Species of Wild Fauna and Flora (CITES).

Also, for the purposes of accuracy and completeness, the collectors should be well qualified to identify, document, and process their collections. Refer to the section of this chapter on field collecting.

Textbox 3.4.   Recommendations for Purchasing Plants

Basic
- Make sure that information about the plants regarding botanical identity, origin, and lineage exists and is accurate before making a purchase.
- Whenever possible, obtain cultivars and hybrids for the core collections from the original source.
- Make sure that plants have phytosanitary certification or have been inspected by botanical garden staff before purchase.
- Only purchase wild-collected plants for the permanent collection that have been obtained in accordance with the provisions of the Convention on Biological Diversity (CBD), the Nagoya Protocol (NP), and national and international laws related to the protection and sustainable use of biological diversity, access to genetic resources, associated knowledge, and benefit sharing.

Exchanges

The exchange of seeds or other propagules between botanical gardens as part of an *Index Seminum* program is unique. The *Index Seminum* is an indispensable means for acquiring and conserving plants if the contributing institutions are rigorous in their participation. It provides a unique conduit for communicating with many of the world's botanical gardens, arboreta, and similar institutions. It has been estimated that the total number of seed lists produced each year is on the order of 600 to 700.[12]

The nature and governance of the *Index Seminum* has transformed with the changing appreciation of people for the conservation of natural resources. The exchange of plants between gardens, whether through the *Index Seminum* program or other means, must now follow the protocols of the Convention on Biological Diversity (CBD) as I have previously made clear. As I have also recommended previously, curatorial staff must apprise themselves of the CBD provisions and implications, and then conduct a

CBD audit of their *Index Seminum* program. In view of the initiatives put forward by the CBD, as well as concerns over the introduction of invasive species, some gardens have dropped their participation in the *Index Seminum*.

Even before the concerns expressed by the CBD were raised, seed and plant imports into the United States have been governed by the U.S. Department of Agriculture (USDA). Plant material imports must be accompanied by USDA import permits and phytosanitary certificates. Information on applying for these documents can be accessed at the USDA website.[13] The importation of small lots of seed acquired through the *Index Seminum* requires a particular permit from the USDA that exempts you from the requirement for a phytosanitary certificate. To import large quantities of seed normally associated with field collections requires a separate permit that should be accompanied by a phytosanitary certificate. Growing concerns about the preservation of genetic diversity, territoriality over natural resources, and the spread of invasive species drive ongoing reviews of permitting processes, and curatorial staffs must stay in touch with the inevitable changes in this legal and ethical area of operation no matter where they are located.

As a conduit for the import and export of exotic species, botanical gardens must also be cognizant of their potential role in the introduction and popularization of invasive plants. This applies to the *Index Seminum* program and other forms of plant exchange. Refer to chapter 2 for information on the risks of exotic species introductions and some strategies for dealing with this problem.

> *Indices Semina* are produced in a bewildering diversity of size, formats, and styles ranging from single, typed sheets to glossy, magazine-style public relations pieces that include color photographs. The first requirement for participation in the exchange is reciprocity. Also, an increasing number of botanical gardens are including Material Transfer Agreements with their seed lists in keeping with the code of the CBD. In signing these agreements, the recipients agree only to use the material for the purpose for which it was supplied at the time of application—research, display, education, etc. If the recipient wishes to commercialize or pass on the genetic material, products or resources derived from it to a third party for commercial purposes then written permission must be sought from the garden. The recipient also agrees to acknowledge the supplier of the plant material in any publication resulting from the use of the material and submit copies of the publication(s) to them.[14]

Best Practice 3.1: Material Transfer Agreements

Many exchanges now occur under the aegis of the International Plant Exchange Network (IPEN) organized by Botanic Gardens Conservation International (BGCI). As part of that network, and stemming from its influence, gardens are requiring that material transfer agreements accompany plant material exchanges. This document sets out the terms and conditions of the transfer. Material transfer agreements should include the complete identification of the taxa involved including the accession number and the IPEN code if applicable. It should also contain the following specifications itemized in the BGCI *Manual on Planning, Developing, and Managing Botanic Gardens*:

1. Material is only provided to institutions working in the areas of research, conservation and education and not to individuals or commercial enterprises.
2. The recipient shall not sell, distribute or use for profit any of the material, its progeny or derivatives.
3. The recipient shall acknowledge (the donor botanic garden), as supplier, in all written or electronic reports and publications resulting from the use of the material, its progeny or derivatives. A copy may be expected to be sent to the donor botanic garden without request.
4. The recipient shall take all appropriate and necessary measures to import material in accordance with relevant laws and regulations and to contain the material, its progeny or derivatives so as to prevent the release of invasive alien species.
5. The recipient may only transfer the material, its progeny or derivatives to a botanic garden, university or scientific institution for non-commercial use in the areas of scientific research, education, conservation and the development of botanic gardens.
6. All transfers shall be subject to the terms and conditions of this agreement. The recipient shall notify the donor botanic garden of all such transfers.[15]

A basic ground rule, implicit in the purpose of botanical gardens, is that all material offered for exchange be of verified identity. At the very least, degrees of verification should be indicated on the list (for more on verification, see chapter 4). This may call for the inclusion of more extensive documentation (e.g., lineage, parentage, etc.) in the list where cultivars and hybrids are concerned. Also, participating institutions should seek to ensure that the seed they offer is viable through viability testing and the use of controlled seed storage techniques that will help preserve viable seeds.

Misidentified, inviable, and genetically mixed seeds are three of the most serious problems associated with the *Index Seminum* program. "Certainly, it is on reflection curious that serious establishments should publish ostensibly scientific lists which are known to contain a high percentage of errors."[16] "The goal of any seed list is to offer viable seed. Any practice which prevents this should be eliminated."[17]

As a whole, botanical gardens choosing to participate in this program should look upon seed exchanges as a critical and selective activity. Here are some important considerations:

- Focus the seed collection on the native flora or comprehensive, specialized collections that are an integral part of the garden's programs.
- Distinguish between seed collected from plants in the wild and that collected in the garden from plants of wild origin.
- For wild collections, provide the station (country, county, etc.), altitude, and habitat of the collection.
- Consider including stocks from research collections.
- Include seed of verified identity *or* segregate the verified and unverified portions of the list and identify each segment as such.
- Process and store seeds in a manner to preserve their viability.
- Seeds that lose viability shortly after ripening should be collected and distributed at the appropriate time. Inform respondents of this tactic.
- Seeds of plants listed in a conservation category should be distributed with discretion or contingent upon follow-up status reports. This will involve permitting and may involve other legally binding agreements.[18]
- Consider including seed handling notes for those taxa with difficult germination requirements.

The basic format for a seed list should include the following information:

- Name, address, and station of the garden including local environmental conditions (e.g., Walter diagram) and contact person
- List of seed available in phylogenetic, alphabetic, or other order with each offering numbered as an ordering key
- Ordering directions and/or a detachable order form
- Expiration date

Seed Storage
A small seed bank is a relatively simple and economical facility to create. In fact, the resources committed to seed collecting and cleaning often far

surpass the expense of properly stored seeds. Work space, a refrigerator, and a storage cabinet are generally all that is required for a modest seed bank operation.

Orthodox seeds may be handled as follows: collect, clean, subject to a tetrazolium or germination test for viability, air dry in paper containers at temperatures below 30°C for a short period of time, and, finally, seal in aluminum cans, glass containers, or plastic-aluminum foil envelopes with a desiccant at 4°C. This is generally all that is required to ensure that your *Index Seminum* collection will remain in good health for the short time it is stored before distribution.

Recalcitrant and other seeds predisposed to germinate immediately upon ripening should be prepared and distributed at the point in time when they are collected. Some temperate gardens have been known to use this approach with seeds of *Quercus* species. For a more comprehensive review of seed and gene banks, refer to chapters 5 and 10.

Surplus Plant Distribution
In order to maximize the capacity of botanical gardens for plant rescue and conservation, safeguard vital germplasm from unforeseen catastrophe, and reduce the waste of resources associated with an unnecessary duplication of effort, coordinated networking is a professional and ethical obligation. There are several professional networking affiliations available for botanical gardens: American Public Gardens Association, Botanic Gardens Conservation International, and the Center for Plant Conservation. These three organizations, in addition to the Plant Conservation Alliance Non-Federal Cooperators Committee (PCA-NFCC), form a consortium of networking opportunities in plant conservation—and other curatorial practices and programs—to empower all gardens toward achieving the regional, North American, and global plant conservation goals outlined in the Convention on Biological Diversity.

Three specific networking opportunities for the exchange of plants are the American Public Gardens Association's Plant Collections Network, the Botanic Gardens Conservation International's International Plant Exchange Network, and the Center for Plant Conservation's National Collection.

- The Plant Collections Network coordinates a continent-wide approach to plant germplasm preservation, and promotes excellence in plant collections management. The Network, originally known as the North American Plant Collections Consortium, accredited its first collections in 1996. Today, Plant Collections Network includes 151 accredited collections throughout North America, stewarded by 83 participating institutions.[19]

- As a collective answer of botanic gardens to CBD provisions, the International Plant Exchange Network (IPEN) was established in 2002. It is a registration system for botanic gardens worldwide to exchange plant genetic resources in compliance with the CBD. The objective of IPEN is to provide a sound basis for cooperation, transparency and communication, taking into account the concerns of both the providers and the users of genetic resources.[20]
- The Center for Plant Conservation maintains a collection [National Collection] of more than 1,600 of America's most imperiled native plants through its network of world class botanical gardens. Our 62 participating institutions safeguard endangered plant material in "ex situ" botanical collections including seed banks, nurseries, and garden displays. An important conservation resource, the National Collection serves as an emergency backup in case a species becomes extinct or no longer reproduces in the wild.[21]

As you can see from these descriptions, the Plant Collections Network is focused primarily on the sharing of germplasm to safeguard against threats to preservation. The other two programs have their focus on the conservation of germplasm and biological diversity. Each of these programs expect participants to conduct themselves by standards of participation that bolster curatorial practice in several crucial areas. All three are critically important networking frameworks for botanical gardens.

Botanical gardens inevitably have surplus plants from their plant collecting, propagation, research, and deaccessioning processes. The U.S. National Arboretum and Longwood Gardens are two examples of institutions that have made a practice of distributing surplus plants and propagules from their collections. Such material could be made available to other gardens through the networks described above. This type of garden sharing is particularly useful among gardens with significant complementary collections, such as those belonging to the American Public Gardens Association (APGA) Plant Collections Network. For those institutions with complementary conservation collections, these exchanges could be construed as a professional and ethical obligation.

Interstate, interprovincial, or interregional exchanges of plants and vegetative propagules must meet the phytosanitary and other requirements for transport across regulatory boundaries. In some cases, this may require that plants be inspected and receive a phytosanitary certificate documenting that the materials are free of pest organisms. Contact your state, regional, and national offices that regulate the transport of agricultural products for details. The next section has more on these subjects.

Textbox 3.5. **Recommendations on Exchanges**

Basic
- International, interstate, interprovincial, or interregional exchanges of plants and vegetative propagules must meet the Convention on International Trade in Endangered Species of Wild Fauna and Flora (CITES), Convention on Biological Diversity (CBD), sanitary, and other requirements for export, import, and transport across regulatory boundaries.
- Material offered for exchange must be of verified identity.
- Seed offered must be viable.
- Distinguish between seed collected from plants in the wild and that collected in the garden from plants of wild origin.
- For wild collections, provide the station (country, county, etc.), altitude, and habitat of the collection.
- Establish and follow quarantine standards for exotic acquisitions.

Intermediate
- Seeds and other propagules of plants, as well as whole plants, listed in a conservation category should be accompanied by a material transfer agreement.
- International exchanges of seeds and other propagules of plants, as well as whole plants, are facilitated through the International Plant Exchange Network (IPEN).
- Include seed handling notes for those taxa with difficult germination requirements.
- Quarantine and hold species judged to have moderate invasive potential for at least five years and remove those that show invasive potential.[1]

Note

1. S. H. Reichard, "Learning from the Past," *Public Garden* 12, no. 1 (1997): 27.

Field Collecting

No facet of botanical garden operation carries the allure and fascination of plant exploring and field collecting. Of course, much of this appeal is derived from stereotypes associated with the exploits of plant collectors from long ago. These collectors most often represented the interests of nurseries, private collectors, or other enterprises with peripheral interests in plants. Their collections found their way to botanical gardens often by indirect means. Today there is a mix of collectors at work ranging from individual enthusiasts, botanical garden personnel, and civil servants in the employ of agricultural or botanical arms of government. Many of these collectors are seeking plants for commercial introduction, comprehensive collections, botanical research, and/or conservation programs.

Botanical gardens can ill afford to embark upon field-collecting activities in an opportunistic or generic "baling hay" fashion as was typical of the old days. Limited resources, conservation restrictions, and shrinking "virgin" territories of relatively unknown taxa limit a generic "bioprospecting" approach and demand that gardens be more focused in their collecting efforts following a strong conservation ethic. Botanical garden staff with responsibilities for building collections must make themselves intimately familiar with the provisions of CITES (www.cites.org) as well as the CBD (www.cbd.int).

CITES provides a legal framework for the regulation of trade in those endangered plant and animal species that are commercially exploited. Any country that is a signatory to CITES has an obligation to maintain an authority that monitors and regulates the trade in endangered species between it and other nations. CITES operates through the issue and control of export and import permits for a number of clearly defined species listed in three appendixes:

- Appendix I lists plant species threatened with extinction, for which international trade must be subject to particularly strict regulation, and only authorized in exceptional circumstances.
- Appendix II lists species that are not threatened with extinction at present, but may become so if uncontrolled trade continues. Trade is permitted of both wild and artificially propagated material provided an appropriate permit is obtained.
- Appendix III lists species that are threatened locally with extinction through commercial exploitation and therefore subject to trade controls within certain nations. International trade in this material requires

an export permit from the country that listed the species, or a certificate of origin.[22]

Botanical gardens are important partners in the implementation of CITES by following its protocols and providing safe storage for plants seized by customs and legal authorities. To the extent that your institution is dedicated to conservation, seriously consider becoming a CITES-registered scientific institution; among other things, this scheme facilitates scientific exchanges between institutions. As a registered institution, you are exempt from the need for permits involving some specimens and exchanges.

Best Practice 3.2: CITES Compliance
Botanical gardens should do the following to be in compliance with CITES:

- Check collections for CITES-listed species and maintain complete documentation;
- Assign clear staff responsibility for CITES matters;
- Always obtain CITES permits and labels when appropriate work with collaborating institutions to compile procedures for obtaining the necessary export and import documents and remember that CITES also covers herbarium specimens, spirit collections, tissues, and DNA samples as well as other specimens/samples of CITES-listed species;
- Disseminate CITES information and/or training to staff, and ensure they understand CITES issues to prevent infractions;
- Prevent any illegally collected plants from coming into the collections "through the back door"; and
- Contact and find out about their national CITES authorities, and consider registering the institution with the Management Authority.[23]

The field collecting impacts and implications for curators of the CITES provisions are clearly expressed in the situation surrounding the collecting and trade in certain flowering bulb plants, such as *Galanthus* spp. Populations of the common green snowdrop, *Galanthus woronowii*, have been under great collecting pressure in the central Asian country of Georgia. All Galanthus species are listed in CITES appendix II, and any collecting efforts must be accompanied by a "non-detriment finding" issued by a scientific authority from the country of origin (Georgia). At a point when the CITES Plants Committee (made up of some botanical garden members) became concerned about the number of snowdrop exports from Georgia, there was insufficient information to issue a finding on the actual detriment to wild populations.

In response, a consortium of botanical gardens including the Royal Botanic Garden, Kew, the Tbilisi Botanic Garden, and Batumi Botanical Garden of Georgia joined with CITES authorities of Georgia to conduct field surveys to determine the population status of *Galanthus woronowii* in the wild and in semi-domestic habitats (gardens, farm fields, etc.). Interviews were also conducted involving local landowners, botanists, traders, and government officials. All this information contributed to the development of a potential sustainable harvest model used to recommend annual export quotas and the management systems needed to meet CITES requirements.[24]

The Convention on Biological Diversity (CBD) is the principal instrument for the international conservation and sustainable use of biodiversity. As of this edition, the United States is still not a party to the CBD, but U.S. botanical gardens are implementing the CBD as a show of support and to facilitate interactions with other gardens in countries that are parties to the CBD. The CBD has three principal objectives: the conservation of biological diversity, the sustainable use of its components, and the fair and equitable sharing of the benefits that come from the use of genetic resources. The key programs established to implement the objectives of the CBD that pertain most particularly to botanical gardens include the Global Strategy for Plant Conservation (GSPC), the Global Taxonomy Initiative, invasive alien species, traditional knowledge, and access to genetic resources and benefit sharing (ABS).

The CBD Global Strategy for Plant Conservation may be the most important access point for botanical gardens to the CBD. It provides an international framework to support and facilitate the conservation of plants at all levels. The GSPC originated and was formulated by a working group of botanical gardens and is composed of five objectives containing 16 measurable and time-limited targets. Objective II and its associated targets is perhaps the most important for gardens and curators engaged in field collecting:

Objective II: Plant diversity is urgently and effectively conserved.

- Target 4: At least 15 per cent of each ecological region or vegetation type secured through effective management and/or restoration.

- Target 5: At least 75 per cent of the most important areas for plant diversity of each ecological region protected, with effective management in place for conserving plants and their genetic diversity.

- Target 6: At least 75 per cent of production lands in each sector managed sustainably, consistent with the conservation of plant diversity.

- Target 7: At least 75 per cent of known threatened plant species conserved in situ.

- Target 8: At least 75 per cent of threatened plant species in ex situ collections, preferably in the country of origin, and at least 20 per cent available for recovery and restoration programmes.

- Target 9: 70 per cent of the genetic diversity of crops including their wild relatives and other socio-economically valuable plant species conserved, while respecting, preserving and maintaining associated indigenous and local knowledge.

- Target 10: Effective management plans in place to prevent new biological invasions and to manage important areas for plant diversity that are invaded.[25]

I urge gardens to go beyond the dictates of target 9 to share in the benefits accrued from collections based on indigenous and local knowledge.

An example of GSPC actions in collections building that botanical gardens may take that contribute to the above targets is participation in the Millennium Seed Bank Partnership. Led by the Royal Botanic Garden, Kew, this is the largest *ex situ* plant conservation program in the world at the time of this writing with many botanical garden partners. Another example is the Zero-Extinction Project led by the Xishuangbanna Tropical Botanical Garden (XTBG) of the Chinese Academy of Sciences. Focused on the preservation and conservation of the diverse forests of Xishuangbanna Prefecture on the Laotian border in southern China, seeds of endangered and vulnerable species will be collected for storage in the XTBG seed bank or grown in the living collections.

The Global Partnership for Plant Conservation provides coordination to support national and international implementation of the GSPC. Botanical gardens should be at the forefront of this strategy and many are, but there is currently a need for more. Gardens should also take guidance from and cite GSPC objectives and targets in their collections-building goals and activities.

Finally, and not the least of which, staff should also become familiar with their own domestic laws and those of countries of concern regarding the collecting and export of plant materials. Do not rely on the word or interpretations of on-site collaborators from foreign countries who may not clearly understand their own domestic regulations. Go through the proper channels and obtain documented "prior informed consent" for your collection activities.

Purpose
With the need for a more comprehensively planned and focused field-collecting program, gardens must have a clear vision of the purposes of such activities relative to their collecting goals. Some gardens may be able to meet all of their plant acquisition goals through the means described earlier in this chapter. Keep in mind, however, that the *Index Seminum* and other exchanges are often padded with a limited number of the most adaptable taxa, often referred to as the "botanical garden flora," and the recirculation of a limited, often unrepresentative, gene pool.

The purpose of field collecting for building plant collections should be made clear by the plant collections policy and plan of your institution. However, these purposes may be summarized in the following generic categories:

- Rescue collecting to preserve genetic diversity
- Collecting for specific, programmatic use, for example, crop breeding
- Gap-filling of collections
- Special research: taxonomy, reproductive biology, and so forth[26]

Regarding rescue collecting, it should be noted that there is a paradigm shift underway in regard to *ex situ* conservation collections. Orthodox thinking and actions for the conservation and protection of biodiversity have traditionally put the highest priority on *in situ* approaches and models. The paradigm established at the 1992 Earth Summit in Rio de Janeiro, and the policy and regulations established there, emphasize the preservation and conservation of ecosystems, habitats, and their component species where they are found: *in situ*. This approach has been manifest in protected areas and reserves managed with integrated approaches. *Ex situ* strategies involving field collecting of plant genetic diversity for preservation at various institutions in support of *in situ* conservation has been part of this conservation paradigm.

The impacts of climate change that are causing shifts and changes to existing ecosystems will render some *in situ* approaches ineffective if the natural habitats for some plants will no longer support them. Given this circumstance, "a number of protected areas may soon no longer harbour the species for which they were originally designated."[27] At the same time, our ability to successfully preserve biodiverse collections *ex situ* has increased. All of this points to a stronger case for a greater role for *ex situ* conservation in the overall scheme of biodiversity conservation efforts. As Pritchard and Harrop point out in their article, "A Re-evaluation of the Role of Ex Situ Conservation," in the BG *Journal*:

> A more fluid regulatory paradigm may need to be identified. Reintroduction of species into the original home ranges will no longer be a desirable outcome of regeneration or captive breeding programmes. Therefore, a critical analysis of what, in some cases, have become sacred ecological cows may be required. This may well result in the distinction between concepts of ex and in situ conservation blurring to the point of disappearing altogether.[28]

Finally, the edification of the curatorial staff is an often overlooked but important purpose of field-collecting programs. To be effective in their positions, curators must have a working familiarity with the relationships and ecology of accessioned plants in their natural habitat.

Planning

Precise definition and good research help control the cost of field-collecting trips and ensure some acceptable degree of success. The garden's collections management plan should be consulted in order to develop useful field-collecting desiderata. The desiderata will dictate that field collections be taxon specific, region specific (multispecies), or both.

Both taxon- and region-specific field-collecting efforts are driven by plant breeding, plant introduction, and taxonomic research programs. Area, region, or multispecies field collecting, however, may also be driven by conservation programs. Species-specific trips may cover wide-ranging areas in order to collect a large gene pool. Region-specific trips may be more geographically cohesive in an effort to collect species adapted to a specific habitat, ecosystem, or climate. These orientations will dictate the manner in which each trip is organized.

> The North America-China Plant Exploration Consortium (NACPEC) has worked to foster partnerships and undertake plant exploration to study and conserve the flora of China for over 30 years. The specific collecting goals of each trip have varied. Many expeditions worked from a large list of target taxa and included collections of a wide diversity of plants opportunistically encountered, while other trips focused on pointed taxa (e.g., Tsuga in 1998 and 1999, Fraxinus in 2008 and 2010, and Acer griseum in 2015). With greater knowledge of the flora of China in hand and limited space in living collections, the work of NACPEC today uses gap analyses of current holdings

and conservation assessments to maximize limited resources for future, more targeted, collection development.[29]

Both types of field-collecting trips require rigorous and thorough planning. As described above for the NACPEC, collaboration and partnerships are critically important for all kinds of field collecting, especially for international trips. As was first discussed at the beginning of this chapter, gap analyses can be very helpful planning tools. The diversity of plants encountered on region-specific trips presents a broad set of challenges.

> The most important [challenge] is the relatively restricted knowledge the collectors are bound to have of many of the species they will be dealing with. This means that it will not be possible to follow an optimal sampling strategy for all the species, interesting and perhaps unique material which an expert would have recognized will be missed and the information on each sample will not be as complete as it might have been. It will therefore be all the more important to tap the large store of indigenous knowledge about plants and the environment maintained by local communities. Furthermore, it may not be possible to collect many potentially interesting species or landraces within the course of the mission because of differences in maturation time. Finally, different kinds of species may require radically different collecting techniques and even equipment. For all these reasons, a multispecies collecting mission will sometimes need to be focused at least to some extent, usually on a "plant category." Examples might include collecting Andean root and tuber crops in Ecuador or forages in the semiarid regions of Kenya.[30]

For taxon-specific trips, types and populations of target species should be well defined. Wild species will often be more difficult to collect than crop species. Locating collectable populations of wild species and identifying the optimal times for collecting are much more difficult. Crop plants, on the other hand, may be located in areas where they are used; propagules may be collected from field plants, farmers, storage areas, or markets. In this regard, and referring to the above quote, it is important to collaborate with and share with indigenous and local groups the benefits accrued from collections.

Existing germplasm collections and their associated documentation could be invaluable to the planning of plant-collecting missions. It is quite possible that the discovery of certain taxa within existing collections may resolve all or a portion of your collecting needs simply by exchange. Also, information derived from existing collections may further justify the collecting trip(s) and provide a better understanding of the area to be covered. Needless to say, part of any planning for field collecting must be to learn as much as possible about any relevant previous work. Collectors will want to gather general information on any relevant past collecting trips and specific information on the taxa collected. To help locate existing germplasm collections, one should begin by consulting databases, directories, and professional organizations that act as clearinghouses for such information. The Botanic Gardens Conservation International databases outlined earlier in this chapter may be extremely helpful in locating those collections. For conservation collections of North American native plants, the institutional collection holders of the Center for Plant Conservation's National Collection are an important resource. Also, in the United States, the U.S. Department of Agriculture–Agricultural Research Service (USDA-ARS) National Plant Germplasm System maintains an up-to-date database of its collections (https://www.ars-grin.gov/Pages/Collections). APGA may be contacted for a directory of its members' collections and anecdotal information regarding recent collecting trips.

There are several organizations that publish international directories of germplasm collections. Perhaps the most widely recognized is the International Plant Genetic Resources Institute (IPGRI), which maintains databases on major groups of crops (https://www.ifpri.org/publisher-source/international-plant-genetic-resources-institute-ipgri). The IPGRI also maintains a separate database of forestry species called TRESOURCE. The Forest Resources Division of the Food and Agriculture Organization also maintains information on forest genetic resources collections (www.fao.org). Finally, the Global Genome Biodiversity Network of the Smithsonian Institution, as "an international network of institutions that share an interest in long-term preservation of genomic samples representing the diversity of non-human life on Earth,"[31] is another resource for locating germplasm collections and information on best field-collecting practices.

General information on past collecting trips may be obtained from published mission reports in the *Plant Genetic Resources Newsletter*, the FAO *Forest Genetic Resources Newsletter*, *CAB Abstracts*, and *Plant Genetic Resources Abstracts* and from the Plant Exchange Office of the USDA-ARS. For a discussion of mission reports, refer to chapter 4. Unfortunately, not all missions are followed up with a report, and not all such reports are published. To locate unpublished reports, contact the institution that sponsored or actually carried out the collecting trip.

Online resources aid information gathering, and we've found the surge of herbarium specimen data uploaded to regional, online floras to be very useful in selecting collection locations.[32]

Herbaria, floras, revisions, monographs, existing germplasm collections and their associated documentation, and their websites and databases, as well as experienced collectors are often consulted for location information about target wild species. Collectors should also research the ecological requirements and habitat preferences of the target species to begin developing a "search image" of the species and collecting sites. The search image is composed of a set of field indicators that should be correlated with the presence of the target species, for example, serpentine outcrops or prominent plant community companions such as certain tree species.[33] Based on the above recommendations, you can see that collectors are in need of good references to help them plan and carry out field-collecting missions. One I have yet to mention that is relevant here and of particularly good use for collection planning is a 2011 update of chapters 8, 9, 10, 13, 14, and 15/16 in *Collecting Plant Genetic Diversity*.[34] This valuable reference has been augmented with a series of chapter updates that you may obtain online. I suggest taking advantage of this material.

At this stage of planning, it should become clear that partnerships and collaborations with local botanists and others knowledgeable in the local flora is important. As we move forward with more detailed discussions of field-collecting logistics, the necessity of local contacts will become more obvious.

Best Practice 3.3: Field Collecting Plan
Once the decision to organize a collecting trip is made and broadly defined goals are established, it is time to consider and document more specific technical and logistical plans. A helpful planning guide for this crucially important part of your field collecting experience is *The Arnold Arboretum Expedition Toolkit*. The following is a synopsis of information from that document. The first step is to establish both a digital and a hard-copy repository for your collecting plan documents, lists, maps, and such. Computer tablets, notebooks, and cell phones are useful tools in the field for keeping and accessing digital field collecting plans, documentation, and references. The repository will then contain the following items and information:

- Collecting team (including collaborators) and outline of responsibilities.
- Collection planning time line.
- Desiderata, organized by location, niche, and taxonomy (indicate type of propagules).
- Sampling strategies (may be included with the above).

- Collecting location(s), important contacts and logistical details.
- A complete trip itinerary.
- Lists of travel and financial requirements.
- All permits.
- List of necessary supplies (consider all target propagules, sampling strategies, and documentation needs).[35]

The collecting team may consist of one person for a very limited, localized and/or opportunistic collection. However, under most circumstances, field collecting is best accomplished with at least two collectors and, for trips with lengthy desiderata, three or more. The principal tasks of field collecting—propagule collecting, herbarium sampling, and documentation—are best allocated to each of three people. When we include on-site collaborators, and the prospect of in vitro, DNA, or other special collections, the team may then be even larger. Consider collaborators that will contribute to the efficacy and expense of field collecting as well as making the results more broadly useful, such as other botanical gardens, herbaria, government agencies, supportive nongovernmental organizations (NGOs), college and university researchers, and local botanical enthusiasts. Appoint or designate a team leader and make the roles and responsibilities of all team members clear in job descriptions for each of them (e.g., leader, recorder, herbarium curator, propagator, etc.). Everyone on the team will be involved in scouting and seed cleaning. Extensive trips involving larger numbers of participants will require a treasurer or accountant to handle the budget, expenses, bookkeeping, and final expense report.

Each of these team members will be responsible for identifying (for acquisition) and organizing their supplies. All team members should maintain a trip journal, the contents of which may then be used to compose the final trip report. If the collecting trip will be collaborating with and obtaining plants of interest and utility to indigenous peoples, the team recorder may need training in special interview and documentation techniques appropriate to communicating with and meeting the needs of those people. There is more on this subject later in this section. Finally, the organizing institution may have the necessary administrative staff to assist with fulfilling other trip-planning and logistical details.

In consideration of the desiderata, the collecting locations involved, and timing, ascertain what is known about the current distribution, population sizes, and reproductive biology of the target taxa in preparation for implementing your sampling strategy. Environmental data will be needed in order to determine optimal timing for a collecting trip. Human uses and interaction with target species may be no less important than other, more typical, types of environmental data depending upon the species and the habitat. Included here would be autecology, taxonomy, population biology, and phenology of the target species. As has already been mentioned, some of this information may be obtained from the documentation on existing institutional collections; bibliographic, factual, and plant collection databases; and herbaria. Finally, and perhaps most important, validate your findings with your site and/or in-country collaborator(s)—this is one of the reasons you have sought them out.

After this information has been analyzed, it might be useful for collectors to put together a list of specific target areas within the overall target region, each with a list of desired taxa to be found there. I am recommending a comprehensive strategy for conservation purposes that requires a high level of familiarity with the taxa on your desiderata. Those detailed recommendations are presented later in this section.

Even with your desiderata in hand, you should be prepared for opportunistic collections involving disjunct taxa, special forms and ecotypes, and collection gap-filling opportunities. Be sure to have access to institutional inventories and double-check your permits before making any such collections (more on permits later in this section). Check your desiderata against information held in the International Plant Sentinel Network for potential biohazard issues (more on the IPSN in a later section) and weed risk assessments available through government agencies such as the U.S. Department of Agriculture APHIS-PPQ Weed Risk Assessment and Federal Noxious Weed list. Consider the collection localities carefully in regard to potential threats and hazards for collectors (difficult terrain, poisonous organisms, etc.) and prepare accordingly. On-site collaborators should be of particular assistance in helping you make these assessments and for offering advice on how to mitigate them. Make sure you have first aid and other medical necessities and documentation.

> To best determine the timeline for collecting, answer the following questions. How much time will be spent getting from the host institution to the collecting site base camp; how much time will be needed if the base camp changes; and how far are the collecting sites from the base camp? By developing a timeline, the leader can calculate the total length of the trip.[36]

Develop a daily itinerary that is generous regarding the anticipated collecting time for each taxon on your desiderata. Traveling to each location in unfamiliar territory and then locating the target taxa may involve more time than even a local resident will anticipate. Be sure to consider alternative sites for taxa of particular interest in case you are unable to locate any plants at the initial collection site, or the plants found there do not possess the necessary propagules. An ideal travel itinerary revolves around a base camp (hotel, homestay, campsite, etc.) location with daily forays to specific collection localities. Base camps also allow you to consolidate the materials you need each day, leaving unnecessary field supplies and previous days' collections at the base camp. Include base camp days for cleaning, documenting, and organizing collections; perhaps one every four to six days. Plan for and identify shipping options for sensitive collections that need special storing or growing facilities back at the botanical garden (alert the necessary garden staff to the possibility before leaving!).

Best Practice 3.4: Collecting Genetic Diversity
The principal ethic in field collecting is to do no harm, either to the site or the population. Of course, not all field-collecting efforts are fulfilling conservation purposes, but regardless of an institution's collecting goals, a consideration for conserving biodiversity must be part of the aim of field collecting given the current extinction crisis. Therefore, I am presenting a sampling strategy recommended by the Center for Plant Conservation in their extremely valuable reference: *CPC Best Plant Conservation Practices to Support Species Survival in the Wild*. Also, the guidelines published by the National Germplasm Resources Laboratory of the USDA-ARS will be valuable.

The CPC guidelines are particularly important for gardens creating genetically representative conservation collections. Based on these guidelines, your protocol should address three important questions:[37]

1. Which taxa should be given priority for collection? Gardens should review their mission, collections management policy, and collections plan for guidance in establishing collecting priorities. For conservation purposes, the potential for reintroduction and research and the degree of endangerment could serve as general criteria for establishing collecting priorities.[38] This, however, is much more easily recommended than applied, and there is further discussion on this subject below.

2. How many sites and how many individuals should be sampled? The CPC recommends that as many populations be sampled as possible, given the resources available, representing the range of variation with no fewer than five populations being the minimum. This approach is crucial for clonal and self-pollinating species where diversity exists between populations but of less importance for out-crossing species. Also, collect from widely dispersed plants within a population; fruiting plants representing a range of phenotypes and fruit set; plants in different microhabitats; and at different times during the fruiting season. For vegetative material, collect tissues from up to 50 unrelated individuals from multiple populations.[39]

3. How much material should be taken from each individual? The CPC recommends an ideal number of 3,000 seeds from a sample of 50 plants in a population, or approximately 60 seeds per plant. Further, if the seed set per plant is too small to meet this recommendation, you may have to collect from a larger number of individuals. If this expectation is too high, collect 10 percent of the seed from the fruiting individuals available that season and plan for return trips limited to no more than five in a 10-year period. Similarly concerning small populations, err on the side of collecting fewer seeds each season extended over several years. If those populations are under threat of extirpation, collect 100 percent of the seeds available, and if fruiting is poor, collect vegetative and tissue samples as well.[40] Chapters 9 and 10 offer more on conservation collections.

At this point, I want to add a few more details in support of the best practices above—and some caveats. Regarding question 1—which taxa to collect—this may involve assessing the degree of endangerment, and collectors must be aware that this condition may be construed to exist outside of the formal designations applied by various conservation and governmental agencies. In other words, "although all endangered plants are rare, not all rare plants are endangered."[41] Collectors should attempt to distinguish between plants of natural rarity, such as edaphic endemics confined to serpentine soils, and rarity based on the effects of human activity. These two types of rare species may require a special understanding of the dynamics of their ecological condition before collectors can establish a rational priority and collecting protocol for them. Complicating this distinction is the fact that various governmental and institutional methods of assigning a degree of endangerment are often inconsistent and irreconcilable.

Collectors may factor into this evaluation unique evolutionary lineages related to genetic distinctiveness, such as monotypic taxa, and relationships to economically and horticulturally important plants. Finally, and of equal importance, gardens must consider the likelihood of successfully propagating and preserving the target species—a general consideration for the collections management policy and plan.

On the subject of question 2—how many sites and how many individuals to collect from—most of our landscape plants, especially those introduced from abroad, represent a very small gene pool. Whether gardens desire horticultural introductions, additional taxa for themed collections, or rare species for conservation purposes, collecting genetic diversity has relevance to all these collection goals and is an integral component of our overall goal to build "complete" collections. Diversity, as I hope you can see from the above best practices, may be partitioned in three ways: (1) within-population diversity, (2) between-population diversity, and (3) ecogeographical diversity. As such, the question of "how many sites" relates to capturing genetic diversity based on the genetic difference among populations of plants.

Sampling populations will involve a comprehensive travel plan and specific curatorial arrangements to adequately document and manage these population collections. The predicted effort involved in collecting genetic diversity at the population level may vary significantly between herbaceous and woody plants. The necessary geographic range to cover in order to capture significant genetic diversity within a genus of woody plants may be quite different than for a genus of herbs.

A population-based collecting strategy should include a consideration of documented ecotypes and geographic and/or reproductive isolation among other biogeographic and life-history characteristics. For endangered species, collections from any imminently threatened populations, such as those in areas under development, should be expedited.

Toward a more complete sampling of genetic diversity, collectors must consider the fraction of diversity embodied among the species of any given population. "Within [among] populational sampling . . . attempts to capture geographical and ecotypic variation, a quantitatively small portion of the overall genome, but of potential adaptive and evolutionary significance. By contrast, sampling design within populations seeks to capture the basic genetic blueprint of the species with a reasonable degree of accuracy."[42]

Individuals of a species, as may be readily observed, have overwhelming genetic similarity. Because of this, collectors may rapidly reach a point of diminishing returns in capturing genetic diversity by undertaking extensive intrapopulation collections. However, for rare species, it would be sensible to collect from among a large number of individuals in the population to ensure capturing the largest degree of remaining genetic diversity.

In regard to question 3—how many seeds and such to collect—also consider that the number of seeds or other propagules collectors should plan to obtain from each individual will depend upon their expectations for survival of the material and the number of individuals they want to establish.

Consider the benefits of repeat visits to collecting sites:

- Collect a diversity of material from among and within populations with different fruiting times
- Overcome collecting shortfalls and difficulties associated with year-to-year variations in fruit set and other phenological phenomena
- Schedule preliminary, or reconnaissance, trips to positively identify target species when characters for accurate identification are present
- Monitor genetic erosion at particular sites or within specific populations[43]

Obviously, repeat visits may be restricted by expense, time, labor, and the availability of local contacts. Overcoming these potential restrictions is part of the necessary commitment and planning required to implement a successful field-collecting program.

Before leaving the subject of sampling protocols for plant collecting, I want to direct your attention to a highly comprehensive resource on this subject for gardens having significant research programs in plant conservation: a 2011 update of chapter 5: "Basic Sampling Strategies: Theory and Practice" in *Collecting Plant Genetic Diversity: Technical Guidelines* (1995) edited by Guarino et al.[44] Using the protocols and equations presented in this update may bring greater precision to the pursuit of research in genetic representation in plant conservation. For example, this is often more important for collecting infraspecific variation within a cultivated taxon for food and agriculture crops than for wild plants.[45]

I want to emphasize again the importance of finding opportunities to collaborate with other institutions having similar collecting needs. Cooperating institutions can often meet all of their needs at substantially reduced costs to each participant. Furthermore, consider partnerships or consortia with grassroots organizations, groups, and institutions in the target region with goals and missions appropriate to the purposes of the field-collecting program. If necessary, collaborate on training programs for local farmers, gardeners, horticulturists, or scientists. Botanical gardens may then rely on a network of trained, local people to locate and collect germplasm to be shared and forwarded to the garden. These may be valuable partnerships from the standpoint of economy, conservation, and the establishment of global cooperation.

As mentioned earlier, region-specific and multispecies trips should be accompanied by a list of target species to be collected from the region. Regardless of the type of field-collecting trip, determine the scope of work to take place in the field as collections are made: seed and/or vegetative collections, herbarium vouchers, photographic documentation, and so forth. From this, a more accurate itinerary may be constructed to govern travel and time spent at each locale. Plan to monitor and adjust this itinerary if conditions dictate changes. Have your lists of target species and your itinerary for any given region checked by local collaborators well in advance of traveling.

Depending upon the length of the collecting trip, the most expensive part will likely be the field time as opposed to the initial transportation cost.[46] In some cases, several trips to the same collecting area may be more economic in overcoming on-site challenges and making successful collections. Preliminary trips to confirm plans and collecting opportunities may also be economically justified to streamline expensive field-collecting time. In any case, when it comes to trip expenses and budgeting, try to arrange to have as much of the expense prepaid as possible. This will help eliminate renegotiated, and usually higher, costs during the trip as well as the need to carry large sums of money while traveling.[47]

Field-collecting plans, whether of national or international scope, should take into consideration all the necessary permits. These may be classified into three categories: entry permits, movement permits, and CBD access and benefit-sharing agreements. The entry permits relate to access to a site and collecting germplasm there. The movement permits relate to transporting germplasm from the site of collection, particularly across borders. The CBD agreements, in the form of prior informed consent (PIC) and/or mutually agreed terms (MAT) may be mandated by the countries you visit or by agreements that you put in place as a participating CBD institution. Regarding the two types of permits, they may be required domestically as well as internationally.[48] Some of these may be in the form of official, governmental, or agency-issued permits, and others may simply be a verbal authorization from landowners.

Plan ahead and allow for what may be long waiting periods to receive the necessary permits. Be aware of multiple jurisdictional authority over any given collecting location and consider reaching out to their representatives in the field as well as those in permitting offices. Field personnel may be able to provide more specific and helpful field-collecting locality and other kinds of information. This is another reason for working with in-country collaborators for both national field collections in unfamiliar locations as well as international collecting.

Send an official cover letter with your permit application(s) to underscore the scientific and conservation purpose of your collecting as well as your guiding CBD-inspired collecting ethics. Also, include a copy of your desiderata, or an amended facsimile, so that the permitting agent may see what you intend to collect and why. In any case, because of the nature of a garden's use of, and work with, field collections, strive to obtain permits without future restrictions on distribution. Acceptable restrictions may come in the form of access and benefit-sharing (ABS) stipulations within the CBD, as mentioned earlier. Regarding private landowners, their requirement for granting permission to collect plants is generally not as rigorous and exacting as governmental agencies. Still, I would treat them in a similar fashion as far as making sure they are completely informed. Your tact may vary slightly in that private landowners may be considered partners in the field-collecting project more so than agencies, and they, or their representative, may choose to join you in the field. A good example of such a partner is The Nature Conservancy.[49]

Movement-type permits authorize collected plant material to be transported across state and international boundaries. An example many people are familiar with involves the transport of certain agricultural products across state or regional borders where there are pest management concerns and quarantine restrictions. This underscores the fact that the required permits likely correspond very specifically to the type of plant and plant material that is being transported. For instance, field-collected orchids are tightly controlled due to serious conservation concerns. As far as plant material goes, cleaned seeds are of much less concern for all kinds of plants than vegetative material. When it comes to international imports of field-collected plants and plant parts, inadequate permitting may have serious legal consequences.

When it comes to export permits for international field collecting, there are no stereotypical or consistent rules and qualifications; each country has its own requirements. Be sure, then, to investigate the regulations of each country you intend to visit. A simple online query: "how to export plants out of China," produced the document, *Import and Export Animal and Plant Quarantine Regulations of the People's Republic of China*. This document was dated 1982 and should be validated before using it as an authoritative guide. Also, review the CITES designations for the plants on your desiderata that may be found in the country of export.

Once you have identified the necessary export requirements for the country where you are collecting, you will need to also identify the import permits required by your home country. In the United States, the necessary information is provided by the U.S. Department of Agriculture (USDA), Animal and Plant Health Inspection Service (APHIS) under plant health and permits. Institutions in the United States often use the USDA small seed lot permit because it allows for multiple collections of small seed lots of approved taxa to be imported without phytosanitary certificates provided they are adequately cleaned. Other countries may offer a similar type of import permit. If, however, you are collecting seed following the Center for Plant Conservation protocols outlined earlier in this section, you will need to obtain a special permit to accommodate collecting greater numbers of seed and provide the necessary documentation to justify that request. Vegetative material will require special permits and may be intercepted at points of entry (usually airports) for sanitary inspections. This necessity may be prearranged with authorities so as to expedite the process and minimize the threat of mortality to your collections. In most cases, seed and other material samples will each require a special identifying label showing import authorization.

Regarding Convention on Biological Diversity agreements between botanical gardens undertaking international field collections and the provider (host) countries and institutions, consider the following:

> For CBD and Nagoya Protocol compliance, botanic gardens need to obtain PIC (if required) and establish MAT with providers. Parties to the Protocol will be issuing internationally-recognised certificates to serve as evidence of users gaining PIC and establishing MAT, and their checkpoints will expect to see evidence of those certificates at later stages.[50]

The requirements for prior informed consent (PIC) and mutually agreed terms (MAT) granting access to "genetic resources" may take different forms, such as a general memorandum of understanding stipulating how the products of your field collections will be shared. The sharing of "results," in the name of the CBD mandate for access and sharing of benefits (ABS), may include the return of plants and propagules to provider institutions; the exchange of knowledge gained about the propagation, preservation, and conservation of subject plants; and educational exchanges of personnel. Some of this exchange may take place under the aegis of the International Plant Exchange Network (IPEN), which provides a system to facilitate the exchange of living plants for noncommercial purposes. These ABS-based relationships and exchanges must be well documented, starting with their initiation at the point of collections building and field collecting. If you are engaged in collecting plant genetic resources for food and agriculture, you must be aware of the requirements outlined in the International Treaty on Plant Genetic Resources for Food and Agriculture adopted by the Food and Agriculture Organization (FAO) of the United Nations in 2001.[51] As you can see, obtaining permits, permissions, and agreements for collecting is an integral part of field trip planning and organization.

Here is a sobering case of the permit requirements for collecting *Hudsonia montana* from the vicinity of the Blue Ridge Parkway in North Carolina. You would need to obtain three different permits. First, you would need a general collecting permit from the U.S. Forest Service as the landowner where the plant is found. You would also need a federal permit from the U.S. Fish and Wildlife Service for collection of an endangered species protected under the U.S. Endangered Species Act. Finally, you would need to obtain a conservation permit from the state of North Carolina, which has a rare-plant conservation law.[52]

An additional twist to this scenario could be added if you expected to export collected material of *Hudsonia montana*. The material would have to be certified as disease free by the USDA. As an endangered plant, it would require additional special permits for export. CITES regulates the export and import of wild species thought to be threatened by international trade. CITES has developed and updates appendix lists of plants that are assigned varying levels of endangerment and protection. Botanical gardens conducting routine field collecting and import/export of plants likely to be listed by CITES should be registered with the secretariat by their governments. These institutions may then exchange herbarium specimens and living plants without first obtaining CITES permits.[53] So, pack your permits—botanical gardens in particular should exemplify a lawful and ethical approach to plant collecting.

There are a number of supplies that you will need to acquire and pack for field collecting. They may be categorized as follows: documentation, propagule collecting, herbarium vouchers, and personal items. Special collections, such as DNA samples, *in vitro* samples, and symbiont organism samples, may require an additional category of supplies. Here is a synopsis describing each of these categories from *The Arnold Arboretum Expedition Toolkit*:

> Field documentation supplies relate to anything needed for the written documents associated with each collection made during the trip. You should have ample collection forms, notebooks for journaling, writing utensils, and any technology including a GPS, camera, and laptop.
>
> Herbarium voucher collection supplies relate to everything necessary to collect and press vouchers. For certain trips, this may also include materials necessary for collecting silica-dried leaf material for later DNA extraction.
>
> Propagule Collection supplies relate to collecting any seed, seedling, or cutting material. This also includes labels for seed bags, as well any special tools used to process seed cleaning such as sieves, bowls, and plates during the expedition.

Personal supplies are those each expedition member will provide to remain comfortable, hale, and hearty. Note that the expedition should supply a first aid kit and other items such as bug spray or sun screen.[54]

I refer you to the *Expedition Toolkit* for a more complete list of field-collecting supplies.

There are a number of ancillary and personal preparations to take note of and complete before embarking on any collecting trips, especially international ones. All participants should be aware of and receive any medical treatments and immunizations that may be required. Also, make sure any insurance requirements are in place and updated. If necessary, also make sure that all team members have a valid passport and visa. Finally, and too often underappreciated, all team members on international trips should apprise themselves of the histories, customs, and rudiments of the language(s) of the people of the countries they will visit.

Fieldwork

After a great deal of thoughtful planning and organization, you have at last arrived at the long-anticipated field-collecting site. There is great enthusiasm and eagerness to get outside to explore, observe, and collect. As difficult as it may be, this is a good time to pause and take stock of your surroundings, team members, and gear. On international trips, take a full day, or even two, to become oriented to where you are and get to know any team members whose company you share for the first time. Also keep in mind that jetlag may hamper the team's ability to start immediately with the necessary attention to detail. Use this time to review collecting goals, time lines, the attributes of target collecting areas, and any new information provided by local collaborators. Also, you may now need to reorganize your supplies to be more accommodating to the upcoming collecting itinerary. All in all, this is the time to begin to develop a multifaceted situational awareness.

Before embarking on the first day's field collecting, begin by reminding the team that your prime directive is to do no harm to the habitat or target populations and plants. Then begin thinking about specifics with an important review of verifying landmarks, search imagery (field indicators of the target species' presence), and the location of collecting sites. Once again, corroborate your field-collecting plans with local authorities and other people who can help verify geographic and logistical information. Something else, although a great deal of research has probably been poured into identifying and characterizing both the target plants and their habitats, it still behooves all field personnel to be alert to unexpected finds in off-target areas. These occasions not only assist in making the necessary collections but also contribute greatly to the collective understanding of that taxon. Any opportunity to contribute to our precise knowledge of the habitat, whereabouts, and other details of the ecology of a taxon is priceless information. In addition, and perhaps most important, these incidents must be carefully documented.

> Also, be aware of certain pitfalls. If you see a target near a temple or sacred ground, always seek permission before collecting. Heed the advice of your local guide in navigating cultural situations. If in doubt, always err on the side of not collecting. The ethically aware collector always knows the limits of their collecting permits, and does not try to bend the law.[55]

Try to establish and follow a collecting routine in the field so as to maximize your productivity and continuity, as well as to avoid confusion. For instance, amid the excitement brought on by successful scouting, it's natural to want to begin collecting immediately when target plants are identified. Such decisions may be premature if the area has yet to be thoroughly inspected to determine the size of the population and the availability of propagules. It is often best to first mark or flag initial finds to make these important determinations before making collections. As a general rule for the sake of thoroughness, once a field collection has begun, complete all the necessary steps for that taxon before moving on to any new collections. This means being very disciplined about all collections being completed as a team and not engaging in individual collecting among team members, which could lead to chaos.

This is a good point in our review of work in the field to consider the potential utility of drones: unmanned aerial vehicles (UAVs) or small unmanned aircraft systems (SUASs). A drone may be of great help in surveying difficult terrain and locating hard-to-find plants—as long as they have access to GPS satellite communications. Drone technology is progressing at a rapid pace leading to drones of great functionality at reasonable cost. Fortunately, that functionality also comes at a lighter weight, which is an important consideration when trudging over harsh terrain and through dense vegetation. The staff at the National Tropical Botanical Garden has employed drones to locate endangered plants in difficult terrain using the following approach:

> [W]e have employed a new system in which a [drone] pilot and observer (with spotting scope) position themselves across the valley from the plants. With an improved vantage point, the observer directs the UAV into areas where plants

occur to collect photos and GPS points. Once the photos are post-processed and mapped, the data can be used to guide seed collection of the species.[56]

In this case, the observers already possessed general information as to where the plants were located, but drones may also be used to survey and locate populations of hidden plants and fruiting individuals. By the date of this printing, it may be possible to outfit and program drones to collect propagules of difficult-to-reach plants. Obviously, this will require a collecting team member with precise skills in operating a drone and its accessories (GPS unit, camera, etc.).

I referred to the utility of GPS (global positioning systems) above, and I want to emphasize how important it may be to collectors in the field. Both the increased availability and ease of use of GIS (geographic information systems) and GPS can make them highly useful, integrated tools in the field if you have access to devices that use them. GIS programs using GPS data will allow collectors to do ecogeographic surveying that will assist in locating areas that are likely to contain taxa of interest as well as map those areas. These tools will combine with helpful locating and mapping programs available online from the Committee on Earth Observation Satellites (CEOS) International Directory Network[57] and DIVA-GIS,[58] a free computer-mapping and geographic data analysis program. Ecogeographical information in GIS layer format is becoming more readily available, and this should prove very helpful to field collecting. GIS use in plant collecting has been most prominent in the search for and studies of crop wild relatives (CWR) and advanced by CWR researchers in government agencies and universities.

I think it is safe to assume that most of your fieldwork will involve seed collecting, since it allows sampling greater genetic diversity and the propagation of a larger number of plants. This means that your scouting and collecting protocols will involve the seed collecting guidelines presented in the previous section on planning. That requires that you determine the size of the plant population and the number of maternal, or fruit-bearing, individuals present. You will then need to determine the seed set by extracting or cleaning some of the fruit and performing a cut test to examine the contents and health of a seed to assess its viability. Prior to packaging seeds, a float test may also be done.

Field-collecting and handling methods are determined by the type of fruit and the storage behavior of seeds. Trees often present the greatest collecting challenges because their fruits may be nearly inaccessible in the canopy. There are four general methods for collecting tree fruits: (1) collecting fallen fruit; (2) removing fruits from the canopy using ground access tools such as pole saws, rifles, and so forth; (3) climbing into the canopy; and (4) accessing the canopy via drone or helicopter.

Collecting fallen fruits is generally not recommended due to uncertainties and risks regarding source, physiological quality, and contamination. However, there are some circumstances where ground collecting is warranted. It is certainly a cheaper, faster, low-risk method, particularly if you are collecting bulk provenance seed as opposed to plant-specific collections. Ground collections may also be justified for rain forests where species are sparsely scattered throughout the community, fruits are hard to see in the canopy, and fully mature fruits must be collected.[59]

To obtain seeds in the canopy from the ground, the following methods may be applied:

- Beating and shaking: useful for trees with weakly attached, mature fruits, for example, legumes (must be well timed)
- Sawing small branches with long-handled or flexible saws
- Shooting down branches with rifles[60]

You should be well trained in tree-climbing techniques and safety before attempting to climb into the canopy to collect fruits. While scouting, generally speaking, dry fruits may be collected and transported in paper bags. However, those that release tiny seeds, such as members of the *Ericaceae*, may be cleaned when collected and the microscopic seeds stored in seamless paper or wax envelopes, or small glass vials. Fleshy fruits may be collected and transported in the field using sealed plastic bags. Here is a useful summary on collecting and handling seeds in the field from *Collecting Plant Genetic Diversity*:

All Seeds

- Avoid damaged seeds and collect them at a maturity that will optimize their drying tolerance.
- Minimize the potential for damage by hand-cleaning seeds.
- Avoid potentially damaging pest control treatments if it is legal to do so.
- Personally insure that seed arrives at the garden without delay.

Recalcitrant Seeds (Exceptional Species)

- Collect close to fruit fall but do not collect from the ground unless recently dispersed.
- Keep seeds aerated and moist in inflated polythene bags. Aerate on a weekly basis by deflating and inflating bags.
- Do not allow seeds collected in the tropics to cool below 20°C or rise above ambient shade temperatures.

- Do not allow more than one month to pass between collection and receipt of seed at the garden.

Orthodox Seeds

- Check meteorological conditions and viability data to determine if seed lots will require active drying with silica gel.
- If possible, keep fleshy fruited seeds in the fruits and the fruits aerated at ambient temperatures.
- Air dry hand-cleaned seeds in a thin layer under shade for three days or more before packing.[61]

The Royal Botanic Gardens, Kew, has established an online version of their Compendium of Information on Seed Storage Behavior: The Seed Information Database.[62] This searchable resource, part of Kew's Millennium Seed Bank project, contains information on the seed storage behavior (recalcitrant, intermediate, orthodox) of an exceptionally large number of taxa from around the world. Searching this database with the names of plants from your desiderata should provide you with some helpful seed-collecting and storage information to assist in planning and conducting your work in the field. For taxa not listed in the Seed Information Database and of unknown seed storage tolerance, there are published predictive models for determining if seeds are likely to be recalcitrant (the most difficult status) and other useful scholarship on this subject under the authorship of M. I. Dawes.

Although I have listed some general recommendations above from Guarino et al. for handling seeds of various types, keep in the mind that many taxa will demonstrate an intermediate response to drying and storage. It has been shown that there may be a continuum of tolerance within an individual species based on its provenance within its range.[63]

Because of the weight of fleshy fruits and their tendency to liquefy and ferment, some should probably be cleaned as soon as possible despite the recommendation above. Large, dry dehiscent fruits with loosely contained seeds may be easily cleaned in the field. Other dry fruits may present so little additional weight in proportion to the seed that it may be easiest to clean these after transport back to the garden. One other factor that impacts the decision to clean seeds in the field is quarantine restrictions. Field cleaning should be done manually and carefully to minimize the impact on seed viability. If the environmental conditions are acceptable, dry fruits may be stored during the trip in permeable containers such as cotton or paper bags transported in a way to facilitate air circulation. Seeds from some fleshy fruits may be cleaned, spread to air dry, and stored in the same manner. Seeds that should be dried below equilibrium moisture content may require treatment with silica gel or other desiccant. In doing so, collectors must have some idea of how much desiccant will be required per seed lot to dry the seeds. A standard recommendation calls for seed and silica gel in a 3:2 ratio by weight. The seeds and the gel should be packed in a sealed plastic container allowing for maximum contact and minimum air volume.[64] Curators and field botanists at the Royal Botanic Garden Edinburgh fold their own seed packets to great success using greaseproof paper and following a reliable origami-type approach.[65]

As was mentioned earlier, seed-handling methods are determined by the seed storage behavior of each species: recalcitrant (exceptional species intolerant of drying) or orthodox (tolerant of drying). As was also mentioned earlier, keep in mind that some species are intermediate between these categories and should, as a conservative approach, be treated as recalcitrant.

Collecting vegetative material of some species may be the only viable means available for collectors. In order to collect genetic diversity, and because vegetative material will be clonal (same genotype), collect from spatially separated individuals with different appearances.[66] Vegetative material may be easily damaged and will deteriorate rapidly after being collected. Species may be collected vegetatively as root cuttings, shoot cuttings, *in vitro* meristems, other perennating organs, or seedlings. In regard to collecting roots and tubers grown as food crops, I refer you to the recommendations in Dansi's "Chapter 21: Collecting Vegetatively Propagated Crops (Especially Roots and Tubers)" in *Collecting Plant Genetic Diversity: Technical Guidelines—2011 Update.*[67]

Root cuttings may be collected and stored in plastic bags containing a moist, sterile medium such as sphagnum moss. Large segments should be collected, which may be trimmed for propagation at a later date. Shoot cuttings should have the leaf area reduced by 75 percent or more, be soaked for several minutes in a solution of soap and 1 percent bleach, and/or dipped in a pesticide to reduce contamination. The cut ends of the shoots may then be sealed with melted paraffin and the whole sealed in plastic bags. Still, root and shoot cuttings will remain viable for a matter of days, not weeks or months, and may require storage in insulated containers such as coolers.

Loss of viability and excessive weight may severely limit a collector's success in obtaining viable vegetative propagules and recalcitrant seeds. An *in vitro* collecting method may be called for under these conditions. If this method is anticipated, one of the members of the collecting team should have training and practice in successfully establishing vegetative material *in vitro*. It will be of particular importance that the *in vitro* plant material be successfully established without the threat of contamination. Therefore, it will be helpful for the responsible team member(s) to have practice in establishing sterile *in vitro* cultures and know what will be required in the field. The necessary materials and supplies must also be identified and acquired. The major considerations are the following:

1. Select the appropriate material—one that will withstand sterilization (shoot tissue, embryos, seeds).
2. Trim it to size—remove damaged parts.
3. Perform surface cleaning to remove soil and pests.
4. Perform surface sterilization of tissues, containers, and tools.
5. Remove sterilant by washing with sterile water.
6. Use plastic containers and minimal, sustaining media with anti-contaminant.
7. Transport in a strong, rigid, and insulating transport container.[68]

The works of V. C. Pence et al. on the subject of *in vitro* field collecting also serve as a useful guide.

Not surprisingly, collections of seedlings and young plants must be well cared for and accommodated, making them unsuitable for acquisition on long collecting trips unless they can be promptly returned to the garden or an accommodating host institution.

Any of the propagules discussed above may come from a plant that can only be successfully grown in a symbiotic relationship with bacteria or fungi. Such a symbiosis most often has a direct impact on the nutritional requirements of the host plants. Success in reproducing the collected propagules and growing on whole plants may be dependent upon collecting and providing the appropriate microsymbiont partner.[69] I'm sure you are familiar with any number of leguminous plants that grow in partnership with a rhizobium that assists in fixing nitrogen used by the host plants. Also, terrestrial and other orchids germinate and grow in association with mycorrhizal fungi that assists in nourishing them. The most important symbiotic partners include the following organisms:

- the root-nodule bacteria or rhizobia that form nitrogen-fixing symbioses with legumes, currently including 50 species distributed among the genera *Rhizobium*, *Ensifer*, *Mesorhizobium*, *Azorhizobium*, and *Bradyrhizobium*.
- *Burkholderia*, which has only recently been recognized as a potential nitrogen-fixing symbiont of legumes, especially mimosa.
- the vesicular arbuscular mycorrhiza (VAM) for phosphorus supply in many plants.
- the actinorhizal associations (*Frankia*) for nitrogen supply in about 200 species, including forages and forestry species such as *Alnus*, *Allocasuarina*, *Elaeagnus*, *Hippophae*, *Purshia*, and *Shepherdia*.

- the ectomycorrhiza (more than 1,000 species of *Basidiomycetes* and *Ascomycetes*, mainly the former) for water and nutrient uptake in many forest species in the families *Pinaceae*, *Betulaceae*, *Salicaceae*, *Myrtaceae*, *Casuarinaceae*, and some *Caesalpiniaceae* and *Diptero-carpaceae*.[70]

In the past, the most typical way of inoculating propagules and whole plants with a microsymbiont involved the incorporation of the field soil from the collecting site during propagation and growing on of the host plant. Now we can add to that protocol the use of microsymbionts from artificial media and enriched cultures.

After determining if the taxa on your desiderata are dependent upon a microsymbiont, check to see if that organism(s) is commercially available. If not, then consider adding to your collecting plans and *permits* the acquisition of the necessary organism(s). It is possible that there is information among the references already cited in this section of collections made of germplasm and their microsymbionts from the taxon you will be seeking. This information may also contain references to the manner in which the microsymbionts were collected and stored in the field. Obviously, you should plan to prospect for microsymbionts in the vicinity of their host plants.

A significant disparity may arise in that the microsymbiont of interest may only be available, or active, other than the time host plant propagules are available. For instance, it may be necessary to collect rhizobium nodules when they are seasonally active and ectomycorrhizal fungi in the spring. This may necessitate two collecting trips or dependence upon a regional collaborator for timely collections. Also, be prepared for a need to isolate and store microsymbionts separately in response to quarantine and other restrictions in the host country before import home. There is a great deal of information on field sampling of microsymbiont organisms in field ecology literature. Begin with a review of "Chapter 26: Collecting Symbiotic Bacteria and Fungi," in *Collecting Plant Genetic Diversity: Technical Guidelines—2011 Update*.[71]

Field Documentation

For a consideration of field documentation, we must back up in the process of field collecting to the point when a determination is made to collect from a population of plants. It is at this point that specimen documentation begins with the recorder assigning a unique collection number to the collected germplasm and any other associated items (herbarium vouchers, photographs, etc.). It is critical that this process keep pace with field collecting and proceed in an orderly fashion. Collection forms, both digital and hard copy, are critical for specifying and organizing documentation. A member(s) of the team should be responsible for taking photographs at each sampling location of the area, the habitat, segments of the plant population, and the variation among sampled plants, including various elements of individual plant anatomy. These images must be identified with the name of the plants and the collection number.

What may be a critical part of field documentation, as well as meeting target 9 of the Global Strategy for Plant Conservation, depending upon where in the world you are collecting and who your collaborators are, is documenting ethnobotanical information on the plants of interest with the knowledge of the indigenous people. To define traditional knowledge documentation, I prefer the definition from the useful reference, *Documenting Traditional Knowledge: A Toolkit*: "TK documentation is primarily a process in which TK is identified, collected, organized, registered or recorded in some way, as a means to dynamically maintain, manage, use, disseminate and/or protect TK according to specific goals."[72] To underscore the importance of indigenous knowledge to field collecting, I turn to Guarino et al. in *Collecting Plant Genetic Diversity*:

- Locating target areas and material.
- Deciding what to collect and how.
- Documenting the collection.
- Assessing the completeness of collections.
- Understanding the origin and distribution of diversity, and the rules of access to it.
- Assessing the extent and threat of genetic erosion.[73]

Documenting indigenous knowledge may be accomplished in either of two ways: (1) *in situ* through special interview techniques with indigenous people or (2) through research with secondary sources such as literature searches for preexisting documentation, interviews of other researchers, and such. Regarding option 1, on-site interviews take the form of a participatory indigenous knowledge journal and are undertaken when preliminary investigations indicate no such documentation exists. The traditional knowledge journal fulfills the following purposes:

> Before scientists can use the knowledge of communities that have oral traditions, they need to assist the community in documenting this knowledge in their own language for their own use. Then the community's knowledge may only be used with a citation. This process avoids the one-sided approach where traditional knowledge (TK) is studied without any benefit to the community.

The TKJ methodology was developed on the premise that farmers and community members have knowledge that is valid for scientific use and that empowering farmers to do their own documentation addresses the problem of knowledge erosion and enables scientists to have access to the information.[74]

In addition, a TKJ also promotes *in situ* conservation by generating a greater interest in preserving the plant diversity of the region on the part of the indigenous people. It also validates their knowledge of and relationship with the regional flora strengthening that lineage for future generations. Finally, this process facilitates an agreement bestowing prior informed consent (PIC) for plant collecting that may stretch into the future. There are special skills required on the part of anyone working with indigenous groups to complete a TK journal project. So much so, that any field collecting and documentation project in areas where little information on the traditional knowledge of indigenous peoples exists may be best undertaken first by specialists trained in the necessary interaction and relationship building.

Regarding option 2 for documenting indigenous knowledge—research—this presumes that a particular group of indigenous people living in a targeted plant-collecting area have already been studied and their customs and ways of living and using plants are documented. In taking this approach, literature searches are the best place to start seeking indigenous knowledge in advance of collecting trips. Anthropological and ethnographic literature will be good sources. Herbaria that anchor a flora for the targeted region may have helpful ethnobotanical information filed and included on specimen labels. The reports generated by crop germplasm collectors over the years also contain ethnobotanical descriptions and citations. Don't forget what may be termed the "useful plant" literature, a description derived from the common title wording of many of these works, as well as titles on edible and medicinal plants. Bibliographic databases can be very useful using some of the key words mentioned above as well as current researchers in the disciplines. Finally, a simple online query brought up the *Center for Indigenous Knowledge for Agriculture and Rural Development* list of indigenous knowledge resource centers.

The question one might ask is why we should be concerned about the traditional knowledge of the community when what we are really interested in is data for the conservation of the materials. The answer is that an understanding of the community's knowledge will provide an understanding of the conservation of plant genetic resources in situ.[75]

In addition to the purpose described in this quote, gathering ethnobotanical information on a collected plant may be useful to the collectors for the successful propagation and cultivation of a plant(s) as well as its use.

Having done your indigenous knowledge homework, when you reach your collecting destination you are ready to seek a cooperative partnership with the indigenous people toward obtaining a prior informed consent (PIC) agreement containing mutually agreed terms (MAT) and an access and benefit-sharing (ABS) government permit. This contact and arrangement may best be handled by a local authority who can expedite the necessary arrangements with the best interests of everyone involved in mind. The resulting documentation becomes an important part of the overall field-collecting record along with your research file of indigenous knowledge collected through prior office work. I will discuss field documentation in more detail in the next chapter on documenting collections. On the subject of indigenous and local partners, ensure that their contributions are not simply documented but also fully credited, and that they share in the accrued benefits from the collected materials and information.

Representative herbarium vouchers should accompany field collections and be identified by the same collection number. An essential reference on herbaria and herbarium specimens is *The Herbarium Handbook* (third edition) by Bridson and Forman.[76] Herbarium vouchers are used for the following purposes:

- To identify collected taxa.
- To demonstrate the range of variation in the population sampled.
- To determine conservation status.
- To identify pests and problems.
- To document the flora of a region.
- DNA samples.

Collectors should identify in advance of their fieldwork those characters that should be included in herbarium specimens of their desiderata. *The Herbarium Handbook* provides some helpful guidance on characteristic sampling on a family-by-family basis. Knowing this, the team member charged with vouchering collections should then take care to collect at least three duplicate specimens that are representative of the sampled population. The choice of duplicate specimens should represent the phenotypic variation within the sampled population, mirroring the same choice of seed lots and other propagules collected. It is possible that multiple specimens will be required to capture both the vegetative and reproductive parts of large-scale plants. I would also advise you to collect extra samples of critical material for identification, such as flowers.

Specimens should be pressed in the field and dried by use of the appropriate field press materials or the Schweinfurth alcohol method. When positioning specimens for drying, be sure that all the identifying characteristics are visible, including the backs of leaves. Some flowers and/or fruits may be bisected to show their internal anatomy (place a nonstick covering on them when being pressed and dried). Specimens from succulent plants will likely need to be carefully dissected or thinly sliced in order to remove the internal fleshy portions while retaining the epidermis and its characteristic identification features for pressing. Very large and bulky specimens, such as palm foliage and fruit, may require drying and storage in boxes. Aquatic specimens that lose their shape out of water may be floated into position on pressing paper and then dried in a press as for other specimens. Make sure that the prepared specimens are clearly labeled with the correct collection numbers before they go into the press by attaching a label to the sample or to the drying sheet upon which it is mounted. Pencil or indelible ink markers are best. Dried specimens may then be removed from the press, bundled, or, in humid climates, packaged in plastic bags and shipped in tightly packed cardboard cartons containing naphthalene or paradichlorobenzene and padded with extra paper. Often, tissues are collected to extract molecular genetic markers important to conservation biology and biodiversity research. How to preserve these tissues both in the field and when taken from *ex situ* collections in the garden is described in chapters 5 and 10.

DNA sampling has become an increasingly important part of field collecting and much of the research stemming from these collections. If DNA samples will be required from your field collections, you will need information on which types of collections (e.g., standard herbarium samples) and which specific organs yield the best quality tissue for DNA sampling. There are two very good references to help you get started with DNA sampling: the Global Genome Biodiversity Network[77] and the article "Organizing Specimen and Tissue Preservation in the Field for Subsequent Molecular Analyses," by Gemeinholzer et al.[78]

At the end of each day, the collecting team will return to base with containers full of fruits, seeds, and perhaps vegetative material. However, that day's work isn't finished; the evenings are a good time to clean, organize, and pack some of your collections as well as double-check the documentation and the contents of your field-collecting digital repository and/or hard-copy files and notebooks. Make sure that all collections and supporting materials are clearly labeled with the corresponding collection numbers. This is also a good time to review identifications since you may have left bulky references behind for the day's fieldwork, and if you have good internet access, you may also use that resource for verifications. Some of your collections may be more thoroughly cleaned for shipment home, especially any dry fruits that have dehisced to easily release their seeds. Fleshy fruits that will require maceration for cleaning may be moistened in sealed plastic bags or other containers to hasten that process. Remove the husks of nuts and examine them closely for signs of predation separating out the viable ones for packing in breathable bags or other containers. Finally, ready your field materials for the next day's work: desiderata, maps, microhabitat information, and so forth.

As a reminder, field documentation of wild collections will be considered in the next chapter.

Textbox 3.6.   Recommendations on Field Collecting

Basic

- The purpose of field collecting for building plant collections is made clear by the plant collections policy and plan of the institution.
- International, interstate, interprovincial, or interregional collections of plants and vegetative propagules must meet Convention on International Trade in Endangered Species of Wild Fauna and Flora (CITES), Convention on Biological Diversity (CBD), sanitary, and other requirements for export, import, and transport across regulatory boundaries.
- The garden's field-collecting activities are focused by a desiderata of collection needs.
- Garden field-collecting trips are organized with technical and logistical collecting plans.
- Representative herbarium vouchers should accompany field collections.
- Field-collecting activities are recorded in a log.
- Wild-collected plants and propagules are quarantined for biosecurity inspection and monitoring upon receipt.

Plant Propagation

A program of plant propagation is critical to the renewal and augmentation of the plant collections. Planned, often cyclical programs to repropagate existing plants in the collection should be standard practice for botanical gardens. This process, including a method for establishing propagation priorities, should be articulated as an integral part of the collections management plan. To be most useful, these programs must be sensitive to gene pool maintenance, particularly the preservation of diversity and minimizing artificial founder effects and hybridization.[79]

To successfully cultivate field-collected propagules of unknown or unfamiliar plants, gardens should develop propagation protocols based upon taxonomic affiliation, ecology, propagation standards, and empirical research. Hybridization programs are an intense and expensive source of new plants justified only by a commitment to specific research goals. For more information on propagation programs, refer to chapter 5. For more information on collections management and hybridization programs, refer to chapter 12.

Personal Collections

Each botanical garden should include a statement of what constitutes personal collecting in the context of that institution. It should be kept in mind that acquiring, collecting, and possessing plants is not itself unethical, and may enhance professional and curatorial skills. Problems may arise when curators and other garden staff collect plants that are also part of the botanical garden collections. Garden employees should not compete with their employers for collections. Here are some other relevant considerations from the American Association of Museums (AAM):

- Museums should have the right, for a specified and limited period of time, to acquire any object purchased or collected by any staff member at the price paid by the employee, excluding objects acquired before the staff member's employment.
- Museum policies should specify what kind of objects staff members are permitted or forbidden to acquire, the manner of acquisition, and variations in this policy for different employees.
- Museum employees must inform their employers about all personal acquisitions, including the circumstances of those collections.
- Museum employees may not use their museum affiliation to promote their own or any associate's personal collecting activities.[80]

Biosecurity

The acquisition of plants by all the means described above are sure to come with issues of biosecurity. The import of wild-collected plants and the movement of plants in and out of botanical gardens makes them sites of particular concern for biosecurity. In a previous time, those issues revolved almost exclusively around pest management focused on insects, diseases, and, to a lesser extent, weeds. Now, however, and after years of import, our concern with weeds and invasive plants has become more acute. Research into invasive species over the last several decades shows that they are second only to habitat destruction as the most important cause of biodiversity loss. A vital part of curatorial practice and collection management is to ensure that the collections held by botanical gardens are not contaminated with invasive plants or insect and disease organisms that, in addition to harming the collections, may then spread to the surrounding environment. Although this subject is also discussed in both chapters 2 and 5, I raise it here to underscore the importance of prevention at the point of plant acquisition.

In consideration of plant acquisitions programs, there should also be a "firewall" of provisions put in place for the biosecurity of the existing plant collections. The first line of defense involves the initial research of taxa to be included on an acquisitions desiderata. This means identifying potentially invasive characteristics in the breeding or other aspects of the plant's biology and known insect and pathogen problems. These investigations may disqualify a taxon from the desiderata at the outset. At the very least, staff authorized to acquire plants may be alerted to possible biosecurity concerns that may be assessed at the point of acquisition (field sites, nurseries, etc.). Field collecting has been the principal topic of this chapter, and it is incumbent upon field collectors to be informed and qualified to identify invasive characteristics and to adequately clean, inspect, and prepare field-collected materials as a matter of biosecurity. These materials should be monitored during the field-collecting period for degradation and signs or symptoms of pests. In addition, some of these materials may be held for governmental inspection by both the exporting and importing countries, another layer of biosecurity before plants and propagules reach the garden.

In regard to gathering information on the potential pests of taxa on your collecting desiderata, there are many online databases and other references that can be helpful. There are two networks to assist curators and their staffs with biohazards involving biological competitors that threaten their collections. The Sentinel Plant Network,

. . . first launched in 2011, . . . is a collaboration between the American Public Gardens Association and the National Plant Diagnostic Network (NPDN). This partnership makes it possible to extend NPDN's diagnostic and "First Detector" training expertise to the diverse collections and public outreach programs of the more than 500 public gardens across the continent. The Sentinel Plant Network contributes to plant conservation by engaging public garden professionals, volunteers, and visitors in the detection and diagnosis of high consequence pests and pathogens.[81]

The International Plant Sentinel Network is being developed by Botanic Gardens Conservation International "to facilitate collaboration among institutes around the world, with a focus on linking botanic gardens and arboreta, National Plant Protection Organisations (NPPOs) and plant health scientists. The aim will be for these institutes to work together in order to provide an early warning system of new and emerging pest and pathogen risks."[82] Each of these networks will provide valuable plant protection early warnings and information to curators and their unexpected synergies may be invaluable in this regard.

The term *sentinel* refers to the large collections of plants held in botanical gardens and the knowledge held by their curators regarding the presence of organisms that have the potential to cause damage. The plant preservation programs, and resulting documentation, serve as an early warning for the presence of potential pest and disease problems heretofore unknown within natural populations of those plants or among plants grown in a particular area. Both of the biohazard networks cited above combine into a broad alert system among botanical gardens, research institutions, and government regulatory agencies. The Sentinel Plant Network can also greatly enhance the capacity of botanical garden staff in pest detection and management through various training and information programs.

Of course, anyone interested in the potential pest and disease problems associated with a particular plant may use their favorite internet browser to conduct a search for this information using the currently accepted name of the taxon. Also, the Center for Agriculture and Bioscience International (CABI) maintains distribution maps of pests and diseases, as well as descriptions of fungi, bacteria, and plant parasitic nematodes. Search "plant viruses" online to see information and databases on these organisms. Finally, the Consultative Group on International Agriculture Research (CGIAR) maintains databases on and links to pest and disease information for food and agricultural crop plants.

Nursery acquisitions may be of equal, and in some cases greater, importance as potentially invasive plants and of concern as points of insect pest and pathogen introduction. Use reliable, local nursery sources with a track record of clean nursery stock whenever possible. Even so, visit these facilities and conduct your own inspections of plants of interest to assess their health as well as overall quality. Among nursery plants, large stock is the greatest biosecurity risk. Plants that have been treated with pesticides should still be inspected as those treatments may mask the presence of pests. Reinspect shipments of plants and their documentation before accepting them and return (or destroy for reimbursement) any of suspicious quality.

We can only expect limited biosecurity protection from field inspections while plants are being collected such that more effective biosecurity procedures and facilities must be put in place. In some cases, this requirement may be imposed by the country or location of export when samples are deemed unsafe for export until placed in an "intermediate quarantine," monitored, and deemed pest free. This is most common with whole plants that may harbor nematodes and soilborne pathogens that are difficult to detect and treat. Other types of vegetative material may be infected with systemic pathogens that are also difficult to detect and treat. Once those samples are considered safe for transfer to the country of import, a new set of permits may be required. Speaking of documentation, all of this biosecurity activity with collected samples must be included in the field documentation for each numbered sample; this includes pest diagnostics, images, and perhaps samples (preserved in appropriate containers, etc.).

Plant quarantine involves the isolation, screening, and treatment of plants to prevent the introduction of pests, pathogens, and invasive species. New acquisitions should be received and initially housed in a designated quarantine area, which may be a closed area of the botanical garden, polyhouse, greenhouse, or a purpose-built facility. In any case, the facility should meet the following requirements:

- Secure and separate location, with regulated access for staff (and visitors), locked doors, barriers, and appropriate signs;
- Controlled drainage;
- Separate, dedicated tools and equipment;
- Plant inspection and diagnostic skills and equipment;
- Facilities for pesticide application;
- Facilities for cleaning and disinfection;
- Facilities for incineration and disposal of infected material; and
- Regular monitoring for pests and diseases; use of traps such as sticky traps, pheromone traps, light traps, and such.[83]

In addition, it is preferable that the quarantine facility be located so that new acquisitions may be received without exposure to the rest of the garden. An initial inspection will take place and be documented (see chapter 4) before plant materials are established for propagation or potted for further growth. These plants are then grown and monitored in the quarantine facility for a period of six weeks or more depending upon the results of monitoring and the overall health and growth rate of the individuals. Growth structures, such as tubers, rhizomes, and corms, as well as potentially invasive plants, may be kept in the facility longer in order to accommodate a full cycle of growth and inspection. Plants are released to the collection and/or for export from the garden only after passing inspection by the designated botanical garden authority.

Persons and support material entering or leaving the quarantine area must be disinfected. Personnel should disinfect their hands, footwear, equipment, and vehicles. Disinfection kits should be available at disinfection stations at entrances and exits to the quarantine facility. These will contain disinfecting sprays, sanitizers, and various applicators and tools for applying these materials. Protective clothing, such as gloves, face masks, and safety glasses should be made available as well.

A final word about invasive species: an assessment of invasive potential may be made through regular monitoring of plants in quarantine, particularly if they enter quarantine with some suspicion of invasiveness. Depending upon the nature of the observed invasive potential, such plants may be taken from the quarantine area and destroyed. On the other hand, they may be considered benign and released from quarantine, accessioned as part of the collection, and monitored as a routine part of collection field documentation without further evidence of invasiveness. If, however, invasive qualities are noted through regular collections documentation, these behaviors may need more specific assessment and documentation for a final determination of invasive qualities that may lead to deaccessioning and careful elimination. This scenario will come up again in chapters 4, 5, and 10.

The pursuit of new acquisitions—well planned and comprehensively undertaken—constitutes the infusion of life into the botanical garden. Botanical gardens must invest in the acquisition of new plants to ensure the viability of their collections for future generations. As noted in the AAM's *Museum Ethics*, "In the delicate area of acquisition . . . of museum objects, the museum must weigh carefully the interests of the public for which it holds the collection in trust, the donor's intent in the broadest sense, the interests of the scholarly and the cultural community, and the institution's own financial well-being."[84]

Notes

1. G. E. Burcaw, *Introduction to Museum Work* (Nashville, TN: American Association for State and Local History, 1984) 57.

2. Burcaw, *Introduction*, 48.

3. Botanic Gardens Conservation International, *IPEN Code of Conduct* (Richmond, Surrey, UK: BGCI, 2018).

4. American Public Gardens Association, *Collections Development Planning Guide*, https://www.publicgardens.org/Programs/Plant-Collection-Planning.

5. American Public Gardens Association, *Core Collections Primer and Collections Prioritization Worksheet*, https://www.publicgardens.org/resources/core-collections-focusing-and-prioritizing-living-collections.

6. Pamela Allenstein, personal communication, October 2021.

7. Michael S. Dosmann and K. Port, "The Art and Act of Acquisition," *Arnoldia* 73, no. 4 (2016): 2.

8. American Association of Museums, *Museum Ethics* (Washington, DC: AAM, 1978).

9. Alberta Museums Association, *Standard Practices Handbook for Museums* (Edmonton, AB: AMA, 1990), 96.

10. Denver Botanic Gardens, *Living Collections Management Plan* (Denver, CO: DBG, 2017), 5.

11. American Association of Museums, *Museum Ethics*, 96.

12. V. H. Heywood, "Role of Seed Lists in Botanic Gardens Today," in *Conservation of Threatened Plants*, ed. John Simonds (New York: Plenum Press, 1976), 225.

13. U.S. Department of Agriculture, "Plant Health (PPQ)," https://www.aphis.usda.gov/ppq/permits/plantproducts/nursery.html.

14. E. Leadlay, ed., *Darwin Manual* (Richmond, Surrey, UK: Botanic Gardens Conservation International, 1998).

15. J. Gratzfeld, ed., *From Idea to Realisation: BGCI's Manual on Planning, Developing and Managing Botanic Gardens* (Richmond, Surrey, UK: Botanic Gardens Conservation International, 2016), 60.

16. Leadlay, *Darwin Manual*, 228.

17. R. A. Howard, "Comments on 'Seed Lists,'" *Taxon* 13 (1964): 90–94.

18. Mark P. Widrlechner, personal communication with the author, March 1997.

19. American Public Gardens Association, "About the Plant Collections Network," https://www.publicgardens.org/programs/about-plant-collections-network.

20. Botanic Gardens Conservation International, *IPEN Code of Conduct*.

21. Center for Plant Conservation, "National Collection," https://saveplants.org/national-collection/.

22. Gratzfeld, *From Idea to Realisation*, 70.

23. Gratzfeld, *From Idea to Realisation*, 71.

24. Gratzfeld, *From Idea to Realisation*, 72.

25. Gratzfeld, *From Idea to Realisation*, 74.

26. L. Guarino et al., eds., *Collecting Plant Genetic Diversity* (Wallingford, Oxon, UK: CAB International, 1995), 32.

27. Diana J. Pritchard and Stuart R. Harrop, "A Re-evaluation of the Role of Ex Situ Conservation," *BG Journal* 7, no. 1 (2010): 4.

28. Pritchard and Harrop, "A Re-evaluation," 6.

29. Anthony S. Aiello, A. Gapinski, and K. Wang, "Collaboration across Continents and Cultures," *BG Journal* 16, no. 2 (2019): 30.

30. Guarino et al., *Collecting Plant*, 37.

31. Global Genomic Biodiversity Network, "About GGBN," https://wiki.ggbn .org/ggbn/About_GGBN.

32. Michael S. Dosmann and K. Port, "The Art and Act of Acquisition," *Arnoldia* 73, no. 4 (2016): 73.

33. Guarino et al., *Collecting Plant*, 38.

34. L. Guarino et al., eds., *Collecting Plant Genetic Diversity: Technical Guidelines—2011 Update*, https://www.bioversityinternational.org/e-library/publications/ detail/collecting-plant-genetic-diversity-technical-guidelines-2011-update/ (Rome, Italy: Bioversity International, 2011).

35. Arnold Arboretum, *The Arnold Arboretum Expedition Toolkit* (Jamaica Plain, MA: Arnold Arboretum, 2018), 8–24.

36. P. Bristol, "Collectors, Start Your Engines," in *Plant Exploration: Protocols for the Present, Concerns for the Future*, ed. James R. Ault (Glencoe, IL: Chicago Botanical Garden, 1999), 41.

37. Donald Falk and T. Holzinger, "Sampling Guidelines for Conservation of Endangered Plants," in *Genetics and Conservation of Rare Plants*, ed. Donald Falk and T. Holzinger (New York: Oxford University Press, 1991), 226.

38. Falk and Holzinger, "Sampling Guidelines," 227.

39. Center for Plant Conservation, *CPC Best Plant Conservation Practices to Support Species Survival in the Wild* (Escondido, CA: CPC, 2019), 3-6–3-7.

40. Center for Plant Conservation, *CPC Best Plant Conservation Practices*, 3-9–3-10.

41. Falk and Holzinger, "Sampling Guidelines," 227.

42. Falk and Holzinger, "Sampling Guidelines," 231–32.

43. Guarino et al., *Collecting Plant*, 40–41.

44. Guarino et al., *Collecting Plant Genetic Diversity: Technical Guidelines—2011 Update*.

45. K. Hammer and Y. Morimoto, "Chapter 7: Classifications of Infraspecific Variation in Crop Plants," in *Collecting Plant Genetic Diversity: Technical Guidelines—2011 Update*, ed. L. Guarino et al. (Rome, Italy: Bioversity International, 2011), 1.

46. B. Yinger, "Objectives and Funding of Ornamental Plant Explorations," *Longwood Graduate Program Seminars* 16 (1984): 29.

47. Bristol, "Collectors, Start Your Engines," 40.

48. Arnold Arboretum, *Toolkit*, 19.

49. Arnold Arboretum, *Toolkit*, 22.

50. Gratzfeld, *From Idea to Realisation*, 77.

51. United Nations Food and Agriculture Organization, "International Treaty on Plant Genetic Resources for Food and Agriculture," http://www.fao.org/plant-treaty/en/.

52. L. R. McMahan, "Advice for the Modern Plant Explorer: Pack Your Permits," *Public Garden* 6, no. 4 (1991): 12.

53. F. Campbell, "What Every Public Garden Should Know about CITES," *Public Garden* 6, no. 4 (1991): 18.

54. Arnold Arboretum, *Toolkit*, 76.

55. Arnold Arboretum, *Toolkit*, 37.

56. B. Nyberg, "Eyes in the Sky: Drones Proving Their Value in Plant Conservation," *BG Journal* 16, no. 2 (2019): 27.

57. Committee on Earth Observation Satellites, "International Directory Network," https://idn.ceos.org.

58. DIVA-GIS, "DIVA-GIS," http://www.diva-gis.org.

59. Guarino et al., *Collecting Plant*, 494.

60. Guarino et al., *Collecting Plant*, 498.

61. Guarino et al., *Collecting Plant*, 451–52.

62. Royal Botanic Gardens, Kew, Seed Information Database, https://data.kew.org/sid/.

63. M. I. Daws et al., "Developmental Heat Sum Influences Recalcitrant Seed Traits in *Aesculus hippocastanum* across Europe," *New Phytologist* 162, no. 1 (2004): 164.

64. Guarino et al., *Collecting Plant*, 438.

65. Sadie Barber and S. Scott, "Botanical Envelopes," *Sibbaldia* 7 (2009): 197.

66. Center for Plant Conservation, *CPC Best Plant Conservation Practices*, 2-8.

67. A. Dansi, "Chapter 21: Collecting Vegetatively Propagated Crops (Especially Roots and Tubers)," in *Collecting Plant Genetic Diversity: Technical Guidelines—2011 Update*, ed. L. Guarino et al. (Rome, Italy: Bioversity International, 2011).

68. Guarino et al., *Collecting Plant*, 514–15.

69. Center for Plant Conservation, *CPC Best Plant Conservation Practices*, 2-7.

70. L. Herrmann et al., "Chapter 26: Collecting Symbiotic Bacteria and Fungi," in *Collecting Plant Genetic Diversity: Technical Guidelines—2011 Update*, ed. L. Guarino et al. (Rome, Italy: Bioversity International, 2011), 2.

71. Guarino et al., *Collecting Plant Genetic Diversity: Technical Guidelines—2011 Update*.

72. World Intellectual Property Organization, *Documenting Traditional Knowledge: A Toolkit* (Geneva: WIPO, 2017), 9.

73. L. Guarino, "Secondary Sources on Cultures and Indigenous Knowledge Systems," in *Collecting Plant Genetic Diversity*, ed. L. Guarino et al. (Wallingford, Oxon, UK: CAB International, 1995), 199.

74. P. Quek and E. Friis-Hansen, "Chapter 18: Collecting Plant Genetic Resources and Documenting Associated Indigenous Knowledge in the Field: A Participatory Approach," in *Collecting Plant Genetic Diversity: Technical Guidelines—2011 Update*, ed. L. Guarino et al. (Rome, Italy: Bioversity International, 2011) 2.

75. P. Quek and E. Friis-Hansen, "Chapter 18: Collecting Plant Genetic Resources and Documenting Associated Indigenous Knowledge in the Field: A Participatory Approach," in *Collecting Plant Genetic Diversity: Technical Guidelines—2011 Update*, ed. L. Guarino et al. (Rome, Italy: Bioversity International, 2011), 3.

76. D. Bridson and L. Forman, *The Herbarium Handbook*, 3rd ed. (Richmond, UK: Royal Botanic Gardens, Kew, 2004).

77. Global Genome Biodiversity Network, "The GGBN Data Portal," http://www.ggbn.org/ggbn_portal/.

78. B. Gemeinholzer et al., "Organizing Specimen and Tissue Preservation in the Field for Subsequent Molecular Analyses," in *Manual on Field Recording Techniques and Protocols for All Taxa Biodiversity Inventories*, ed. J. Eymann et al., ABCTaxa 8 (n.p.: Belgian Development Cooperation, 2010), 129.

79. E. Guerrant and Linda McMahan, "Practical Pointers for Conserving Genetic Diversity in Botanic Gardens," *Public Garden* 6, no. 3 (1991): 20.

80. Burcaw, *Introduction*, 48.

81. International Plant Sentinel Network, "Home," https://www.plantsentinel.org/index/.

82. American Public Gardens Association, Sentinel Plant Network, https://www.publicgardens.org/programs/sentinel-plant-network/about-spn.

83. Gratzfeld, *From Idea to Realisation*, 132.

84. Burcaw, *Introduction*, 11.

CHAPTER FOUR

Documenting Collections

"The significance of a [plant] in the collection lies not in itself alone but also in the information relating to it."[1]

"In order to be able to interpret and communicate knowledge effectively, a museum must first have detailed and accurate information about the objects in its collection. Museums can provide an efficient service only if their information resources are readily available and if their records are revised as a continuing process."[2]

Perhaps no single facet of a botanical garden so thoroughly distinguishes it as a living museum, separate from parks and other such facilities, than the documentation it maintains on its plant collections. Without proper documentation, museums and botanical gardens have a limited story to tell and little reference value. Moreover, these institutions are ethically and legally obligated to maintain basic information about their collections according to accepted professional standards.[3] At very small institutions, the curator may be directly responsible for documentation. However, at most botanical gardens this essential part of the curatorial program is usually the purview of a registrar, curator of records, or plant recorder.

It is important for all botanical garden personnel to realize that information is both a resource and a product. Imagine, for example, that the garden director and education coordinator would like to see the garden's lone specimen of *Lithocarpus henryi* moved to be a central part of a display of *Fagaceae*. Based on the garden's catalog information for this taxon, you find that it is intolerant of the full sun exposure in this location, is poorly adapted to trans-planting, and is currently under stress caused by recent severe winter weather. With this information, your recommendation would be to leave it where it is. But something else is revealed: this investigation has made it clear that a single plant (and accession) of Lithocarpus henryi is insufficient, both for educational programming and preservation. The documentation the garden compiled on Lithocarpus has been a resource in helping you make a decision that benefits the collections, programs, and efficacy of the institution.

Imagine further that a taxonomist makes an arrangement to examine your specimen of Lithocarpus henryi and your records on the plant for a revision. The source, phenological, herbarium, and other information you maintain on the tree is a product of the garden for use by the taxonomist. In other words, it is a product that someone wants. The taxonomist will take your information and synthesize it to generate new information. Garden staff do this all the time as well. As alluded to in the first quote at the beginning of this chapter, the botanical garden's second greatest treasure is the informa-tion it keeps about its collections.

A Generic Documentation System

Each individual institution has documentation needs that are specific to its mission, programmatic goals, and collections. Therefore, it is difficult to describe a specific "recipe" for documentation that will neatly fit the needs of each botanical garden. However, no matter what the idiosyncrasies of a particular garden's information needs are, there is a basic structure that all gardens might find useful. I hope that the generic system described here will serve as a useful outline to help gardens establish standardized documenta-tion systems or reorganize existing systems, which then may be augmented with the details of their specific documentation needs.

The value of a documentation program depends upon what information is documented and how it is documented. Randomly organized and stored information, no matter how accurate and pertinent, is useless. To be useful, documentation must be systematized—it must have a structure. Therefore, a program of documenting collections should be adequately considered and carefully planned, including priorities for the information to be documented and carefully observed rules and procedures for preparing records.[4] The plan-ning and analysis of a documentation system should be complemented by an analysis of the scope of the collection and the staff, time, and resources avail-able to do the job. These factors will affect the amount of detail designated for the system to handle. Above all, information in the system should be accurate and current. Facts and opinions coexist comfortably and practically when distinguishable. The system itself must contain standards for format, terminology, and timeliness. For example, is it acceptable that new acquisitions be documented within 14 days of receipt?

Botanical gardens often make the mistake, particularly during the computerization of records, to try to capture as much detail as possible. In many cases, this approach inevitably leads to incomplete documentation. "To be effective, a system does not have to be complex. In fact, the system should be as simple as the complexity of the collection allows."[5]

Some gardens may be well advised to start with the rudiments of a complete system and expand it as their capacity increases. Others may find it more useful to build a comprehensive system, begin documenting at a general level of detail, and become more specific over time.

The major component of a documentation program is a system for handling information. The scope of the system will include all collections-oriented documentation including registration, cataloging, indexing, information retrieval, and collections control data.

The documentation system is driven by inputs and should provide a means for organizing this input in a way that supports the program and facilitates meaningful outputs. This is accomplished by staff who design, develop, implement, and maintain the system using manual and/or automated means. These days it is possible for garden staff to simply select and purchase a suitable system that is available as commercial software. A system's success depends upon the accurate and efficient recording and retrieval of information reliably keyed, or connected, to the plants in the collections.[6] The information must be easy to retrieve, interpret, sort, and reorganize. In other words, one should be able to find information on the collections simply and easily. One should also be able to find items in the collections with similar ease and feel confident that all collections are documented and that all documentation represents existing collections. Those of you with experience know that this is easier said than done.

Key uses of the documentation system include but are not limited to the following:

> *Care and control of collections* by providing mechanisms and sources to help locate items, manage internal movements, undertake inventories and respond to audits, improve security and reduce the risk of loss, and maintain details of preservation.

Facilitating the use of collections by supporting publications and educational programs, providing sources for research, and supporting the development of displays.

Preserving information about items in the collections or of interest to the garden by providing facilities for its long-term storage and access.[7]

The system described here is one of information management. The focus of this model will be the stages and procedures for processing and storing collections information. I divide the procedures into two general categories: those pertaining to the collections attributes and catalog information, and those pertaining to control of the collection. Catalog procedures are used to manage catalog information—that is, intrinsic information deduced by an examination of plants (accessions) in the collection (description, condition, phenology, etc.) and extrinsic information acquired through summative research (taxonomy, natural history, etc.). Control procedures are used to manage inventory control, location control, retrospective inventorying, and stock checking (or stocktaking)—is the plant in the collection, where is the plant, is it documented?

Inventory control concerns the development and maintenance of a com-prehensive, numerically ordered inventory of the collections in the care of the museum, as an essential basis for their management, use and detailed cataloguing.

Location control concerns the maintenance of methods for tracking the location and movements of these collections while they are within the museum . . . providing a link between the inventory and the collections themselves.

Inventorying is the process of developing inventory and location records at the time of acceptance of a collection or when reprocessing an inadequately documented collection.[8]

"Stocktaking" procedures may be employed in any segment of a botanical garden's collections management operation to monitor the implementation of the collections management policy and demonstrate accountability to others. In the case of documentation, this amounts to internal auditing pro-cedures to evaluate the effectiveness of the documentation system.

Catalog and control procedures are applied at all stages during the docu-mentation process. For planning, implementing, operating, and auditing a documentation system, it is useful to identify its various documentation stages. These stages may be determined by the general nature of the data collected and records created throughout the documentation system. Each of the documentation activities that take place at each stage builds on the activities of the previous stage and represents a progression in the complete documentation of a taxon or plant. At each stage of the system, documentation is created, stored, modified, and/or enhanced. Roberts identifies the following documentation stages in *Planning the Documentation of Museum Collections*:

- *Pre-entry stage*: documentation procedures that take place before the actual entry and acceptance of plant taxa or other objects into the garden, for example, field notes, hybridization data, purchasing information
- *Entry stage*: documentation procedures that take place when plant taxa or other objects enter the garden but before they are accessioned, for example, evaluation data, quarantine data, propagation data
- *Accession stage*: documentation procedures that occur at the point of accession (processing a taxon into the garden's collections through assignment of an accession number linking taxon and documentation)
- *Registration stage*: documentation procedures that occur after acquisition but before development of individual catalog records, for example, basic documentation centered on the circumstances of acquisition, accession, obvious extrinsic qualities of the taxon at the time of accession, and necessary control data regarding the disposition of the plant within the institution
- *Catalog stage*: documentation of individual accessions derived from collection use and research aiding the museum with long-term management and program delivery, for example, verification, phenology, cultural adaptation, stability, and so forth
- *Output stage*: making use of the information accumulated during the previous stages to produce publications and indexes and to respond to collection queries
- *Exit stage*: documentation procedures that occur when accessions leave the garden[9]

If we view the documentation system from the point of view of a plant entering and moving through the system, the volume of records kept by a garden will become more diverse and comprehensive from the "pre-entry stage" through the "catalog stage." Once documentation on an accession has proceeded through the catalog stage, one may hope that the accession is thoroughly documented to that point in time.

Figure 4.1 is a diagram of a generic documentation system illustrating the principal stages and examples of the activities and processes that occur at each of these stages. Neither the activities nor the processes are all inclusive. Also note that the quality of documentation at any stage of the system is subordinate to the quality of the previous stage. The quality of the details held in files associated with the documentation that occurs during the entry stage of the system directly affects the quality of the subsequent files created during the accession stage. Keep in mind that a myriad of different file and record types along with different formats may be created and used in association with each of these stages based on the needs of each garden.

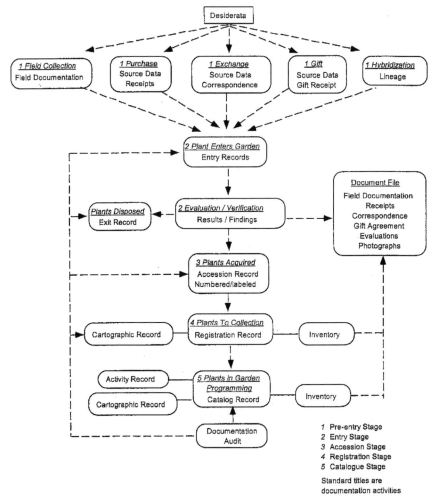

Figure 4.1. Documentation System Chart of Activities, Processes, Stages
Source: D. A. Roberts, *Planning the Documentation of Museum Collections* (Duxford, Cambridge: Museum Documentation Association, 1985), 31.

Some of you might feel that the documentation process may be more simply broken down using fewer representative stages. "Pre-entry" and "entry" stages and the types of documentation established during them are not often practiced or recognized at botanical gardens. The documentation procedures conducted during the "accession," "registration," and "catalog" stages are often viewed as synonymous portions of a single stage (accessioning) of documentation activity. In addition, most gardens are now using "off the shelf" software programs as the basis for their documentation programs. These systems are created with a particular architecture and organization conceived by the software developer, hopefully with experience and input from botanical gardens, that likely differs somewhat from what is presented here. In any case, I hope that, after further reading, you will consider the documentation stages in figure 4.1 valid and helpful in understanding, planning, implementing, and using a documentation system. It is the goal of this chapter to define, describe, and help establish a standard documentation model, one that is consistent and aligned with museum practice.

Typically, documentation systems contain a series of main files that correspond to specific types of documentation and several index files based on specific themes. Automated programs make it possible to structure the relationship between files in a myriad of ways. The major components of the main files are records that summarize all the relevant information about a plant in the collections. Each of these records has data categories configured as text, integers, or other relevant data bits.

Documentation systems call for a series of activities and processes that will take place at the described stages. Most of these activities and processes will revolve around the creation and updating of records. Some of these activities will only occur once in the documentation of an item (such as the accessioning process), while others may occur repeatedly or even on a continuous basis (such as inventorying). The information that results from these activities is held in various manual and/or automated files that will likely include separate entries about accessions in the collection.

The procedures and activities of any documentation system are implemented through a series of manually and/or electronically processed operations. Roberts has broken these down into convenient categories for system planning and auditing:

- *Displaying* sorted, indexed, or otherwise retrieved information on various media
- *Editing* information to correct errors or make changes to information based on current research

- *Entering* information into a computer-based system
- *Inverting* new records for index files
- *Maintaining* and storing secure copies of files
- *Manipulating* information into a standard format
- *Merging* new records with existing
- *Modifying* the records by the addition or elimination of data categories
- *Recording* information for a new record or supplementing an existing record
- *Retrieving* information
- *Sorting* records and information into standard sequences
- *Transferring* information between files or other systems[10]
- *Validating* information

Before outlining the activities and types of documentation that take place at each documentation stage, I offer a general recommendation about putting in place documentation processing steps and protocols. So often unexamined assumptions about how things will proceed and be accomplished are proved wrong after what might be an extended period of needless difficulty and error; unfortunate conditions that may then become embedded in the program over an unexpectedly short period of time. The sooner a clear and effective protocol and documentation routine is established, the sooner and more likely it is that collection documentation will be accurate and up to date. For new gardens and new or updated documentation programs, this may be expedited by pulling the documentation team together—all those staff who contribute to a particular process stream—to outline procedural steps, choreograph the necessary actions to complete these steps, and conduct practice runs of those steps to assess outcomes. For instance, the entry stage described here may be planned, choreographed, and practiced with the curator, recorder, and propagator in a mock-up fashion or to document real acquisitions. This work will help you avoid inefficiencies, errors, and gaps in the system that may become embedded in the operation if they persist for long, making it more difficult to correct down the road.

Pre-entry Stage

This is the stage when information is recorded and collected relating to plants and other items that *might* be added to the garden's collections. Field collections of propagules and whole plants are good examples, with field notes a primary source of documentation. Also, this stage may include vendor and other source information from purchases, files of desiderata, and a general correspondence file. The work that occurs at this stage concerns gathering information for potential use in registration and catalog records and collection control details about the taxa such as field-collection numbers.

Documenting Field Collections

For *ex situ* collections from the wild to be successfully used, they must be supported by quality documentation. Field documentation, also known as "passport data," may consist of two types: conventional scientific data and indigenous knowledge.

> It is at least partly on the basis of such data that samples in different collections can be recognized as duplicates, that appropriate conditions for regeneration, characterization and evaluation can be identified, that material now extinct in the field can be reintroduced to the area where it was originally collected, and that users of conserved germplasm are able to make an initial decision regarding the suitability of the material for inclusion in breeding, introduction or screening programs.[11]

Field-collecting trips are not usually blessed with an excess of time and resources. Unfortunately, this may set up a conflict between collecting propagules and collecting data. As is made clear in the above quote, the value of wild *ex situ* collections is heavily dependent upon documentation. Adequate trip research, preliminary trips, and the use of data collection forms will help mitigate conflicts and expedite data collection.

Whether you use a paper form, data logger, or laptop computer for recording data, a data descriptor list is needed to ensure that data collection is standardized and complete. Descriptor lists should contain certain essential descriptors, facilitate the incorporation of data into the permanent record, and adhere to and promote standardization. Good sources of standardization for guiding the development of descriptor lists are the International Plant Genetic Resources Institute (IPGRI), Botanic Gardens Conservation International, Center for Plant Conservation, and the reference *Collecting Plant Genetic Diversity: Technical Guidelines* and the 2011 update of this reference.

In the past, field data was typically collected on acid-free, long-lasting paper forms formatted to be easy to fill in, update, and correct. Now, field collectors more commonly use the convenience and organizing capability of electronic devices for field collecting notes. Still, hard-copy field notes recorded using preprinted data collection forms are a useful intermediate technology, and if put to use, collectors should strive to limit them to a single page that will help reduce the possibility of lost data commonly associated with multiple forms. These forms are bound in a collecting book or stowed as individuals in a ring binder or folder/clipboard unit. The form should be designed in such a way as to program the recording of data. Multiple-choice or binary descriptors that only require checks or tick marks are preferable to data fields requiring free-text descriptions. Nevertheless, some flexibility should be built into such forms with the provision of "other" categories and/ or blank spaces for some free-text entries.

More commonly, field collectors are using a number of electronic devices for data collection, including smartphones, computer tablets, laptop computers, and other data loggers. Laptop and tablet computers are more useful in this regard than data loggers because they accept free text (depending upon software and program setup).[12] Field laptop computers are outfitted for greater resilience to field conditions and may be preloaded with specific software applications for more rapid, complete, and accurate field documentation, such as BG-BASE. Smartphones are often used as data assistants in the field with field applications networked to larger software programs. The quality of smartphone photographic technology has improved dramatically, but in order to avoid bottlenecks and conflicted use issues, it may be preferable to have a camera available for taking specimen and other photographs. Portable power banks are available, some with solar charging capability, to assist in keeping electronic devices fully charged in the field. Keep in mind that there is still the possibility for technical, reliability, power, and transport problems that make paper backups a good idea. Recording devices or the recording mode of a multifunctional device are very useful for documenting indigenous knowledge about plants.

Many collectors will also maintain a field notebook, or diary, along with the "official" collecting data. The notebook may include a general account of daily operations, orientation and travel information regarding how a site was reached, information about valuable contacts, and a synopsis of the day's collections including a list of all the field numbers. The notebook can serve as a quick reference to what was collected on any given day and an easy check on the accuracy and sequencing of collection numbers. The field notebook will also be a primary source of information for a published "mission report."

The following basic set of field data descriptors from Moss and Guarino compares favorably with the field data collecting forms also used by the Arnold Arboretum, among others:

Sample labeling
  Expedition identifier
  Name(s) of collector(s)

Collecting number (or collector's number)
Collecting date
Type of material [seed, pollen, etc.]

Sample identification
Genus, species [infraspecific designator]
Vernacular name
Herbarium voucher number
Identification numbers of other associated material [e.g., pests, photographs, etc.]
Status of sample [wild, weedy, cultivated]

Sampling information
Number of plants sampled
Sampling method [random, formula, etc.]
Collecting source

Collecting site localization
Country
Administrative unit [county, province, etc.]
Precise locality
Latitude of collecting site
Longitude of collecting site
Altitude of collecting site[13]

Each of these categories deserves further explanation:

Expedition identifier. Collecting trips should be identified by a unique or codified name that may be used to keep track of samples and all other material associated with that particular collecting trip. This is particularly important for trips mounted by several institutions.

Name(s) of collector(s). This descriptor, along with the collecting number, provides a unique identity to each sample. Both the name(s) of the collector(s) and the collecting number should stay with each sample.

Collecting number. The collecting number should be assigned starting with 1 and be given to subsequent samples in consecutive order continuing through additional collecting trips. Renumbering samples in accordance with each trip may lead to duplicate and confusing collecting numbers. If both vegetative material and seeds are taken from the same plant, they should receive the same collecting number. Other material collected in support of the plant material, such as pest samples, should be given the same number as the plant sample. The collecting number is not only recorded as part of the field data and in the field notes but also should be marked on the propagule container and recorded on a tag placed inside the container—a total of four records of the number.[14]

Collecting date. Generally recorded in the format day/month/year, this information is useful for the collector in keeping track of activities, determining the timing of future collecting trips, and evaluating the possibility for genetic bias in the sample based on timing.

Type of material. This information is particularly useful in determining current and future handling and cultural requirements for the samples. It is also useful for gene-pool analysis.

Sample identification. Each taxonomic rank is usually accorded its own data field on the collecting form. Be prepared to identify specimens in the field by equipping yourself with the appropriate floras, annotated checklists, and other identification aids (refer to the previous chapter for more specific recommendations). However, be prepared to provide provisional identification if necessary using the appropriate annotation (e.g., *Clethra* cf. *pringlei*) or an arbitrary designation if no determination is possible. Include a field on the record form for notating references used in identifying plants.

Vernacular name. Recording these names may be important to a determination of plant identity and the documentation of ethnobotanical data. Vernacular names in languages other than English may present problems in transliteration. Whenever possible, the vernacular names should be recorded in local script.[15] Many dictionaries will provide standards for transliteration, and these may be useful references for inclusion on the trip.

Herbarium voucher number. This should be the same as the sample number.

Identification numbers of associated material. Other specimens, such as pests, mycorrhiza, and so forth, should be assigned the same number as their associated plant sample. Photographs should be identified by a roll and frame number.

Sample size. Make a determination about the size of the sample collected relative to the population encountered. Also, indicate what sampling method was used, for example, random, systematic, and so forth, and include the details of the method. This information will be useful in assessing questions of genetic diversity, genetic erosion, and other factors related to the genetic quality and gene-pool maintenance of the sample and source populations.

Source. Indicate from what source the samples were collected. This may be categorized on the record form in a multiple-choice format. Moss and Guarino suggest the following set of choices:

Undisturbed natural habitat
Disturbed natural habitat
Weedy habitat
Farmer's field, plot or orchard
Landscape or garden
Store or other vendor
Institute, experiment station
Threshing floor
Other[16]

Source location. It is important to record the geographic location from which the sample was collected. One example that includes country and appropriate administrative units would be as follows: "US/WA/King Co/ MtBaker-Snoqualmie Nat'l Forest." The precise locality may be recorded in a statement beginning with a major highway, roadway, or town such as: "North Bend WA, 10 miles northeast on Hwy 18, 15 miles east (rt.) on Forest Service Road 522, 2 miles south (rt.) on Hinman Trail, rock outcrop 100′ east (lt.) of trail." This location should be well marked on a map of at least 1:250,000 scale or larger. A global positioning system (GPS) receiver is particularly useful for this purpose. GPS receivers must be calibrated to correspond to certain map data. This information should be included in the map legend and may be programmed into the receiver (see the "Cartographic File" section later in this chapter for more information on GPS). Altitude may be determined with the aid of an altimeter that is reset and checked as often as possible. Latitude, longitude, and altitude may also be derived from useful maps. Other details and specifics regarding location should be described fully in the field notebook. Finally, collectors should consider using site numbers so that multiple collections from single sites will not require redundant descriptions of the same site.

Documentation equipment. The following is a list of basic field equipment for gathering and recording data:

Collecting forms or other data storage equipment as described above
Field notebook
Pencils and other markers
Maps
Checklists and other identification aids
Binoculars
Hand lens
Altimeter

Compass
Clinorule or clinometer
Tape measures
GPS receiver or other capable device
pH kit
Color chart (RHS)
SLR camera, tripod or monopod, flash, and film[17]

You may compile a longer list containing more technical equipment and materials including ecological profiling software and a drone, just to name two.

The subject of documenting the knowledge of indigenous people within the target collecting area first appeared in the previous chapter. This includes a prior informed consent (PIC) agreement with the accompanying mutually agreed terms (MAT) and the access and benefit-sharing (ABS) government permit. That documentation becomes an important part of the overall field collecting database and mission report. For more details and references on the subject, I refer you back to the section on field collecting in chapter 3.

Mission reports serve as a useful summary of the goals, methods, and results of field-collecting trips. The Agricultural Research Service of the U.S. Department of Agriculture requires that mission reports be filed with their Plant Exchange Office within 60 days of trip completion. Their reports contain the following information:

a. Catalog of collections: a record of all collections including information on the sample data collection sheet. This should be in electronic form.
b. Narrative report: three to five single-spaced typewritten pages (more, if necessary). The narrative should be as short as possible and avoid details presented in the catalog. Include significant observations likely to be of interest to germplasm users or other explorers who may visit the same areas in the future. Provide a list of contacts (domestic and foreign) with complete addresses and indicate how they contributed to the mission and how they might contribute to future missions in the same country.
c. Page-size map showing itinerary: identify principal points on the itinerary and most important collection sites.
d. Information on any threats to genetic resources in the area visited.[18]

Mission reports should be published to make them accessible to other professionals and the general public. Refer to chapter 3 for a list of publishers of such reports.

Entry Stage

Botanical gardens should establish procedures for documenting all incoming material, distinguishing between temporary and permanent holdings, and itemizing individuals of a group. They may have both temporary and permanent collections as well as several plants of a single acquired taxon. They have the additional challenge of controlling and tracking propagules of uncertain viability and unique documentation requirements.

The curatorial staff must structure entry procedures and protocols that will take into account the matrix of decisions that must be made to serve the documentation requirements of new plants: temporary, permanent, or unknown collection status and propagules versus plants. Many gardens will simply enter all acquisitions in an accession register regardless of the plant's status, intended use, or the predicted length of stay within the collections. This approach creates a more cohesive and simplified documentation process. It also means that every acquisition will have some documented history within the institution however brief its acquisition or use. However, this will also require more frequent edits in the accession record, increasing the risk of errors and omissions.

Many other types of museums will establish an entry register separate from the accession register unless the item was acquired specifically for the permanent collection. The item will be labeled with a unique entry number that will be used to establish a record in the entry register and attached to an entry file containing any supporting documentation such as receipts, passport data, and so forth. Gardens will likely retain such acquisitions in a greenhouse, nursery, research holding area, or quarantine facility for examination, evaluation, or eventual accession into the permanent collection.

The initial documentation, whether in a separate entry register and file or the accession register and registration record, should include certain catalog and control categories. The basic information should include the following:

Entry number (could be a modified version of the accession number)
Date and method of reception
Source (name, address, telephone number, and other data)
Identification as received
Condition
Field data if appropriate
Location

Accession Stage

At this stage, plants are formally incorporated into the botanical garden collections as a permanent acquisition by the assignment of an accession number. This is a critical part of the documentation process and the documentation system if the garden is to maintain proper and effective control over its collections. An accession, then, is a group of clones or members of a line received from the same source at the same time.

Acquisitions are entered into an accession register with an accession number. This is a unique number that will stay with the plant and its supporting documentation as long as it remains in the collections. The accession register is a permanent, official record of all the garden's holdings recorded in sequential order by accession number. Additional documents and records pertaining to an accession become part of the registration record described later.

The accession register may be a bound volume or loose-leaf notebook with preprinted data fields or a computer file. It is a closed record with few annotations made unless the accession is removed from the collection. The type of information recorded in the accession register closely mirrors that described for the entry register.

The accession number is usually a compound sequence of digits denoting year of accession, order of receipt within that year, and, in some cases, number of items per accession. For example, the 100th accession of 1996 may be notated as 96.100 or 96-100. Old collections may use more digits to further segregate centuries, such as in 996.100. Plants of an accession may be designated with an additional set of digits or letters such as in 996.100.1 or 996.100.A. A group of clones or members of a line may be given the same number indicative of their genetic similarity. Propagules of a taxon are most often assigned a single accession number as a batch. The successful propagules may then be distinguished as above with a digit, letter, or other data-symbol extension to the accession number.

Accession numbers are not only used to link all documentation about plants but also must be attached to the plants themselves. There should be a very specific and rigorous specification in the documentation manual for when and how plants are labeled and who is responsible—the longer the interval between accessioning and labeling, the more opportunity for mislabeling and confusion. Labeling plants—the label size, materials used, information displayed, and position of the label on the plant—is a subject of considerable debate.

Gardens have differing needs regarding how their accession labels function beyond the basic need to attach accession numbers to plants. The basic accession label may contain only the accession number. Accession and interpretive needs may be combined in a single label that includes the accession number, plant name, and other bits of information of greater interest to garden visitors. Nevertheless, it is often helpful to the staff to include the botanical name of the plant on the label in addition to the accession number.

Using the appropriate nomenclature can be a vexing problem. The World Flora Online and the International Plant Names Index are valuable resources for checking the validity of plant names. Gardens should have copies of *The International Code of Botanical Nomenclature* and *The International Code of Nomenclature for Cultivated Plants* on hand to guide them in naming plants. Complicating the matter of nomenclature is the plethora of trademark names that now obscure the lineage and correct nomenclature for many cultivars. Gardens must be wary of the validity of certain fancy, marketable cultivar names and diligent to check their registration. Many popular garden plants are being marketed with their trademark names portrayed as cultivar names. In some cases—the patented "blue hollies," for example—these errors have persisted into the horticultural literature. Botanical gardens cannot ignore the use of trademark names in commerce and should make it their business to correctly interpret and distinguish them from registered cultivar names.

Some basic labeling considerations are the following:

- The label must contain the accession number in readable form.
- The label material must be reasonably durable.
- The labels must be adjustable and/or removable and must not harm the plant.
- The placement of the label must be consistent.
- The method of attachment and placement must deter label/plant separation.

Plant labels may be made from laminated paper, engraved plastics and metals, embossed metals, photosensitive metal coatings, and other types of materials.

In addition to standard text, some institutions are placing bar codes on their accession labels to augment and facilitate their automated documentation processes. As of this writing, bar code technology has progressed much more slowly than anticipated. Nevertheless, it is receiving greater attention as the technology has become more accessible and practical. Bar codes allow for more accurate, consistent, and direct collection of inventory, evaluation, and other types of data collection in the field with handheld data loggers, portable computers, and, most commonly, smartphones. I want to emphasize that the key advantages of bar code technology are accuracy and speed.[19] Transcription and other types of manual data recording errors are reduced. Bar codes must be fixed to accession labels by weatherproof and UV-resistant means, and they are vulnerable to vandalism just as are the labels themselves.

In order for bar code technology to work within the documentation system, it must contain the plant accession number, and it must be read by a capable scanning device that is compatible with the documentation software program. The principal problem in making bar code technology useful since it was raised in the first edition has been compatibility. Avoid using intermediate technologies to interpret and organize the bar code scans as this may nullify the speed advantage of barcoding.

The early technology used linear bar codes that could be difficult for scanning devices to read quickly in the field. Also, the bar codes themselves were easily damaged by scratching, nicking, and smearing. Two-dimensional bar codes are now more commonly used, and the technology for scanning them has been incorporated into all sorts of devices. Of the two-dimensional bar codes available, the most common are QR, or quick response codes, and DM, or data matrix codes with Reed-Solomon correction capability for better field application.[20] Data matrix codes are of particular use for botanical garden accession labels because they may be printed in very small formats (e.g., 3 mm x 3 mm). DM bar codes may also be effectively printed in various ways to accommodate differing background colors, among other things.

In their article, "Barcodes Are Dead, Long Live Barcodes!" for *Sibbaldia*, Havinga and Ostgaard list three basic garden documentation functions of bar code technology:

1. A barcode is scanned and the record is displayed so that details can be checked and updated.
2. A barcode is scanned and only the plant status (e.g., dead, alive, etc.) can be updated directly. Additional details are not displayed.
3. A barcode is scanned and the record is added to a list, after which the next barcode can be scanned, enabling a list to be compiled quickly.[21]

The last function of list making is particularly useful for batch functions involving record updates. Stocktaking and inventories are vastly sped up using labels with DM bar codes. Also, areas of the garden where the status of plants changes rapidly, such as the propagation and nursery area, benefit from the ease of bar code labeling and technology. Bar code technology is most effective for gardens with large collections and, therefore, a large volume of documentation and a need to manage that documentation more efficiently.

Another technology that is emerging for tracking accessions, especially trees, are radio frequency identification (RFID) transponders. RFID is a system in which radio frequency communication is used to exchange data between a mobile device equipped with memory and a host computer. A wide variety of electronic devices such as televisions, radios, and wireless telephones use radio frequency technology to transmit or receive information. RFID operates much the same as current laser-scanning bar codes: information is stored within the label and retrieved from a reader through a scanning process. From there, a computer makes sense of the information, such as by linking it to a product or, for our purposes, a plant accession.[22]

One scenario using RFID technology involves the use of a TIRIS-type transponder (4 millimeters in diameter and 23 millimeters long) embedded in the trunk of a tree using a type of hollow "nail." The transponder is queried, or "interrogated," using a personal digital assistant (PDA) carried in the field by a user. The amount of information that may be accessed through this system is partly governed by the amount of power the components are capable of generating. A greater power requirement means bulkier equipment. One will have to strike a balance between the ideal field information requirement and the bulk of the equipment required to deliver that information. That balance must include the ability to display text large enough to easily read with a screen that is bright and visible in the prevailing environment in addition to having the capability to deliver the right kind of information. While walking, the user must be able to carry the device in one hand and interact with it using the other hand. RFID applications are now compatible with smartphone technology. For this application, smartphones need to be outfitted with additional hardware and software.

Similar to bar code systems, there are benefits of an RFID tagging system: more information is available to curatorial staff in the field; information may be updated from the field and/or reconfigured to provide new information to field users; and the RFID tagging device, unlike labels printed with bar codes, will not likely be lost, stolen, or vandalized. Liabilities of the system are the initial and replacement costs of the transponder, compatibility with existing documentation software and electronics, and use with smaller plants. At the time of this printing, the promise of RFID technology has yet to be realized due to the liability limitations listed above. Still, when RFID technology proves more practical, it may be a very helpful documentation technology for botanical gardens. Also, those gardens using bar code technologies should find it possible to incorporate and use RFID technology without too much difficulty. Keep watch on the forestry industry for new information on the use of RFID technology that may be applicable to botanical gardens.

Registration Stage

At this stage, new records and files are established using attribute and control data associated with the accession's current status within the collections of the garden. These records are also more item specific than the accession records. They are commonly referred to as "registration records." In reality, they are a composite of some data taken directly from the accession records and more recently collected data from other sources. The registration records may be composed of several hardcopy or electronic files and indexes containing different types of documentation, for example, an accession file with specific data related to an accession; a document file with supporting documents such as receipts and letters; a cartographic file showing map locations for plants of that particular accession; and a source file with specific information about plant sources. Some of this data was established and filed during the entry or accession stages as entry records and/or in an accession registry.

Accession and other data are copied or shared at this stage to a permanent file system that comprises the bulk of the documentation system. This file system may be sets of file cards or bound registries but is now most often stored in computer databases with thorough backups. Additional details relating to the broad categories of information established in the entry and accession registers are recorded for *each specimen* of an accession. In other words, the original data categories, such as "source," may be expanded by several fields of new source data. To reiterate, records established during the registration stage become more "item" specific. Additional attribute information relating to the identity, description, or historical data may also be added. For instance, accession records may contain the name of an accession as it was received, while registration records usually contain data regarding the verification of that identity. Registration records begin to add greater breadth and depth to the documentation of your collections.

Still, this record should be considered basic and centered on the circumstances of acquisition, accession, obvious extrinsic qualities of the accession, and necessary control data regarding the disposition of the plant within the institution. More detailed information derived from collection use and research will become part of collections documentation during the cataloging stage. *However, I cannot emphasize enough that each garden must determine the scope of their records based on the anticipated or actual use of their collections.*

The following subsections list and describe the types of files generally found in registration records.

Accession File

Accession files usually consist of the following data fields stored on formatted file cards, record forms, or computer databases:

Accession number
Plant name
Verification
Accession date
Material received (seed, scion, etc.)
Condition
Source (immediate and original) and lineage
Location

The accession number codes for the year and order in which the item was received along with the item number within a group, for example, year (996), order (100), item (01). The plant name will be recorded as it was received unless verification dictates otherwise. The level of verification may be indicated within a range from none to fully verified by a recognized authority. A record of how the accession was received and in what quantity and number it was received is often included. The source of the accession is listed with the most basic of information and a pointer, flag, or link to a source index file with more detailed information. The location may be indicated with a coordinate number, grid location, or landmark description and a pointer, flag, or link to a location index file with more accurate information.

To facilitate the exchange of data about collections, it is important that gardens adhere to a set of data standards. Botanic Gardens Conservation International (BGCI) has developed an international transfer format (ITF) for just such a purpose. This format was designed primarily for the exchange of computerized data but provides a useful standard for hard-copy records as well. See table 4.1.

The importance of facilitating data exchange between botanical gardens cannot be overstated. The use of data-compatible documentation software programs has made data exchange possible and more convenient. However, BGCI has advanced this possibility even more with a set of searchable databases outlined in the previous chapter: GardenSearch, PlantSearch, GlobalTreeSearch, and ThreatSearch. The following is an excerpt from the BGCI webpage:

Table 4.1. Botanical Gardens Conservation International ITF (v. 1.0)

Field	Field length/type notations
Accession number	12
Accession status (active, dead, etc.)	1
Accession material type (seed, cutting, etc.)	1
Family	22
Genus name	22
Specific epithet	40
Infraspecific rank	7
Infraspecific epithet	40
Vernacular names	Text
Cultivar epithet	40
Identification qualifier (aff. or cf.)	9
Verification level	1
Verifier	20
Verification date	8
Provenance type (wild or cultivated)	1
Donor	20
Country of origin	40
Primary subdivision (state, province)	100
Locality	Text
Collector	20
Collector's identifier	12
Collection date	8

Source: Botanic Gardens Conservation International, 1998.

BGCI's PlantSearch database is the only global database of plant species in botanic gardens and similar organizations. BGCI's GardenSearch database is the only global source of information on the world's botanical institutions. BGCI's ThreatSearch database is the only comprehensive global database of conservation assessments of plants. GlobalTreeSearch is the only comprehensive global list of tree species and their country-level distributions.[23]

I will discuss more on data sharing in a later section.

Herbarium File

A herbarium file is more correctly a *herbarium* and consists of labeled herbarium specimens arranged in a taxonomic or other practical order. Specimens in a herbarium are pressed, dried plant parts mounted on special sheets of buffered or acid-free rag paper and stored in insect-proof steel cabinets. Herbaria are facilities most commonly used by taxonomists. Herbaria conserve the type specimen—an original specimen to which a particular name was first applied—of each taxon described by botanists. Each sheet is given an accession number matching that of the plant that the specimen represents and has an accompanying label with collection data. A simple flag field should be included in the accession file to indicate the presence of a herbarium record. Such specimens, when properly established and cared for, conserve many structural and chemical characteristics of plants and last almost indefinitely.

Similar in purpose to a herbarium, a spirit collection consists of plant parts preserved in full dimension in a container of specially prepared fluid. This type of documentation is particularly important for the identification of certain plants, such as the Orchidaceae. For this family, orchid flowers are preserved in vials containing a preservative fluid consisting primarily of ethanol or denatured alcohol.[24]

For more on herbaria, see chapter 8. Also, the bible for herbarium creators and curators is *The Herbarium Handbook* (third edition, 2004) by Bridson, and Forman.

Image File

Photographic files generally consist of black-and-white or color prints and/ or color slides and/or image files. These files further document the existence of the accession and its condition and may include photographs from the source location. Photographs may also be helpful in identifying the accession. All photographs should be labeled with the accession number of the plant shown and include a scale. Black-and-white prints should include a gray scale. Digital images should be accompanied by a metadata file of information on each image. As in the other related files, a simple flag field should be included in the accession file indicating the presence of a photographic record. The file should include a photograph of the entire plant along with more detailed photos of flowers, foliage, fruit, and so forth. The files should be stored primarily by name in alphabetical or taxonomic order and secondarily by accession number.

Document File

This is a standard file of documents, papers, correspondence, and other printed material that document information and data about the accession. Each file may be cross-referenced with its accession by the accession number. The files may then be stored primarily by name in alphabetical or taxonomic order and secondarily by accession number. In this way, you may be able to access similar taxa in one section of the file regardless of their accession numbers. Botanical gardens should be cognizant of proper archiving techniques to preserve document files and make them more readily accessible.

Cartographic File

In the narrowest and most basic sense, this file is critical to the botanical garden's inventory and for location control of plants in its collections. However, in the broadest and most elaborate sense, cartographic files in the form of mapping systems have evolved from line-drawn maps to computer-aided design (CAD) products and, finally, into geographic information systems (GIS) that serve to vastly improve curatorial practice, among other aspects of botanical garden programming. More specifically and in one form or another, maps contribute to collections documentation in the following ways:

- Pinpoint the location of an accession.
- Provide a useful spatial perspective on collections.
- Illustrate change and show patterns across the landscape and the collections.
- Make the collections more broadly accessible to visitors and users.[25]

All but the smallest botanical gardens will find that GIS is the most useful and powerful cartographic tool for garden map making and geospatial data management. Unless you are fortunate to have a GIS-experienced staff member, I suggest you consult with a GIS expert to help you envision and organize a GIS data model that will meet all of your projected needs, and then select the software that supports that model. For example, they will help you identify all the potentially important attribute data you wish to include in the database that will contribute to making maps to assist in documentation, curatorial practice in general, and other garden programs such as research. The power of GIS isn't simply in making a detailed map,

> GIS can also be thought of as a method to visualize data from a variety of sources in ways that reveal relationships, patterns, and trends. By combining the query and statistical analysis capabilities of a database with the visualization and geographic analysis benefits of maps, GIS helps to answer questions and solve problems by presenting your data in a way that is quickly understood and easily shared.[26]

One of the best places to begin is the American Public Gardens Association / ESRI partnership.

> Our current partnership emerged in 2012 from a collaboration between Esri, the UC Davis Arboretum & Public Garden, the Zoological Society of San Diego, and the Missouri Botanical Garden with the shared goal to increase

the capacity of public gardens to conserve plant diversity through the use of geographic information systems. By providing ex situ collection and landscape managers with access to leading-edge information management tools and training resources at low-cost, we hope that it will support your efforts to increase awareness and concern for plant conservation through the sharing of rich scientific and interpretive information with your staff, researchers, and visitors.[27]

For greater accuracy, the GPS approach, although slower and more costly, relies on rugged field equipment for this purpose by makers such as Trimble and Topcon that offer a range of GPS receivers and antennae as well as compatible software. Another drawback to GPS map making is the less reliable performance of receivers under heavy tree canopy and close to large structures. Under these circumstances, you may need to employ a hybrid approach that also uses surveying equipment such as a laser range finder. The most accurate base-mapping approach is surveying, which will also take the most time and money. Surveying depends upon survey markers, or monuments, as positioning standards for surveying other elements on the ground. These should be referenced to a fixed survey marker associated with the U.S. Geological Survey (USGS) or other authoritative mapping agency. There is robotic total station surveying equipment that allows for a single person to conduct the work. The makers Trimble and Topcon also offer total station surveying equipment as well as GPS receivers. The data results from surveying will also need processing using software for this purpose available from the total station suppliers.[28]

Survey areas, or grids marked with monuments, should be coded with a numbering system for efficient use and identification in the documentation system. These codes may be based on the relative position of any grid or area to a known baseline or other starting point. For a grid system developed from a baseline, the grids may be numbered in consecutive order from one end of the baseline to the other. Grids on either side of the baseline may be given a number extension based on their distance to either side of the baseline. With this system, the grid number 4-2E represents a particular grid that is the fourth one from the starting point of the baseline and the second grid east of it at the fourth position.

Apply identifiers to grids and areas consistently. For example, grids may be numbered according to the position of their southeast corners. The grid or area codes and locator data within the grids, such as coordinate numbers, should be recorded in the appropriate location field or fields in the accession record and may also be included on accession labels.

In deciding what features will be included on maps, you may discover that aerial photography reveals all the major base features for mapping necessary for using the heads-up approach to making a base map. Then, plants and objects obscured from the heads-up mapping technique may be mapped in the field using one of the other two approaches. With a hybrid approach such as this, you may choose to rent some of the necessary equipment to accomplish this work, such as GPS receivers, unless you know through prior planning that you will be using them as a matter of course in maintaining cartographic files—in which case you will be purchasing them. GPS may be efficiently and conveniently used by one person traversing the garden mapping plants with a GPS receiver. Many geographic information system software programs allow you to import GPS data directly for automatic mapping of plants and other objects.

The subject of drones first came up in regard to field documentation. They may also have a place in garden surveying, in map making, and in collecting other kinds of data. Using drones for aerial surveying and mapping requires specialized equipment and software as well as drone-operating experience. Currently, there are drones configured for agricultural surveying that may be retrofitted for use in botanical gardens. They are expensive and may be better rented or contracted with an experienced user.

The cartographic file, when effectively managed with a GIS, can play a pivotal role in the use of collections documentation for a wide range of curatorial and other garden programs.

Source File

All field data on wild-collected plants should be placed in or copied to this file. The following information should also be included here:

- Name of source (verified spelling)
- Name of contact person if necessary
- Address, telephone number, and other important contact numbers
- Chronological list of accessions from that source
- Accession numbers of all accessions received
- Method of acquisition
- Purchase information, for example, receipts and so forth

The sources may be filed alphabetically by name of source or accession. Some of the material stored in this file may be copied and cross-referenced with the document file. Most documentation software contains a screen or page for source information.

Catalog Stage

> The purpose of a catalog is to enable intellectual access to the objects in a museum's collection. The catalog record goes beyond the registration record with the addition of information about the object that usually requires analysis, research and related expertise. Each object in the permanent collection must have a registration record; however, because of the time and research required to catalog, many objects may not have a complete catalog record. The registration record and the catalog record are usually the same physical record, with catalog information being added to the basic record sometime after registration of the object.[29]

To register a plant is to record the data necessary to identify it in the collections and to document the means by which it was acquired. To catalog a plant is to record all the scholarly information about it.[30] Over time, the catalog records will contain the bulk of the data on any given accession—the master record. In effect, the entry, accession, and registration stages provide information that establish and legitimate a plant in the collection. At that point, now that we've identified what it is and where it came from, it is the catalog data that provide the bulk of the information that the curatorial staff, garden and other educators, and researchers will use in connection with that plant. This is a dynamic record, unlike the accession record, that continually changes to reflect the staff's growing knowledge about the collections. Research, preservation, and education programs reveal, synthesize, and use information about the collections, all of which is documented during the catalog stage. This package of comprehensive documentation may be referred to as the catalog record. Another way to conceive of this, if you are a BG-Base user, for example, is that many of the collections files such as "locations," "names," "plants," and "verifications" are considered catalog records.

Catalog records, or files, are usually an extension of the previously established registration records. In some cases, this simply means the completion of several additional, previously blank, fields on a set of record forms or cards, or within a computer database. In fact, then, the registration and catalog records may be merged together on the same storage tool, whether it be a large cabinet of hard-copy files or on a computer.

Catalog records are created using many different approaches. In some cases, the registrar may create an agenda for catalog documentation that follows a chronological or a subject/theme order. In other words, it might be decided that the correct identity of all the accessions for the year 1999 will be more thoroughly verified on a chronological schedule according to their order of receipt. On the other hand, this process could be scheduled according to taxonomic collection priorities beginning with the most important group to the garden, for example, *Quercus* sp. The operations of other garden programs will also produce data worthy of documentation in the catalog record, such as horticultural and collection preservation data on pest problems. The curator, or other person responsible for supervising collections documentation, must be responsible for making the necessary institutional connections to coordinate the receipt of important collections documentation produced by other departments. This internal networking is often one of the greatest challenges for gardens when it comes to documentation, particularly catalog records.

Catalog records often include, but are not restricted to, the following types of data files:

- Naming information, including authority, synonymy, parentage, and other types of generic information associated with this named taxon
- Verification including process documentation
- Phenological and other developmental data
- Horticultural, propagation, condition, and preservation data, including tracking of horticultural, preservation, and propagation treatments—an activity file
- Conservation data
- Control data, including location, labeling, and distribution information
- Historical data
- Miscellaneous descriptive data

Catalog records, whether hard copy or electronic, must be configured to provide for easy access and cross-referencing to other related records typical of relational database software used today.

Activity Files

This file documents the activities associated with collections management programs such as collections care and preservation, conservation, propagation, and other programs that may effect a change in an accession. This may become quite a large set of files necessitating a separate filing system altogether. Files may be created and named according to the activities they document. They may also be composed of a series of data fields that reflect the steps followed in each activity. One can easily imagine a file structure for propagation activities constructed in this way. Tracking propagules and plantlets is a particularly troublesome part of the documentation process in the propagation area. Propagules and plantlets should be clearly labeled. This may mean that each taxon is propagated in an individual container and each plant receives its own individual propagation or accession label.

Collection care and preservation activities are often quite extensive, leading to a large file or database. The details of the activity will remain within the activity file, while the results of the activity will also be recorded elsewhere in the catalog record as summary data. This activity file may also be closely tied to the cartographic records since many of these activities are programmed, scheduled, and tracked on a geographic basis. Maps are a convenient medium for actually seeing where certain activities have taken place and particular practices have been applied—a tailor-made function for GIS.

Activity files will require substantial cross-referencing with other documentation containing relevant background on the accessions subject to the activity, a built-in part of most of today's documentation software. As mentioned earlier, the results of certain activities may be duplicated in another part of the catalog record as synoptic or summary data. Needless to say, activity files are very dynamic and constantly expand as new activities take place and new data is obtained. The use and management of activity files may be integrated with and assist in publishing procedures manuals for those activities.

It is my hope that, as you are considering the various curatorial activities that contribute to the building of a comprehensive catalog record in your garden's documentation system, the need for internal networking among curatorial staff becomes obvious. Certainly the horticulture staff, involved as they are in all manner of collections monitoring and preservation, may contribute greatly to catalog documentation. This is a difficult internal relationship to foster and, because of it, there is often a backlog of necessary catalog updates that may go back years. Some gardens remedy this situation by conducting periodic data capture projects. These are often multipurpose operations that include verifying identifications, auditing documentation, and updating the catalog files.

Best Practice 4.1: Royal Botanic Garden Edinburgh Data Capture
In their article, "The Data Capture Project at the Royal Botanic Garden Edinburgh," in *Sibbaldia*, Franchon, Gardner, and Rae characterize the situation many gardens face and the importance of catalog data:

> Great priority is given to making detailed field notes and the process of documentation is often continued during the plants' formative years when being propagated. However, for the large majority of plants this process often stops

once the material is planted in its final garden location. The Data Capture Project at the Royal Botanic Garden Edinburgh is an attempt to document specific aspects of the plant collections so that the information captured can be of use to the research community even after the plants have died.[31]

The Royal Botanic Garden Edinburgh (RBGE) data capture project proceeded in three phases: (1) identifying priority groups of plants for data capture; (2) collecting the necessary data; and (3) processing and evaluating the specimens and the data. Phase 1 prioritizing and selecting data capture groups is planned and conducted according to a set of priorities dictated by the collection. In this case, priority was given to groups of wild-collected plants followed by single accessions of cultivated plants that have cultural or historical significance. Further priority is given to selected families or genera based on a clear indication in the collections policy, impending research projects, conservation concerns, availability of expert attention, or plants that are short-lived. Included in this first phase is a complete literature review on the group to identify the most up-to-date taxonomic, ecologic, and other information used to help characterize the families and/or genera of study.

In phase 2 of the RBGE project, all the necessary steps are taken to collect samples and data to verify the identity of plants in the study group. This includes herbarium vouchers, photographs, DNA samples, and any other information helpful for characterization of the plants (more on verification later in the section "Retrospective Documentation and Inventory"). In the final phase, the samples are described, herbarium vouchers are mounted and filed, identities are processed and verified, field descriptions are categorized, and the accumulated data are documented in the system.[32]

The result is a much more complete documentation file on the studied accessions with upgrades in information to several areas, including the catalog record. The subject family or generic group is documented in a way to provide highly valuable outputs—described in the next section—and make their use in other garden programs possible.

Output Stage

This is the stage when the garden's documentation and collection data—particularly the catalog data—are synthesized, reordered, and processed to produce valuable outputs for collections management, education, and research programs, including public access to the collections. The use of computers and garden documentation software has quite possibly made the greatest impact at this stage in the sorting, reordering, and other outputs in response to sophisticated queries.

Printed versions of the entire catalog record may be compiled and reformatted as a comprehensive reference publication to the collections. These publications are more commonly encountered in other types of museums, although their use in botanical gardens may be equally justified. Indexes based on particular collections features, characters, or details may also be compiled as a specialized directory. An index is usually produced for internal use but may also be used to guide public access to particular aspects of the plant collection. One may ask, given the near completely digitized world in which we now live, why print anything? David Rae asks and answers this question in his article titled "The Value of Living Collection Catalogues" for *Sibbaldia*:

> With the increasing use of the World Wide Web and internet some might query the need for a printed "Catalogue of Plants." After all, a printed catalogue is relatively expensive and out of date from the moment it is published. A web version, on the other hand, can be updated easily and regularly and is accessible world-wide at any time. There are, however, some important advantages in having a printed version. Not least of these is the fact that it helps focus staff on important preparatory activities such as detailed stocktaking of the living plants and ensuring that the most up-to-date names are being used.[33]

Knowing how documentation work may pile up and backlogs result, I can see how working to publish a catalog can be an important motivating influence on updating documentation. Nevertheless, these publications are also an important point of access to the collection for our constituents: public visitors, educators, and researchers. Finally, keep in mind that printed and digital versions of a collections catalog may be complementary.

Simpler, more common outputs are inventory lists, labels, and maps. Focused, indexed maps, or map layers, may be produced by a computer and would allow for detailed tracking, locating, and access to the collections. As an example, the education program may find this particularly useful in structuring themed tours of the collections. Documentation output is of particular value to researchers who often wish to query the system extensively for answers to questions generated by their studies of the living collection.

Exit Stage

This is a purely control-oriented stage that deals with the loan or deaccessioning and disposal of accessioned plants. Loans, however, are not as common among botanical gardens as they are among other types of museums. The process and procedures associated with deaccessioning and disposal are just as important as those associated with acquisitions and accessions. "A museum's collections management policy should establish minimum criteria for reviewing proposed deaccessions of various classes of material. Acceptable reasons for removal should be listed and guidance should be given as to when outside opinions or appraisals should be sought."[34]

There are many reasons a garden will deaccession plants from their collections: questionable authenticity and/or uncertain identification, deterioration of the plants, or simply that the plant is no longer relevant to the programs and purposes of the institution. On deaccessioning, the collections management policy of the Arnold Arboretum states:

> *Deaccessioning* is the process of removing a living specimen from the collection, but does not include the removal of any records related to that accession. Deaccessioning decisions are made by the Curator of Living Collections, in consultation with the Living Collections Committee.[35]

The deaccessioning process should be clearly documented. The accession register, registration, and catalog records should be marked or flagged with a "deaccession" notation. The method of disposition should also be recorded in the appropriate field or documented in the appropriate document file. Registration and catalog records may be refiled in a deaccessions file. Digital records carry a "deaccessioned" field that is appropriately cross-referenced. Accession numbers are not reused once a plant is deaccessioned.

Documentation Standards

For the sake of accuracy, continuity, and universality, botanical gardens should employ and adhere to rigorous standards in their documentation systems. These standards will apply to the following:

Operational procedures
Data input
Data processing
Data output
Definition of files

Data categories (e.g., international transfer format)
Terminology conventions

Most of the information presented in this chapter is meant to help gardens establish standardized documentation systems. Specific system standards pertinent to a particular garden's documentation needs should be established during the planning phase of any documentation system. These standards may be categorized as those pertaining to the external functions and uses of the system and those pertaining to its internal use and operation.

External standards would include the international transfer format described earlier in this chapter. Another such standard would be the use of internationally recognized country abbreviations in field and other records.

Internal standards would include, among other things, data categories that may be used for cross-referencing between records and files or the building of index files. Also, and of particular importance, the syntax of the information to be recorded should be standardized, especially if it will appear in more than one record or file location.

Recognizing that most, if not all, botanical gardens will purchase documentation software with an architecture that, in effect, establishes standards, I still urge curators and registrars to give significant preliminary thought and planning to standards in documentation. Certainly this type of cognitive familiarity with the process will help you make rational choices about equipment and software your program needs as well as greater clarity about workflow planning and implementation, allowing you to use what you have to document the collection well and to be adaptable to necessary change.

Documentation standards may be established and reinforced by a documentation procedures manual. This is simply a written set of instructions prepared by the garden to help those responsible for the establishment and operation of an efficient collections documentation system. This manual may also be used as a training resource for all new collections documentation personnel. Such a manual will also serve to reassure garden supporters and the public that the garden has assumed responsibility for documenting its collections. Finally, the preparation of a documentation procedures manual serves as a useful vehicle for reviewing the efficacy of the entire system as well as particular records and files.

Zimmerman recommends that the following types of information be included in a documentation procedures manual:

- Definitions of key terms (i.e., registration, accession, cataloging, deaccessioning, etc.)

- Definitions of records maintained at the garden (i.e., pre-entry, entry, registration, catalog, etc.)
- General descriptions of registration, cataloging, deaccessioning, and other procedures to be followed at the garden (i.e., text, outline, point form description, or flow chart of each phase)
- An explanation of how to prepare and manage records supported by examples
- A description of accession numbers, collections labeling, and other means of associating records with plants
- A description of the references and authorities followed by the garden for plant classification and so forth[36]

The following paragraphs are an extract from the *Plant Records Procedures Manual for Longwood Gardens Employees* and may serve as a useful example pertaining to plant name changes.

It is occasionally necessary to change the names of accessioned plants in order to correct misidentification or to keep our names current with those published in horticultural literature. These changes are usually initiated by Plant Records personnel[;] however, if any employee has reason to believe the name of a plant should be changed, this information should be brought to the attention of the Taxonomist or Assistant Taxonomist.

Semi-monthly lists of names to be changed will be distributed to all Foremen, along with new brass labels reflecting the name changes. The Foremen are responsible for keeping Section Heads informed of changes in their sections, and seeing that Section Heads receive the new brass labels. Each label will have a paper tag attached indicating the obsolete name. Section Heads are responsible for removing the obsolete brass labels and attaching the new ones. One brass label for each appropriate location will be sent automatically with the change list. If additional brass labels are needed, they should be ordered by the Section Head, using the Label Request Form. Section Heads are also responsible for ordering new display labels to replace those that have become obsolete due to name changes. Section Heads should also make corrections on their Plant Status cards to reflect name changes. Any questions on name changes should be directed to the Assistant Taxonomist.[37]

Computer Information Systems

Documentation lends itself very well to the automation provided by computer information systems (CIS). These systems can be valuable tools in the documentation process due to their capacity to speed the creation, organization, and retrieval of records.[38] With computer-processed documentation, gardens have the capability to sort or sequence records, to display or print data, or to select specific records for each item of information processed. These activities may be combined in any order.[39] The ease and power with which this work is done is based on how well gardens have planned the use of computer information systems: computerization of their records, what software and hardware they have subsequently chosen for this work, and how the data is recorded during the computerization process. Thus, the proper use of CIS can significantly improve and enhance many documentation functions.

Although it is certainly implicit that any kind of documentation is accomplished these days with the aid of a computer, it is still important to recognize that the application of these devices to an inherently flawed, incomplete, disorganized, or unused system will not rectify these problems. Computer information systems applications are no substitute for the necessary analysis, planning, and organization that must be applied to the development and operation of good documentation systems. *Therefore, before computerizing documentation, gardens should reconsider what records they keep and how those records are processed and used—a similar process to that used to establish a documentation program. Keep in mind the standard computerization axiom, "garbage in, garbage out."*

If gardens have well-developed internal CIS programming capabilities, they may choose to develop their own software and computer networks for documenting their collections as well as other garden operations and programs. It is more likely, and advisable, that botanical gardens acquire and utilize one of the currently available software packages for collections documentation. Acquiring the right software and hardware from among the material available off the shelf now becomes a process of intelligent shopping *supported by a thorough knowledge of documentation needs.*

A rather dated but still valuable resource for gardens as it applies to making choices about software platforms for collections documentation is *A Guide to the Computerization of Plant Records* by Richard A. Brown, published by the American Public Gardens Association (APGA). This guide can assist garden staff in evaluating and assessing their program needs and the ways in which the available software may meet those needs. Included in this publication is a list of "25 sure-fire activities that can lead you down the path to failure" when automating records by Bonnie Canning, contributing editor of *Administrative Management* magazine. I would like to paraphrase that list here by way of creating a planning and analysis checklist.

- Do not simply copy a system used by someone else without thorough analysis of how it meets your needs.
- Vendor-designed systems most often meet their sales quotas, not your needs.
- Create a staff or user committee to assess needs and examine possibilities.
- Be cognizant of the planning, acquisition, start-up, and training time required to complete the process.
- Complete budgets should account for all of the above in addition to hardware, software, materials, technical support, maintenance, and workspace modifications.
- Provide adequate staff training and create a new procedures manual, or update an existing one, to support the automated processes and to establish standards, conventions, and guidelines.
- In assessing your automation needs, include the conversion of existing files.
- Choose reliable, flexible, and expandable hardware and software.
- Prioritize the implementation process in phases.
- Plan and budget for adequate system backup.
- Minimize the number of vendors used to help centralize product accountability and support.
- Consider how your automated data will be validated and allow for this process.
- Prioritize the data that will be captured by the system.
- Evaluate your need for cross-references, multiple-value fields, and linked files to streamline data input, data editing, operation time, and other system functions; seek out systems that will accommodate them.
- Automation will only amplify the problems associated with poorly planned and disorganized manual systems.

With these recommendations in mind, proceed with three major steps recommended by Brown that will help you think about how a given software program aligns with those recommendations:

1. Understand both the basic and specific purposes of your documentation system and its records.
2. Identify the system users (staff, board, visitors, etc.) and their needs, including how they use and apply the data documented by your system.
3. Select a computer system (hardware, software, support, work environment) that will satisfy the purposes of your documentation program and the needs of its users.

Before committing to a particular software program or upgrading to a new one, put together a team or committee of core users to study and implement the change. As they proceed with the relevant analysis, be sure to have all of the committee's work documented and see to it that the perceived needs that a software system must satisfy are articulated in writing. In fact, before making any commitments to a particular system, write up your requirements in the form of a request for proposal (RFP). Regardless of how much work you may have put into assessing your needs, writing up this assessment may significantly expand and change the committee's understanding of the garden's needs.[40]

Small gardens may make good use of both word processing and spreadsheet software programs for documentation. One of the liabilities of using those more rudimentary programs may be a lack of compatibility with relational and other types of more complex databases that are commonly used for documentation, networking, and reference. There are now several widely used and adaptable relational software programs designed for botanical garden documentation programs and use on desktop computers. These programs are often packaged with training and technical support options, technical upgrades, and compatibility with handheld devices for use in the field such as tablets, notebooks, and smartphones. Another added benefit of most of them is compatibility with mapping programs and GIS to handle cartographic files, making a complete and powerful computer information system for documentation.

Before rushing out to purchase any off-the-shelf documentation software program, you should still implement a process to probe your specific documentation needs and create a documentation plan to meet them to ensure that you are acquiring what you need and not simply what the vendor wants you to buy. In addition to knowing what you need, many of the current garden software package capabilities are designed according to a traditional model of collections development, management, and use for such institutions. Those gardens that pursue a specialized purpose or theme, such as historical and landscape gardens, may find that these software programs are much more comprehensive or complex than they need and that their specific documentation needs are better served by other, less complex and less expensive software alternatives.

Below is a synopsis of commonly available botanical garden documentation software programs. The functionality of these programs changes with new versions, and I encourage you to look at the current information on these programs online available from the developers for their updated list of modules and functions.

BG-BASE

Of the current, widely used programs, BG-BASE was the first to become available in 1985. BG-BASE is a relational software program (like the others described here) with seven modules consisting of living collections, preserved collections, conservation, education, propagations, ArcGIS Connector, and HTML/web. Users and administrators can customize or restrict particular modules on a user-to-user basis. This program has the largest number of data fields and is compatible with the AutoCAD-based BG-MAP as well as the ArcGIS Public Garden Data Model for creating cartographic files and map making. Of the available programs, BG-BASE may be the most difficult to learn because of the number of modules, fields, and its overall breadth.

BRAHMS

BRAHMS, the Botanical Research and Herbarium Management System was developed in 1990 by the Department of Plant Sciences at the University of Oxford. Like BG-BASE, it is a large, relational software program with the breadth of utility to serve a number of natural history institutions and programs via select modules. There is a module for botanical gardens that provides for most, if not all, a garden's documentation and other needs as well as accommodating networking with other natural history institutions, such as herbaria. BRAHMS is packaged with links allowing users to map data to a preferred platform, which ranges from Google Earth to ArcGIS. Also like BG-BASE, learning the more advanced functions of BRAHMS is somewhat challenging.

IrisBG

IrisBG was developed in 1996 by Compositae AS in collaboration with the University of Oslo Botanical Garden and is now used by a large number of gardens. Like the other systems listed earlier, IrisBG is a relational software system of integrated modules mostly presented in form and tabular views with high functionality in collections documentation and other botanical garden needs. Of the three described here, it is probably the easiest to learn. IrisBG has a mapping module that exports to Google Earth and ArcGIS. As for BG-BASE, the ArcGIS Public Garden Data Model will synchronize well with IrisBG for cartographic data, mapping, and map making.

All three of these programs can export data in a range of formats for various types of analysis, interpretation, and application. They also allow for comprehensive data storage and tracking in the area of collections care and preservation with rapid data entry functions for mobile devices. These programs allow for a wide range of data sorting and matching outputs that will greatly assist a range of research-oriented queries.

As important documentation database software products were being developed and put to use, digitizing and managing cartographic files with practical software solutions was lagging behind. Now, however, there are some practical and powerful cartographic software programs available for use in collections mapping and geo-spatial forms of collections analysis and research. The first software program designed specifically for botanical garden mapping and map making is BG-MAP. This AutoCAD-based program was designed to link with BG-BASE forming a rudimentary geographic information system (GIS) to map collections and respond to queries requiring both attribute and geographic data to provide powerful curatorial and collections management information.

In addition to BG-MAP, and as I mentioned earlier in the section "Cartographic Files," the American Public Gardens Association and ESRI have come together to promote and facilitate the use of GIS in botanical gardens. This partnership relies on the ArcGIS Public Garden Data Model that is based on Esri's ArcGIS software. The three collections documentation software programs I described above may be linked to ArcGIS. However, as of this writing, the ArcGIS Public Garden Data Model facilitates synchronization with BG-BASE and IrisBG.

As I hope you can see by now, a very important element to consider in developing a computer information system is the selection of attribute database and cartographic database software that is compatible. This relationship should be clear when you consider the number and kinds of queries you make of both your geographic and attribute databases. For example, you might want to know where in the garden are all of the plants damaged in last winter's cold or all of the plants of Mediterranean nativity. These queries involve the melding of attribute and geographic databases, a function at the heart of GIS. The goal in obtaining and using a GIS is to provide at least a rudimentary linkage between the garden's attribute database and a graphical display system.

The real power of cartographic databases lies in two areas: (1) the combined use of topographical and thematic map data as described in the discussion of cartographic records and (2) spatial analysis and modeling for the purpose of tracking specialized data and projecting changes in monitored phenomena (e.g., the spread of disease). Both CAD and GIS software have varying capabilities in this area. In fact, CAD software does not have a spatial analysis capability. CAD will convert maps and images into digital data for selective retrieval and display. David Cowen explains the utility and limitations of CAD software in the following excerpt from his article in the journal *Photogrammetric Engineering and Remote Sensing*:

> In essence, CAD systems handle geographic data in the same manner as photographic separations are used for the production of topographic maps. Different types of geographic features are placed on individual layers that are then combined and printed with different colors and line styles to generate the final product. Although the concept is the same, CAD systems provide much more versatility in terms of display functions than do their photographic counterparts, and are particularly beneficial for editing and updating.
>
> While offering major improvements over photo-mechanical methods of map production, CAD systems have severe limitations when it comes to analytical tasks. In particular, it is difficult to link attributes in a database to specific geographic entities and then automatically assign symbology [symbols] on the basis of user-defined criteria.[41]

In other words, GIS systems may be used to make new maps based on themes derived from your plant records, for example, provide a map highlighting all the plants in the collections that suffered cold damage more than once in the last 10 years. CAD systems, by themselves, are graphic systems for storing and retrieving topographic data to produce maps. Unlike GIS systems, they cannot automatically create thematic maps based on values and attributes stored in the plant records database regarding, say, plant care activities, conservation status, or phenology. Since Cowen's article first appeared, the basic differences between CAD and GIS programs have not appreciably changed other than the fact that each is now easier to use and more affordable.

As mentioned earlier, CAD-based systems do not have the capability of GIS to perform analytical functions. For instance, they can't overlay two or more thematic map layers in a database to find structural relationships between the themes, and they can't use previously derived relationships to locate areas that meet specific criteria. GIS software can also statistically analyze the extent of different classes on a thematic layer or correlate the extents of the classes on two different thematic layers. GIS software can use thematic data to create a model of the real world. These models can then be used to simulate real-world occurrences, making feasible a whole range of studies based on garden changes, trends in collections preservation or conservation activities, and the examination of the possible consequences of other future decisions. Finally, GIS software is often a complete system containing both attribute and cartographic databases that seamlessly interact, such as ArcGIS.

The differing functions of CAD and GIS software programs are only meaningful within the context of a garden's programmatic needs. Day-to-day documentation needs will likely be adequately fulfilled by certain CAD systems. Many types of public gardens simply have need of easily generated and edited maps showing where plants and other features in the garden are located. GIS programs may be more suitable for those gardens with more comprehensive curatorial research programs that place greater analytical demands on the documentation system. Regardless of the cartographic software you choose, you will also need to decide how best to map individual plants and features in the field. Most gardens are using either GPS or total station technology for this purpose.

Shopping for computer hardware, software, and the associated materials and services is made more effective if it is done systematically so that all the available products may be easily compared. The bottom line is to ensure that the system will address your garden's particular needs within the confines of a particular budget. A checklist will help you with the process of selecting the right system for your garden.

Retrospective Documentation and Inventory

Inventorying large collections, whether they be stock items in a department store, 19th-century paintings in the collections of the Metropolitan Museum of Art, or woody plants in an arboretum's collections, has been likened to painting the Golden Gate Bridge—nonstop. The ongoing documentation of new plants and edits to location files and fields represents this process. Even the most comprehensive, organized documentation programs succeed or fail by the accuracy and efficiency of their inventory control processes. Compiling and maintaining accurate inventories is the heart of the collections control portion of a garden's documentation system. Retrospective documentation is a means to improve the standard of documentation of an existing collection through a retrospective inventory, verification, cataloging, and indexing program.[42] "If the museum has any doubts concerning the effectiveness of its approach to inventory and location control, it is recommended that detailed consideration be given to the adoption of formal control procedures, a retrospective inventory program and regular internal documentation and collections audits."[43]

An inventory is essentially any part of the records and cartographic database showing that an accession is in the collection. Manual documentation systems may contain a separate inventory list or record that is often attached to the maps comprising the cartographic records. Computer information systems should have the capability to compile an inventory of the collection from the existing databases. Invariably, however, there are lapses in the system that may lead to lost, inaccurate, or disconnected documentation and a consequently inaccurate and incomplete inventory.

Periodic assessments, or audits, of the documentation system will help reveal problems in the accuracy and completeness of your documentation. The Museum Assessment Program (MAP) administered by the Institute of Museum and Library Services in Washington, D.C., may be a useful resource in helping you assess the efficacy of your documentation program and the accuracy of the information it contains. If you discover problems, the next step will be to ascertain their extent. Problems at the level of individual plants or with smaller collections may be rectified through a rather simple process of backtracking, source verification, and other means. Serious problems will likely warrant a complete retrospective inventory and verification process.

I've already alluded to the importance of embedding the collection of catalog data in other collections work, such as preservation activities. As documentation becomes a more automated part of other activities involving direct contact with the collection, the inventory of the collection should become more complete and accurate as a matter of course. Still, and as I noted earlier, it may be necessary at times to conduct a retrospective inventory. Such work checks that accessions in the system actually exist in the collection and can be tracked. This is also referred to as "stocktaking" and means checking that the plant and its label exist. Using either hard-copy maps and printed inventories or portable devices with access to this information, field personnel locate the accessions and their labels recording any discrepancies. Regarding field processes for retrospective inventories, see the following example from the Arnold Arboretum of Harvard University.

Best Practice 4.2: Field Inventory Processes at the Arnold Arboretum
The Arnold Arboretum *Plant Inventory Operations Manual* begins with a succinct list and explanation of what is to be done during a collections inventory:

- all accessioned plants to be routinely verified as being present or absent;
- the specific map locations of accessioned plants to be verified for accuracy and precision;
- the identities of accessioned plants to be coarsely verified for accuracy; and
- the primary means of collections control in the field (records labels) to be assessed for accessibility, accuracy, and presentation.

To meet these objectives, the Arboretum fields expert curatorial staff able to conduct inventories as well as troubleshoot an array of taxonomic, cartographic, and horticultural puzzles.[44]

The importance of maintaining an accurate inventory is the raison d'être of the documentation program and can't be overstated. The manual states: "[B]ecause curatorial and horticultural staff rely heavily upon inventories and maps of the living collections in their daily work, failure to maintain these at their fullest can result in an incomplete and inaccurate inventory. This may in turn jeopardize the future preservation of the accession in question."[45]

The manual specifies that inventory field checks will be conducted every five years in specified locations by the plant records manager. The Arboretum is subdivided into five operational regions, hence a single region will be inventoried each year. The staff of the Arboretum recognize that there may be ad hoc inventories conducted at any given time in areas and with collections that are undergoing special or focused programming use. This five-year rotation may also be modified based on the occurrence of natural or other kinds of disaster events that impact regions and collections within the Arboretum.

The inventory process proceeds in eight steps using a two-person team:

Step 1—Notify the Horticulturist: inform the horticulture staff of the impending inventory work and obtain their advice and feedback on the process.

Step 2—Gather Essential Materials and Tools: computer, GPS receiver, cell phones, measuring devices (tape measures, LaserAce hypsometer, Haglof clinometer), references, pruners, temporary labels.

Step 3—Preliminary Review of Records and Maps: this will speed the field checking process.

Step 4—Conduct the Fieldwork: Two-person team working on corroborating plant identity and location with the documentation as well as the condition of the accession labels. Also, assessing overall plant condition for the catalog record. There are established standards and forms for all the procedures that take place at this step.

Step 5—Complete Office Tasks: entering or amending record and map data, filling out activity requests, label making, consulting.

Step 6—Follow-up Field and Office Tasks: confirm absence of plants not found, attach new accession labels, authenticate maps.

Step 7—Summarize Work with Horticulturist: share inventory results and summary report and activity requests.

Step 8—Archive Results: double check that all inventory data has been properly archived in the system.[46]

Finally, the Arnold Arboretum *Inventory Manual* contains a helpful resource list and appendixes for its users with specific details and instructions for conducting various aspects of the inventories.

The subject of collections evaluation will be taken up in greater detail in chapter 8, "Research and Curatorial Practice."

This is a good point at which to take up in greater detail the subject of collections verification. Gardens striving to develop comprehensive and credible living collections have an active and ongoing verification program. Keep in mind that it is quite easy to become sidetracked into a full-blown taxonomic verification project. In many cases, this may be a warranted step to take if a documentation audit reveals problems in the system of a magnitude to justify such action. Routine verification can be considered a regular and ongoing part of the catalog stage of documentation.

To begin with, verification shouldn't be considered an all or nothing proposition. Gardens with active collections acquisitions programs operate with collections verifications at various levels of authentication; it's a dynamic process in keeping up with acquisitions and authoritative reclassifications. Documentation systems should allow for an indication of various levels of verification—from none to the most authoritative verification—by an expert on that particular taxon, with variable levels of verification in between. A common level of verification (other than "none") is most often achieved by curatorial staff using a current, authoritative reference, such as a monograph.

This process may begin by confirming the name given to the plant when it was received by the institution, or, if unknown, undertaking to determine the plant's identity. This is followed by checking to make sure that the name is valid. Those accessions with an unverified identity might be given the notation "affinity" or "aff." A good example might be any number of plants acquired through field collections when careful recognition and verification may not have been possible. In this case, the taxon name should be recorded in the field records and later transcription into the documentation system with the designation "aff.," such as "*Plantaginaceae*, aff. *Veronica* . . ." until it receives some level of verification. Given the difficulty many gardens encounter in keeping up with verifications, working to move accessions beyond the category of "none" or "aff." is an important first step.

Galen Gates, director of plant collections at the Chicago Botanic Garden, makes the following recommendations for verifying the identity of collections:

> For an institution to establish a reputation of veracity or scientific accuracy, a methodical system to authenticate its holdings is critical. Three methods are currently used to verify collections:
>
> 1. The classic phenotypic approach of comparing morphological traits with previously authenticated herbarium specimens and scientific literature;
> 2. A molecular approach of comparing DNA and other relevant chemotaxonomic material with other authenticated samples; and
> 3. The newer approach of digital imagery of exterior features taken during peak bloom periods that are then compared with known specimens or with literature.[47]

In an effort to gain more control and efficiency when verifying the identity of plants in the collection, the Royal Botanic Garden Edinburgh (RBGE) decided to organize the process around groups of genera. The staff began by selecting an important generic group that needed verification and then building a complete herbarium collection representing that group. Completing the herbarium collection on the target genus may be the most time-consuming step depending upon the status of the existing herbarium collection at the start of the project. In some cases, the staff at RBGE worked over the course of a year or more to amass all the necessary samples to serve as a complete reference for verification. Identifying and bringing together the necessary reference literature is another important piece of the background work that was done before the actual verifications began. While all this background work was being done, the routine, somewhat ad hoc, collections verification process continued unabated.

An example of how this genus-based processed worked is illustrated with *Acer*. The following is an excerpt from the report by Rob Cubey and Martin F. Gardner in *Sibbaldia*:

> Two hundred and forty-two accessions of *Acer* were verified and the exercise resulted in the confirmation that the vast majority (197) of the accessions were correctly named. However, it resulted in name changes for 35 accessions through identification or re-identification and a further ten accessions had their names changed due to synonymy. Two accessions were found to be dead and four accessions could not be verified due to a lack of diagnostic characters.
>
> The collection and drying of the 302 specimens that were gathered took around ten days in total. The review of the taxonomic literature took another

day, and the curation of the existing reference material in the herbarium took a further ten days. The actual verification of the 302 specimens once the background work had been completed took three weeks.[48]

The RBGE genus approach to verification demonstrates an efficient and accurate means of verification when the necessary staff and materials can be brought together at one time to accomplish the work. It requires a dedicated approach that has great benefit for increasing the programmatic value of the institutions collections.

Documentation programs should be regularly audited to verify a standard of operation regarding collections control and curation. A program to audit documentation will include the monitoring and evaluation of the overall effectiveness of entry, acquisition, inventory, location, and exit control procedures to ensure that they are effectively and consistently applied. The process will also ensure that the documentation staff are familiar with documentation priorities.[49] Such an audit may be structured to have a secondary, but also very important, impact in helping to determine if the collection is developing according to the collection policy.

An audit may involve a broad evaluation of the entire documentation program or a random audit of selected portions. In either case, the audit must be objective and should be conducted by an individual or a team of individuals independent from the documentation staff. Again, the Museum Assessment Program may be a good resource for this work the first time through, and thereafter the garden should make a more permanent arrangement to handle a regular audit.

Roberts suggests three methods that may be used as the nucleus for an audit program:

- A review of practical control documentation *procedures* to verify that the procedures are being correctly implemented;
- A check on the control and attribute *records* to verify that they are being correctly maintained; and
- A *physical check* or *stocktaking* of the collection to verify that items in the collections have not been misplaced or misappropriated and that accurate documentation has been prepared concerning their acquisition and location.[50]

Conducting regular audits will reduce, and may eliminate, the need for a full-scale collection verification by providing an opportunity to identify and correct documentation problems before they affect the entire collection.

However, if the results of an audit suggest it, a retrospective inventory and verification may do several things:

- Improve the standard of control over an existing collection for which inventory and location records have never been developed
- Establish new records
- Reconcile collections and existing records
- Verify correct identification

Before taking on such a project, a proposal should be drawn up articulating the extent of the documentation project, the staff time required, the inventorying and verification process, and the effect on other collections activities. The proposal should then have the full support of the senior staff and others affected by the undertaking. Before being fully implemented, the project processes should be fully tested on a small but representative portion of the collection.

Data Sharing

Sharing collections and collections information may be one of the most pressing issues in plant conservation for botanical gardens. To assist in this endeavor, there are several documentation and data-sharing tools available. The data management software programs described earlier all have data-sharing capabilities. For instance, BG-BASE has a multisite searches webpage. Botanic Gardens Conservation International has several collective and publicly available databases—GardenSearch, PlantSearch, Global-TreeSearch, and ThreatSearch—that have been discussed previously in this chapter and in chapter 3. In chapter 3, I used them in a hypothetical example to demonstrate their utility in searching for a particular group of plants and a particular taxon regionally, nationally, and globally in making decisions about collections development.

GardenSearch is, as the name implies, an online, searchable directory of the world's botanical institutions. This includes not only botanical gardens but also seed banks, zoos, and network organizations. The database is complete to the extent that all participating institutions keep their profiles up to date (you may have to query specific institutions for the most up-to-date information). Each entry includes general information as well as information about collections, specialty programs, facilities, and staff expertise. PlantSearch is a detailed extension of GardenSearch focused on inventories of institutional collections: "In addition to hundreds of living plant collections around the world, PlantSearch includes taxon-level data from gene and seed banks, cryopreserved and tissue culture collections. Find out which plants (threatened, rare, alpine, medicinal and more) are maintained in botanical collections around the world. . . . You can find out how many botanical collections report holding each taxon."[51] It is the only comprehensive global database of plant species in *ex situ* collections and indicates which of these are considered threatened based on its linkage to the global IUCN (International Union for Conservation of Nature and Natural Resources) Red List.

GlobalTreeSearch is also a searchable database of trees and their global distributions. The database may be searched by taxon or country, and it contains more than 60,000 species and their range maps. A search for *Lithocarpus* brought up 340 species showing the countries where they are found. ThreatSearch is a global database showing all threat assessments for a specific taxon. The threat data is compiled from the National Red List; the Royal Botanic Gardens, Kew; NatureServe; and the CNCFlora. A search of *Lithocarpus henryi* indicates that this species is not listed in any threat category. These BGCI databases are incredibly useful, and I encourage all botanical institutions to contribute to them.

Other, more comprehensive data-sharing *and* collection-sharing opportunities exist under the auspices of the American Public Gardens Association's Plant Collections Network and the Center for Plant Conservation's National Collection. "The American Public Gardens Association Plant Collections Network coordinates a continent-wide approach to plant germplasm preservation" based on multi-institutional, shared family, generic, and ecological/geographical collections. "Plant Collections Network is a long-term collaboration between the American Public Gardens Association and the USDA–Agricultural Research Service."[52] The Center for Plant Conservation's National Collection, like the APGA network, is a multisite collaboration of 62 institutions preserving more than 1,600 imperiled native American plants as living collections of plants and seeds.

Preserving Documentation

No less an effort should be mounted to preserve documentation than would be directed at the plant collection itself.

Responding to disasters—regardless of what form they take—comes down to two key principles: redundancy and dispersal. It is important to keep multiple copies (redundancy) of data, software, and servers in different physical locations (dispersal).[53]

In other words, the documentation system should be located in facilities with environmental controls suitable for the long-term maintenance and preservation of paper documents and computer hardware. These facilities should also be outfitted with fire protection equipment and be easily accessed by fire department personnel.

All documentation should be maintained in duplicate with the duplicates preserved at another location. Duplicate hard-copy documents on microfiche and store these in a suitable location away from the records office. Computer databases should be duplicated on disk or tape on a daily basis and also stored at a secondary location.

Originals and duplicates of documentation should be carefully stored and preserved. Paper and other hard-copy documents should be appropriately archived to maximize access, utility, and long-term preservation. Computers should be placed in a location and used in a fashion to maximize their performance. Database entries and changes should be saved regularly throughout the computer session, and storage media such as USB memory sticks, external hard drives, network attached storage, and other means should be clearly labeled, filed, and protected.

Archives

The concern botanical gardens have for documenting their collections and preserving that documentation is, in effect, a type of archiving.

> [Archives] provide the intellectual infrastructure that documents the history of a public garden's endeavors and are the repository of its permanently valuable records. The records held by the archive serve as confirmation of decisions made and actions taken; they capture the work, creativity, and thinking of both individuals and departments or units within the institution; they give insight and evidence about the culture of the organization.[54]

The person at museums charged with creating and managing the archive is the archivist, a profession not too different from a botanical garden registrar. As such, I think it makes institutional sense for the registrar at most botanical gardens to have some familiarity with archival practice, particularly as it pertains to preserving collections documentation, but also for preserving other documentation of importance to the institution. In this expanded effort, they may need the advice and consent of other botanical garden professionals outside the area of collections management.

Documentation outside the area of collections management will need consideration as a separate archival unit and, therefore, will necessitate planning, policy definition, and an outline of practice specific to that unit. It's likely that this will involve the curating and preservation of historical documents, images, and other things unlike those connected to the plant collection. The registrar will need some basic training and/or support in how to preserve unusual types of paper, inks, photographic materials and the adhesives that are used with these elements.

Preservation of archived material involves steps that may be viewed as a combination of what is done for plant collection records and herbarium vouchers. This includes environmental conditions such as temperature, humidity, and exposure to UV radiation, as well as infestations of various pest organisms. Some helpful archivist resources include the following:

- National Archives: Archives Library Information Center: Archives and Records Management Resources
- Society of American Archivists: Standards and Best Practices Resource Guide
- Historical Society of Pennsylvania Resources for Small Archives
- Elizabeth Yakel, *Starting an Archives* (Lanham, MD: Scarecrow Press, 1994)
- David W. Carmicheal, *Organizing Archival Records* (Lanham, MD: Rowman & Littlefield, 2018)

For additional information about preserving documentation, consult with an archivist at the local historical museum, your computer vendor, and officials from the local fire department.

In summary, I have described a template for a generic documentation system that is broken down into specific documentation stages. Within each of these stages, curatorial personnel carry out certain activities and processes that document collections. The collections are documented by the acquisition, creation, modification, preservation, and use of data stored as records in various electronic, hard-copy, and/or document files.

Textbox 4.1. **Recommendations for Documenting Collections**

Basic
- The garden acquires and preserves documentation on its plant collections in the following categories:
  - _ Accession number
  - _ Plant name
  - _ Verification
  - _ Accession date
  - _ Material received (seed, scion, etc.)
  - _ Condition upon receipt
  - _ Source (immediate and original) and lineage
  - _ Deaccession date

- The garden establishes, preserves, and updates documentation on its plant collections in the following categories:
  - _ Garden location
  - _ Current condition
  - _ Preservation and other activities directed to the collections

- The garden labels all of its accessions with their accession number.
- Electronic documentation uses a convertible or widely accepted software platform.
- Documentation data is accessible to the public.
- The garden preserves its documentation by duplication and responsible storage.
- A documentation audit is performed every five years.

Intermediate
- The garden builds and maintains separate pre-entry and entry records of acquisitions before they are accessioned.
- The garden builds and maintains detailed catalog records on its collection.
- Field collections are fully documented.
- Collections are documented in a herbarium and verified.
- Documentation procedures are described in a procedures manual.
- The garden uses shared or standardized data fields useful for networking documentation between institutions.

Advanced
- Collections are photo documented.
- Hard-copy documentation is properly archived.
- Field-collection mission reports are published.

Notes

1. Carl Guthe, *Documenting Collections: Museum Registration & Records*, American Association for State and Local History Technical Leaflet 11 (Nashville, TN: American Association for State and Local History, 1970).

2. Wright Report, Great Britain, in D. A. Roberts, *Planning the Documentation of Museum Collections* (Duxford, Cambridge: Museum Documentation Association, 1985), 1.

3. American Association of Museums, "Stewards of a Common Wealth," in *Museums for a New Century* (Washington, DC: AAM, 1984), 46.

4. E. Orna and C. Pettitt, *Information Handling in Museums* (London: Clive Bingley, 1980), 11.

5. K. A. Schmiegel, "Managing Collections Information," in *Registrars on Record*, ed. Mary Case (Washington, DC: American Association of Museums, 1988), 55.

6. C. Sawyers, "Where to Start: Plant Records," *Public Garden* 4, no. 1 (1989): 41.

7. Roberts, *Planning the Documentation*, 25.

8. Roberts, *Planning the Documentation*, 98.

9. Roberts, *Planning the Documentation*, 27.

10. Roberts, *Planning the Documentation*, 36.

11. H. Moss and L. Guarino, "Gathering and Recording Data in the Field," in *Collecting Plant Genetic Diversity*, ed. L. Guarino et al. (Wallingford, Oxon, UK: CAB International, 1995), 367.

12. Moss and Guarino, "Gathering and Recording," 370.

13. Moss and Guarino, "Gathering and Recording," 369.

14. Moss and Guarino, "Gathering and Recording," 375.

15. Moss and Guarino, "Gathering and Recording," 376.

16. Moss and Guarino, "Gathering and Recording," 382.

17. Moss and Guarino, "Gathering and Recording," 373.

18. National Germplasm Resources Laboratory, *Plant Exploration Guidelines for FY1997 Proposals* (Beltsville, MD: U.S. Department of Agriculture–Agricultural Research Service, 1995), 3.

19. Reinout Havinga and Havard Ostgaard, "Barcodes Are Dead, Long Live Barcodes! Improving the Inventory of Living Plant Collections Using Optical Technology," *Sibbaldia* 14 (2016): 134.

20. Havinga and Ostgaard, "Barcode," 134.

21. Havinga and Ostgaard, "Barcode," 135.

22. Sean Hoyt et al., "A Tree Tour with Radio Frequency Identification (RFID) and a Personal Digital Assistant (PDA)" (paper presented at IEEE IECon '03, Roanoke, VA, November 2–6, 2003).

23. Botanic Gardens Conservation International, "BCGI Databases," https://www.bgci.org/resources/bgci-databases/.

24. J. Atwood, "Spirit Collections," *Public Garden* 12, no. 1 (1997): 35.

25. David D. Michener, "Collections Management," in *Public Garden Management*, ed. D. Rakow and S. Lee (Hoboken, NJ: Wiley, 2011), 263.

26. Alliance for Public Gardens GIS, "Guide to GIS for Public Gardens: Botanical Gardens, Zoos, and Parks," March 2013, 1, https://publicgardensgis.ucdavis.edu/sites/g/files/dgvnsk6621/files/inline-files/Guide-to-GIS-for-Public-Gardens-March-2013.pdf.

27. American Public Gardens Association, ESRI partnership, https://www.publicgardens.org/members/member-affinity-programs/esri-gis-software.

28. Alliance for Public Gardens GIS, "Guide," 52–54.

29. Alberta Museums Association, *Standard Practices Handbook for Museums* (Edmonton, AB: AMA, 1990), 111.

30. M. C. Harty et al., "Cataloguing in the Metropolitan Museum of Art, with a Note on Adaptations for Small Museums," in *Museum Registration Methods*, ed. D. Dudley and I. Wilkinson (Washington, DC: American Association of Museums, 1979), 222.

31. Natasha Franchon, Martin Gardner, and David Rae, "The Data Capture Project at the Royal Botanic Garden Edinburgh," *Sibbaldia* 7 (2009): 77.

32. Franchon, Gardner, and Rae, "Data Capture," 78–80.

33. David Rae, "The Value of Living Collection Catalogues and Catalogues Produced from the Royal Botanic Garden Edinburgh," *Sibbaldia* 6 (2008): 1.

34. M. Malaro, *A Legal Primer on Managing Museum Collections* (Washington, DC: Smithsonian Institution, 1985), 146.

35. The Arnold Arboretum, The Living Collections Policy (2007), 5.

36. C. Zimmerman, "Preparing a Collections Management Procedures Manual," *Dawson & Hind* 15, no. 1 (1988/1989): 26–28.

37. R. Darke, ed., *Plant Records Procedures Manual for Longwood Gardens Employees* (Kennett Square, PA: Longwood Gardens, 1991), 2.

38. American Association of Museums, "Stewards," 46.

39. R. A. Brown, *A Guide to the Computerization of Plant Records* (Swarthmore, PA: American Association of Botanical Gardens and Arboreta, 1988), 6.

40. L. Sarasan and J. Sunderland, "Checklist of Automated Collections Management System Features, or How to Go about Selecting a System," in *Collections Management for Museums*, ed. D. Andrew Roberts (Cambridge: Museum Documentation Association, 1988), 57.

41. David J. Cowen, "GIS versus CAD versus DBMS: What Are the Differences?," *Photogrammetric Engineering and Remote Sensing* 54, no. 11 (1988): 1552.

42. Roberts, *Planning the Documentation*, 114.

43. Roberts, *Planning the Documentation*, 98.

44. Arnold Arboretum, *Plant Inventory Operations Manual*, 2nd ed. (Cambridge, MA: Arnold Arboretum, 2011), 2.

45. Arnold Arboretum, *Plant Inventory*, 2.

46. Arnold Arboretum, *Plant Inventory*, 6–17.

47. G. Gates, "Characteristics of an Exemplary Plant Collection," *Public Garden* 21, no. 1 (2006): 29.

48. Rob Cubey and Martin F. Gardner, "A New Approach to Targeting Verifications at the Royal Botanic Garden Edinburgh," *Sibbaldia* 1 (2003): 22–23.

49. Roberts, *Planning the Documentation*, 128.

50. Roberts, *Planning the Documentation*, 128.

51. Botanic Gardens Conservation International, PlantSearch (database), https://www.bgci.org/resources/bgci-databases/plantsearch/.

52. American Public Gardens Association, "About the Plant Collections Network," https://www.publicgardens.org/programs/about-plant-collections-network.

53. Carissa K. Dougherty, "Disaster Recovery Planning: An IT Perspective," *Public Garden* 31, no. 3 (2016): 7.

54. Sheila Connor, "Public Garden Archives," in *Public Garden Management*, ed. D. Rakow and S. Lee (Hoboken, NJ: Wiley, 2011), 353.

CHAPTER FIVE

Preserving Collections

Collecting wisely and preparing good documentation would hardly make sense if the collections were then allowed to deteriorate or disappear. Caring for collections is part of the definition of a museum.[1]

In competition with more visible public programs and popular special exhibitions, which offer immediate, tangible rewards to the museum and for which funding is more often available, the less glamorous, behind-the-scenes activities can too easily be pushed aside. In the pressure to serve the public's immediate enthusiasms, it is perhaps too easy to slight its ultimate best interests.[2]

The first obligation of botanical gardens—the adequate management of the plant collections—is strongly predicated on the preservation and care of those collections. But what is meant by "preservation"? The use of the terms *preservation* and *conservation* among botanical gardens seems almost interchangeable and often vague. For the purposes of this manual, preservation is defined as any action taken to stabilize and prolong the life of collections and other botanical garden features. Conservation, on the other hand, may be considered any action taken that contributes to the survival and *recovery* of a species, taxon, habitat, or landscape. Conservation, then, may be interpreted as primarily a restorative practice. There are elements of science and art to both of these undertakings. The preservation and conservation of plants are primarily scientific endeavors. The preservation of exhibits, displays, and landscapes is primarily an art.

The preservation activities of botanical gardens extend beyond the collections of living plants to other segments of the institutional collections, which may include seeds, tissues, herbarium specimens, landscape features, and other objects. The primary focus of this section will be on the plant collections themselves.

Preserving plant collections is, at its root, a horticultural endeavor carrying with it a required breadth and depth of knowledge appropriate for the cultivation of diverse collections of plants. As described at the beginning of this manual, plant preservation is most often the direct responsibility of the horticulturist or collections manager who reports to and collaborates with the curator. Textbooks and other references to basic horticultural and landscape maintenance practices abound, hence these subjects are not covered here. However, preserving the plants themselves is not the whole of our obligation; we must also preserve their genetic and programmatic integrity. To adequately preserve plant collections, we must expand the scope, meaning, and perspective commonly associated with plant care and horticultural practice to make them consistent with the needs of botanical collections. In addition, we must document our preservation activities and use this documentation to improve our preservation efforts. It is this complex of preservation requirements that adds a uniquely curatorial dimension to an otherwise horticultural activity.

Horticulture practice as it is applied to botanical collections takes on a whole new level of comprehension and rigor. Although, as I mentioned above, it's not my intention to deliver a primer on basic horticulture practice, it is worth a quick review of what constitutes horticulture practice to put it into a botanical garden context. Horticulture practice may be subdivided into five broad technical areas:

- Controlling the plant's environment;
  - Microclimate: site selection, exposure, soil
  - Soil chemistry and amendment
  - Water supply

- Directing plant growth;
  - Physical: pruning
  - Biological: grafting
  - Chemical: growth regulators

- Controlling biological competition;
  - Weeds and invasive species

- Vertebrate / invertebrate predation
- Pathogens and parasites

- Plant propagation;
  - Seed germination
  - Vegetative rooting
  - Grafting
  - Tissue culture

- Plant improvement.
  - Plant selection
  - Plant breeding

These technical areas comprise all of the informed decisions and practical skills that constitute horticulture practice, a unique blend of science and art. The horticultural decisions necessary for work to proceed are based on a blend of scientific understanding in botany, plant ecology, and other disciplines. The skills required to perform the necessary tasks stemming from these decisions may be considered an art form; grafting is a good example. The decisions made and actions taken having to do with each of these technical areas must be done with greater acumen and effectiveness for botanical garden applications. The principal reason for this has to do with the tremendous diversity of plant life, and their requisite horticultural requirements, in the charge of the curatorial staff. The areas of plant propagation and plant improvement are often segregated from the others for special focus and application.

When we consider the importance of recognizing and understanding the environment necessary to successfully cultivate a particular plant, along with the decisions and practices required to provide that environment for what is usually a tremendous variety and diversity of plant life grown at botanical gardens, this alone is a daunting task. In consideration of this challenge, I am reminded of research on the successful cultivation of an extremely difficult group of plants: tropical rheophytes (which can be defined as aquatic to semiaquatic plants that grow in fast-flowing water) in highly controlled environments with warm temperatures, specialized soils, and careful irrigation regimes.[3] Institutions dedicated to growing such plants may also accept the challenge of growing plants from tropical cloud forest environments with their own difficult and exacting requirements for growth. Botanical garden curators and horticulturists must be very familiar with the environmental contexts that support such plants and the specific factors for growth that

their habitats provide. Then, they may be able to successfully devise horticultural regimens for controlling the environment that will allow them to
grow and preserve these plants for education, research, and display. These
decisions, and the necessary horticultural knowledge, come to bear early in
the curatorial program of a botanical garden when decisions are made on the
nature of the plant collection and the acquisition of individual taxa.

The same may be said for the application of the other four broad areas of
horticulture technology as they are required in a botanical garden context.
For example, an extensive understanding of the natural history, growth and
phenology, reproductive biology, and other botanical aspects of the plants
in a garden's collection are necessary to direct their growth appropriately
and successfully through physical, biological, or chemical means for optimum development and longevity. Also, a wide range of plants acquired
by a botanical garden from many sources over the course of one year may
harbor—and once in the garden, may also attract—a wide range of biological competitors in the way of various pest organisms. Targeted monitoring
and sampling regimens may need to be devised and a variety of management
protocols tested and evaluated. Curators in receipt of a limited number of
seeds and other propagules from exotic plants acquired in the field and/or in
exchange, must research, devise, and follow a clear set of propagation protocols to minimize losses on the way to the successful growth of these plants.
And finally, plant improvement is the raison d'être of many botanical garden
research programs, and they embody their own set of specialized horticultural
techniques involving plant selection and/or breeding. Coming back to my
initial point on this subject, horticulture practice in a botanical garden must
be of the highest caliber within the broadest context, an important segment
of a curator's expertise.

Of equal importance to all of the above, and something else that distinguishes horticulture practice in botanical gardens, is the necessity to
thoroughly document all this work as a means to research and perfect our
preservation efforts and educate the public.

Sustainable Practice

"An overarching tenet for collections preservation programs is that they
be sustainable. For botanic gardens, as leading environmental organisations, we face an even greater challenge not only to improve our own
environmental performance but also to ensure that we develop our
institutions as models of sustainability."[4]

Every facet of a botanical garden operation should be scrutinized for sustainable practice, and a good place to start is the American Public Gardens Association Sustainability Index.[5] The Public Gardens Sustainability Index "is a suite of attributes intended to inspire gardens to advance their own garden sustainability programs and operations to further the mission of their institution while connecting to local, national, and global sustainability efforts."[6] The "Environmental" segment of the index is most apropos to sustainable horticulture technologies for collections preservation.

Another valuable resource is *The Climate Toolkit for Museums, Gardens and Zoos*.[7] The toolkit is divided into eight sections with one dedicated to "Landscapes and Horticulture" for collections preservation. These two important resources, as well as what is presented here, serve primarily as general guidelines for sustainable practice. The curatorial staffs of each garden may work within these general recommendations to craft an approach specific to the unique circumstances of their institutions landscape setting and plant collection.

Controlling Biological Competition

From the standpoint of collections preservation, many would agree that controlling biological competition—pest management programming—is perhaps the area of greatest challenge for developing an efficacious *and* sustainable preservation program. For the collections manager or horticulturist, cultivating and preserving healthy accessions is of principal concern. If they follow the basic tenets of good horticulture they will most often be spared the disappointments associated with unhealthy plants. However, and as we all know, even the most ardent horticulturist may be plagued with pest problems. Regardless of the type of pest problems encountered, curators and horticulturists, like any good managers, will need to devise and implement a flexible and diverse strategy for managing the problem to obtain the most suitable outcome. Comprehensive pest management is pest identification and control based upon a sound knowledge of management, biology, and horticultural principles. Curators should possess a sound knowledge of the key pest organisms typically associated with the collections in their care.

Pest management involves the correct identification of potential pest organisms, a determination of pest status (is it a pest or a guest?), and the development of strategies for controlling pests in which all available techniques are evaluated and consolidated into a unified and strategic program. If we substituted the words *issues* or *problems* in place of "pest," we would have a working definition of good management that could be applied in any context. Keep in mind that the term *pest* has no ecological validity. People place an organism in a pest category, and some species may be considered pests in one context and beneficial organisms in another.[8]

Pest management, like all good management, is derived from a flexible, diverse, and integrated approach. With this understanding, the phrase "integrated pest management" is somewhat redundant. Also, good managers do not offhandedly dismiss any potential management strategy without careful analysis. Pesticides present one possible pest control strategy that must be carefully considered. Once the broad legal definition (in the United States) of "pesticide" is fully understood, one that includes many sustainable alternatives, then their potential utility in pest management programs takes on a new, and more essential, quality. On the other hand, certain pesticides present such serious hazards that this aspect alone makes them worthy of special consideration within any pest management operation. A sustainable approach to managing biological competitors is effective pest management that may be practiced over time in a manner that does not result in unacceptable environmental, human health, or economic problems.

An important facet of this is being able to objectively evaluate organisms to determine their potential and/or actual pest status. This step will allow you greater flexibility in managing a pest in a sustainable manner. Of course, even before determining pest status, a pest manager must positively identify the organism. You must then recognize that organisms with the potential to be pests may become so through a gradual process influenced by environmental conditions and the actions of collections staff members. Their pestiferous status may gradually escalate from a minor condition to a major one. In other words, organisms of concern to pest managers may fluctuate in their pest status over a gradient from benign to significant. As it turns out, many pests may be effectively and sustainably managed by simply suppressing their population.

Sustainable management of biological competitors, be they insects, pathogens, or weeds and invasive plants, involves a process of sampling and monitoring. Monitoring is an area of pest management that often lacks rigor among practitioners who are having difficulty developing and using a sustainable program. For one thing, monitoring is not simply observing the status of accessions, pests, and other phenomena in the course of carrying out other operations. Rigor must be applied to regularity, continuity, objectivity, documentation, and analysis in the monitoring process. Good management decisions are dependent on the quality of the information acquired by the monitoring portion of the pest management program. Carelessness, skimping, and other breakdowns in the monitoring process will be reflected in poor pest management decisions and outcomes. Monitoring and decision making are a very synthetic portion of the pest management process—one builds on the other. Monitoring and sampling are core elements to a sustainable pest management program. The following are some advantages:

- Provide early warning to prevent problems and reduce costs.
- Determine the specific cause and severity of problems.
- Target specific treatment areas.
- Determine the most effective and economical timing and method of treatment.
- Provide the opportunity to use more sustainable and safer control methods.
- Evaluate control efficacy.
- Document a pest problem history that is useful in predicting and avoiding future problems.
- Contribute to improved plant collections, display, and landscape quality.

For more on managing invasive plants, see chapter 3, "Building Collections." The monitoring and sampling requirements for sustainably managing biological competitors can serve an important dual purpose for botanical gardens: collecting developmental and characterization information on accessions for their catalog record, as well as any other research-oriented data collection.

The specific requirements of a sustainable, effective pest management program are often underestimated in the short term and overestimated for the long term. One principal cause of failure in attempting to plan and implement a comprehensive, sustainable pest management program is that it is simply not given adequate priority within the overall preservation management program. This is particularly true when you consider the monitoring aspect of the program. Sustainable pest management cannot be separated from plant management, and this interrelated nature exists in the day-to-day operation. In short, pest management must be embedded in and integrated with ongoing cultural practices for the care of the collections.

Valuable assistance for controlling biological competitors may be found within the Sentinel Plant Network (SPN). The mission of the SPN is as follows:

The Sentinel Plant Network contributes to plant conservation by engaging public garden professionals, volunteers, and visitors in the detection and diagnosis of high consequence pests and pathogens.[9]

The SPN is a partnership between the American Public Gardens Association and the National Plant Diagnostic Network as well as botanical gardens and scientific agencies from around the world. The IPSN does the following three things:

- Provide public garden professionals with training and diagnostic support to better monitor and protect their collections.
- Facilitate greater collaboration about invasive pests and pathogens among public gardens and with other organizations through improved databases and communication protocols.
- Enhance garden outreach efforts to educate their communities on the impact of high-consequence plant pests and pathogens and engage individuals as First Detectors.[10]

I suggest that all botanical gardens enroll in the Sentinel Plant Network.

Controlling the Plant's Environment

In addition to managing biological competition, another important area of focus for sustainable practice is controlling the plant's environment. Greenhouse and other covered growing environments are the greatest challenge as consumers of energy, water, engineered media, fertility supplements, and growing containers. The greenhouse structure itself—site orientation and positioning, the structural elements, glazing, and HVAC equipment—need careful evaluation and selection for efficiencies, sustainable energy inputs, and customized controls toward those ends. The source and type of energy to power the greenhouse must be carefully considered for its impact on climate change and resource conservation. The components of engineered growing media—as well as all soil amendments—should be evaluated for sustainable sourcing, production, and reuse. The same may be said for growing containers. Water quantity and quality (source and discharge) is a major part of any greenhouse sustainability audit. Closely connected to sustainable water use is greenhouse fertility management, including choice of materials, their application, and reductions in discharge.

Many of the same sustainability considerations for greenhouses applies to outdoor growing areas. This begins with a careful consideration for taking full advantage of prevailing microclimate conditions for successful collections preservation—right plant, right place. To the extent that we must manipulate the plant's environment further with alterations to the existing soil—including structure, texture, fertility levels, and moisture content through irrigation—there will be opportunities for sustainable practice regarding systems, material choices, and applications. As with greenhouse equipment and materials, a consideration of cradle-to-cradle certification and application—a measure of safer, more sustainable products made for a circular economy—must be included.

The park-like context of many botanical gardens often includes large areas of turf. These are traditionally notorious consumers of energy, horticultural materials, and water. Consider the use of low maintenance turf selections, graduated turf care maintenance levels based on visitor use and exposure when scheduling mowing intervals and other maintenance inputs. Finally, regarding turf care and other plant care equipment, replace all those with internal combustion engines with electric substitutes. Where this isn't possible, consider using less polluting fuels or simply reverting to manual power.

The concept of sustainability isn't confined in botanical gardens to a concern for preserving the life of plant collections. Another important component of sustainable practice in botanical gardens, one that is addressed throughout this text, is preserving genetic diversity within our collections and collections research.

Preservation Policy

With so much invested and at stake regarding effective and sustainable collections preservation, a botanical garden's preservation commitments and basic standards should be outlined in the collections management policy. Larger institutions, or those with highly comprehensive preservation programs, may establish a separate preservation policy, manual, and plan for their collections. A preservation policy would establish preservation priorities and horticultural standards. It would also specify monitoring and reporting requirements and reinforce linkages with the documentation program. Some standard practices outlined in such a policy would include the following:

- Develop a preservation management system to guide and govern the application of all collections preservation practices. See to it that it can be successfully linked to the collections documentation program.
- Use preventive, proactive, and hygienic preservation practices that will enhance the health and vigor of collections, minimize debilitating stresses and pests, and preserve genetic diversity.

- Ensure that preservation programs include adequate provision for regular monitoring and documentation of the condition of collections and the results of preservation practices.
- Use preservation practices that are environmentally sensitive, sustainable, and instructive to the public.
- Ensure that the permanent collections are not subjected to potentially damaging garden programs, projects, research, or other type of access.
- Ensure backup power supplies and adequate warning systems for environmental control equipment affecting collections.
- Establish a program of propagation renewal for all collections based on regular evaluations and collections priorities.
- Ensure that the garden completes and provides access to comprehensive design and program statements for exhibits and displays to help guide their preservation and care.
- Implement risk management techniques to evaluate the insurance needs of the institution and purchase the appropriate insurance policies.

Collections Preservation Plan and Procedures Manual

Collections preservation standards may be itemized and delineated in a collections preservation plan and procedures manual. As I described in the previous chapter on documentation, this is simply a written set of instructions prepared by the curatorial staff to help guide those responsible for the establishment and operation of an efficient collections preservation management system. This manual may also be used as a training resource for all new collections preservation personnel. Such a manual will also serve to reassure garden supporters and the public that the garden has assumed responsibility for the effective and efficient preservation management of its collections. Finally, the preparation of a collections preservation procedures manual serves as a useful vehicle for reviewing the efficacy of the entire system.

The Arnold Arboretum's *Landscape Plan* serves as a good example of a collections preservation plan and procedures manual. The introduction clearly articulates the purpose of the *Landscape Plan*:

> The Plan's goal is to maintain exemplary standards of horticultural care and management that provide optimal growing conditions for our collections, enhance the ecological health of the Arboretum environment, and present a landscape of outstanding quality for visitors, students, the surrounding community and other key constituencies.[11]

Notice how this introduction touches on three key components of collections preservation: provide optimal growing conditions for the collections (preservation), enhance ecological health (sustainability), and preserve landscape quality (appropriate context).

The plan and management activities are organized within a geographical framework that divides the landscape into 7 regions that are then subdivided into a total of 70 management zones. The regions are broadly defined by topography, landscape type (collections, natural areas, etc.), and management needs. The zones are contiguous areas within each region that may be defined by an obvious character, such as the location of a specific collection, landscape feature, or specific management requirements. The zones are mapped with a number (e.g., 1.1 = region 1, zone 1) and receive management priority and intensity designations of "high," "moderate," or "low." Priority refers to a zone's relative importance for management within a given year, while intensity indicates the nature of the collections and landscape care.

There is a narrative provided in the plan for each zone that includes a profile, account of the special needs of that area, and discussion of arboriculture conditions. What follows this general introduction are bulleted annual care lists of specific preservation and management tasks grouped by seasonal requirements—these are impressively detailed. There are three textboxes showing weed problems and management options, other pest problems (insects and diseases) and their options, and curatorial notes regarding collections development and field-checking requirements. This last set of features underscores the importance of articulating programmatic crossover areas in policies, plans, and manuals to underscore the interactive and synergistic nature of all collections management components. It is also a good reminder to horticultural staff of the curatorial implications and connections of their work. Finally, there is a list of long-term tasks that are required periodically as well as a list of capital projects in the works for that zone.

The thorough and informative nature of this plan doesn't end there. There are 11 appendixes containing additional and detailed information on other areas of collections and landscape preservation:

Appendix A: Mowing Operations
Appendix B: Snow Removal Operations
Appendix C: Infrastructure and Hardscape
Appendix D: Curation and Plant Records Office
Appendix E-1: Noxious Weeds
Appendix E-2: Insects
Appendix E-3: Diseases

Appendix F: Secondary Paths
Appendix G: Drainage
Appendix H: Arborist Calendar
Appendix I: Horticulture Team Assignments[12]

Appendix D, "Curation and Plant Records," contains information on inventories, labeling, and resources connected to these subjects.

To prescribe all of the preservation needs of the Arnold Arboretum as they were identified in 2011 required a 395-page document. Obviously, the preservation needs of large plant collections and complex landscapes is a comprehensive and highly important part of the collections management program, and documenting those needs may be a lengthy process. However, once completed, it is an invaluable living document that will then only require modest updates as the collections develop and new information and technologies arise.

Best Practice 5.1: Arnold Arboretum's *Landscape Plan*

- Preservation plan activities and procedures are organized within a geographical framework.
  - The property is subdivided into named "regions" corresponding to topography, landscape type, or a broad category of preservation activity (e.g., Hemlock Hill); and
  - Regions are subdivided into preservation-specific "zones," contiguous areas within each region with common specifications for preservation that serve as the basic unit for organizing preservation activities and procedures (e.g., *Abies* collection or Japanese Garden).

- Preservation areas are named, clearly mapped, and labeled within the plan.
- Preservation zones are assigned a priority based on the required level of management intensity (high, moderate, low). These priorities are indicated on the map.
- All the necessary preservation procedures and activities for each zone are described and scheduled in a special section. These sections make up the bulk of the plan and include the following:
  - A synoptic profile of the specific zone;
  - Special preservation priorities are identified;
  - Arboriculture requirements are listed;
  - An itemized annual care plan by season;

- A description of key pests; and
- Other curatorial needs and plans including long-term and capital projects.

- The plan is augmented with important resource information in an appendix (e.g., tree preservation calendar of activities, detailed key pest information, infrastructure maps, specific curatorial procedures, etc.).

Some collections and specialized botanical garden displays and exhibits may require special instructions for their preservation that are outlined in a subsection of the procedures manual or in a manual all their own.

Textbox 5.1.   Recommendations for Preservation Policy

Basic
- The botanical garden has a written collections preservation policy statement in its collections management policy that outlines its fundamental standards and requirements for collections preservation and care.

Intermediate
- The botanical garden has a separate, written collections preservation policy that outlines all of the requirements and standards for collections preservation and care.
- The botanical garden has a collections preservation procedures manual for staff and volunteer training and reference.

Preserving Genetic Diversity

To reiterate an earlier point, preserving the plants themselves is not the whole of our obligation; we must also obligate ourselves to preserve their genetic integrity and diversity. Preserving genetic diversity is most relevant for botanical gardens that direct their research efforts to plant conservation. Conservation collections are most often established and held as "gene banks." Gene banks, as you will see later, may contain whole plants or various types of plant propagules.

It is important to recognize that the preservation requirements of display plants are quite different from gene bank collections. Preserving genetic diversity is the raison d'être of gene bank collections. As an example, the preservation of field gene banks of conservation plants must be meticulously documented. Their reproduction must also be closely monitored and controlled. Gardens must carefully evaluate their capability and capacity to adequately preserve conservation collections—whether they be of a single clone, an ecotype, a variety, or an entire species.

Let's take a closer look at how cultivating wild-collected plants in botanical gardens can change their genetic and phenotypic status and what this means for preserving them, especially for use in conservation programs. With an emphasis that goes back several decades for botanical gardens to build collections of well-documented wild-collected plants, there has been scrutiny and criticism of these collections becoming genetically impoverished, hybridized with closely related horticultural plants, and maladapted as a result of cultivation. These changes are anathema to the purpose of *ex situ* collections for reintroduction and reinforcement, collections that should be kept genetically and phenotypically as they are in nature.[13]

Superficial phenotypic changes, or plasticity, that are responses to environmental changes (e.g., plants in shade with larger leaves) are of little concern compared to phenotypic change connected to genetic change. This can happen in cultivation and propagation in a process known as "domestication." Four common changes associated with domestication are changes in seed dispersal and dormancy mechanisms, increase in size, and synchronization of phenological timing. Phenological changes like these that occur in cultivation are called "unconscious selection."[14]

Genetic changes of concern for *ex situ* collections of wild plants are genetic drift and inbreeding depression. Genetic drift results in the loss of genetic diversity. The smaller the *ex situ* population, the greater the impact of genetic drift. Inbreeding depression can lead to the loss of fitness and vigor in a population of plants. Like genetic drift, the impact is more pronounced in small populations and among taxa with particular breeding mechanisms, such as outbreeding and self-incompatible species. Both of these conditions can lead to the accumulation of deleterious mutations. The bottom line is that these genetic phenomena threaten the integrity of genetically diverse *ex situ* collections.

The best remedy for these genetic preservation issues is population size and separation. Recommendations from several sources point to ideal *ex situ* collections of 500 individuals to combat the forces of genetic drift with frequent introductions of new germplasm. Regarding inbreeding, the recommendations are less (100 individuals), but they will also likely require controlled pollination. The placement of these populations in the landscape and within field gene banks must be carefully considered for separation. One final consideration is that the more like the natural conditions under which these plants grow in the wild that the *ex situ* environment is, the more likely that selective forces will operate to maintain genetic diversity consistent with plants in the wild.[15]

For preserving genetic diversity, *ex situ* collections must be built from an adequate number of genetically representative samples. As a reminder on how that begins, I refer you back to chapter 3 on building plant collections and plant acquisition. Gardens must consider and accommodate the most appropriate methods of preserving genetic diversity and how to retain it within collections once they are amassed.

Textbox 5.2.   Recommendations for Preserving Genetic Diversity

Basic
- Genetically diverse collections receive the highest priority of care available to the collections.
- Genetically diverse collections are rigorously documented and labeled.
- Vegetatively propagate "backup" clones to minimize the chance loss of any one clone and the resulting loss of genetic diversity within the collections.
- Remove volunteer seedlings of out-crossing taxa from the collections and do not use seed from such taxa as a means to renew the collections.

Intermediate
- Botanical gardens wishing to preserve genetic diversity above all other goals must be committed to acquiring and preserving a sufficiently large sample of a taxon.

Advanced
- Garden breeding programs must carefully follow guidelines in the literature designed to maintain the highest possible genetic variability in the progeny.

Plant Collections

This section covers the principal preservation requirements for collections of living plants, the bulk of concern for botanical garden preservation programs. At the end of this section, there will be a consideration of special collections of whole plants with a subsection on field gene banks and one on historic collections. Later sections will outline the preservation requirements for collections of seeds (gene bank) and tissues (tissue banks). Final sections will cover the particular preservation requirements for plant production and propagation programs; exhibits, displays, and landscapes; and lastly, a consideration of collections risk management, emergency preparedness, and insurance.

Having reviewed basic horticulture technologies, it is beyond the scope of this text to outline the preservation needs of a representative number of taxa important to botanical gardens. In fact, much remains to be discovered on this subject, providing rich fodder for botanical garden research programs. Nevertheless, botanical gardens and arboreta are challenged to conduct their preservation programs effectively and efficiently. The approach we as curators, collections managers, and horticulturists use to preserve our plant collections must be comprehensive, integrated, systematic, and preventive. Although our preservation practices will differ based on differences in our plant collections, environmental conditions, collections use, and so forth, we must manage these practices effectively.

Botanical gardens, like other large landscape facilities, are facing escalating labor and materials costs as well as social and environmental pressures for greater sustainability. Curators and others must have information from which they can make sound management and preservation decisions. The basic components from "A Management Approach to Maintenance" by Richard Harris form a useful template for structuring a preservation management program:

- Involve collections personnel in developing the program.
- Be based on and work from a collections and landscape inventory: what are we preserving?
- List preservation tasks: how much pruning? pest management?
- Describe preservation tasks: for example, pest monitoring.
- Establish preservation standards: for example, all trees will have 24-inch turf-free mulch rings at their bases.
- Set task frequencies: tree rings mulched yearly.
- Schedule preservation tasks: tree rings mulched in early winter.

- Implement preservation tasks.
- Monitor preservation activities and schedules.[16]

The Arnold Arboretum's *Landscape Plan* described earlier represents a comprehensive approach to preservation management. To the above set of recommendations, I would add one more component that is implemented on a periodic basis depending upon the results of program monitoring. It is known by the acronym DMAIC: define, measure, analyze, improve, control. When we combine this element with monitoring, we have identified an improvement cycle that should be seen as an ongoing process of the collections preservation management system. The DMAIC process is used to improve, optimize, and stabilize an element within our preservation management system.[17] In other words, it is a periodic auditing and problem-solving mechanism that is invoked in response to issues identified through monitoring preservation practices.

This program will be effective only if it is comprehensively linked to the documentation program and does not become top heavy in planning and descriptive detail.

A computerized preservation management program must be compatible with, and contribute to, the garden's documentation program. It may be attached to the documentation program as an "activity file" and be linked to the cartographic database. This configuration would then contribute to and be a part of the catalog record. The current array of available documentation software programs described in chapter 4 aid the documentation of preservation activities.

Preservation activities, as they apply to any given accession, begin at the point of acquisition and/or propagation in the botanical garden nursery. Plant propagation and production are discussed in a later section. Acquisitions of whole plants must be screened for the presence of biological competitors—pest organisms—and a proclivity for invasiveness in a quarantine area that is usually associated with the nursery. For more on quarantining, refer to chapter 3, "Building Collections." Quarantine facilities should have their own procedures manual outlining their operational protocols.

The horticulture staff of a botanical garden, in order to make informed decisions about collections planting and preservation, must be intimately familiar with the natural conditions and limiting factors of the garden site. A site plan and inventory of the natural features of the garden may be part of its founding documents. If not, such an inventory should be undertaken so that the range of horticultural practices necessary to effectively manipulate and control the environment for successful cultivation and preservation of the developing collections is possible. This will involve a study of topography, slope, aspect, and microclimate; watershed attributes including water courses, springs, seepages, wetlands, and drainage; soil attributes including profiles, texture, chemistry, and fertility levels; natural plant communities and their succession including noteworthy taxa such as indicator and endangered species. This work may require the assistance of an environmental consultant.

To the extent that garden staff are familiar with and fully understand the horticultural implications of the natural conditions of the botanic garden site and the potential impact of climate change on those conditions, the better prepared they are to contribute to decisions about collections development and plant acquisitions adaptable to those conditions. Naturally, gardens should seek to minimize their requirements for manipulating plant-growing environments to help ensure the successful and sustainable preservation of accessions.

Acknowledging that it is unlikely a botanical garden site will be ideal for all the accessions in a botanical garden's collection, there is one particularly important element of controlling a plant's environment that must not be neglected: the soil. Good soil management will help mitigate and offset many other, more difficult to control limiting environmental factors. Within this category of activity, perhaps the most impactful element is the maintenance of organic matter. As I am sure you all know, organic matter content improves soil structure and tilth, pore space and drainage, fertility and nutrient-holding capacity. All but the most narrowly adapted plants benefit from the organic matter component of the soil.

Of course, before undertaking any program of soil amendment and modification, a soil test must be performed. Based on the results of that test, soil may be prepared and amended prior to planting with various types of composts to boost both the organic matter content and fertility levels. Where time allows, cover crops may be grown on the land before it is planted with accessions. Various types of rye grass and legumes may be used for this purpose and then tilled into the soil using a chisel plow for the best incorporation. These same cover crops may be interplanted between the rows in field gene banks (outlined below) and planted as ground covers in display beds if they are aesthetically acceptable. Displays and exhibits may be mulched with composts and other organic matter with a low carbon/nitrogen ratio to improve and maintain organic matter content and fertility. Regular soil tests should be performed to monitor the organic matter content and soil chemistry as plants grow and soil conditions evolve.

Tree Preservation

When it comes to controlling the plant's environment for successful collections preservation, an area that is often fraught with controversy for botanical gardens is the disposition and removal of native trees. The curator and horticulture staff are well advised to consider the matrix of native trees associated with a forested site to be as integral and substantial a part of the natural infrastructure as the prevailing weather. Any wholesale removal of native trees to make way for plant collections, especially in public spaces, is certain to raise the ire of surrounding communities. The apparent irony of such actions by a purported tree preservation- and conservation-oriented institution is not lost on the public.

Of great importance to the preservation of existing native trees and the establishment and preservation of tree collections is the emerging understanding of forest ecologists on the interconnected and symbiotic relationship of forest trees through mycorrhizal networks. Our growing awareness of these networks and their impact on tree communities reveals a new and more compelling mandate for forest preservation as a critical element of environmental sustainability.[18] Such networked communities of native trees within botanical gardens are assets not only for environmental and native forest preservation but also may serve as sites for the enhanced preservation of introduced forest trees and forest-adapted plant collections that may be accepted into the established network.

> A critical mechanism by which new recruits [native seedlings and planted accessions] could benefit from mycorrhizal fungi is through integration into "common mycorrhizal networks"; such networks are formed when the mycelium of an individual mycorrhizal fungus connects multiple plants simultaneously, and in several ecosystems they have been shown to have key roles in nutrient transport, seedling establishment and inter-plant signaling.[19]

The implications of forest mycorrhizal networks—centralized around mature "hub trees" as the predominant resource distribution and signaling nodes—for natural area preservation cannot be overstated. This new understanding of forest ecology is giving rise to a new paradigm for forest management that also has implications for the preservation of tree collections.

Trees are the green infrastructure of the botanical garden landscape and form the bulk of the plant collections biomass. The preservation management program described earlier may be customized and refined to focus on tree preservation. Gardens may subdivide their tree collections into preservation categories as a means to determine and track their preservation requirements. Typically, this is done according to age and location, but unusual, rare, and endangered trees will factor into the creation of preservation categories and priorities as well. Young trees, for example, will require different types of preservation measures than older trees. Tree hazards, of course, will be considered as part of this scheme. Gardens may then determine the degree of attention any tree will receive based on the priority it is given in the collections. Arboreta and botanical gardens with significant tree collections and groves of native trees should join ArbNet, an interactive community of arboreta, to stay abreast of tree preservation issues, practices, and research of particular concern for those institutions with a programmatic emphasis on trees.

Tree preservation certainly demands special attention in any collections management system. For some institutions, a separate tree care or arboriculture management system may be warranted. The Arnold Arboretum's *Landscape Plan*, described earlier, includes special notations on the arboriculture needs of the various landscape management "zones" within the Arboretum and details a several year, rotating schedule of arboriculture work in a lengthy appendix.

The University of Maryland Campus Arboretum and Botanical Garden uses a tree management plan that outlines the arboriculture framework and standards of care, including all basic tree care practices, tree protection measures on construction sites, documentation, and inventory requirements.[20]

Important tree collections, as well as noteworthy and historic trees and tree groves, should be given consideration as to their succession over time. Some of this work may occur under the banner of "collections evaluation" that specify cycles of propagation in response to evaluation outcomes. I will offer more on this subject later in this chapter and in chapter 8. A botanical garden's foundational trees—those trees that make up the forested character of any landscape, be they part of a developed collection or part of the natural matrix of trees—should receive successional planning; this may be of particular importance for urban gardens.

Each botanical garden will identify and/or devise their own goals for tree succession. These may be governed principally by the historic presence of a dynamic native matrix of trees that exist in networked groves as described above or by impressive groves of tree collections brought together following research and educational themes such as taxonomy and ecology. In any case, a tree succession plan should be guided by a clear set of goals consistent with the institution's mission, collections policy, and so forth.

The succession plan for Madison Square Park in New York City serves as a useful example. The developers of the plan make a simple statement that may serve as the overall vision and goal for the park's succession plan: "[T]rees in Madison Square Park should maintain and enhance its woodland quality."[21] In addition there are three stated goals:

1. Increase the diversity of species in the Park overstory, choosing where possible natives such as might have occupied Manhattan before European occupation . . .
2. Maintain a solid cadre of major species—the elms, oaks, and London planes at least—to constitute the principal overstory trees of the Park.
3. Increase the diversity of the understory.[22]

A set of subsidiary, aesthetic goals are identified that speak to diversifying the branching and textural patterns of the forest canopy. There were also two tertiary goals added to address two specific areas of concern involving poorly performing tree groups.[23]

The next step involved updating an 18-year-old tree inventory to include a condition rating for every tree based on eight factors: root structure, root health, trunk structure, trunk health, branch structure, branch health, twigs, and foliage. A numerical rating of 1–4 is assigned to each factor with a tree in excellent condition receiving a score of 32. The scores were divided by this total to obtain the condition class rating in a percentage, with any tree receiving less than 60 percent considered problematic. These percentages are deemed most useful for an approximation of tree condition ratings. From these percentages, a longevity estimation was applied to the trees using three categories: >20, <20, <10 years. Trees that had been removed since the last inventory were also documented.[24]

Taking the example of succession planning from Madison Square Park, we may add a component that is having an increasingly important impact on succession planning and tree replacement, not to mention collections development as a whole: climate change. There is a growing body of research on the expected impact of climate change on the natural history of various biomes, ecosystems, and habitats of Earth, including the ecology and plant geography of regional floras. This information may prove very useful for estimations of the changing requirements for plant growth in certain areas and the influence of those changing requirements on collections preservation, including succession planning and tree preservation. This subject is also taken up in chapter 11 on species recovery programs.

Longwood Gardens uses climate change modeling in making tree management decisions for ongoing preservation and succession planning, a great deal of which comes from the U.S. Department of Agriculture (USDA)

Forest Service.[25] These models are based on carbon-driven climate change scenarios that describe varying impacts on native forest trees depending upon which scenario is used. This information is used by Longwood arborists to predict the future limiting factors and mitigating practices that may be needed to preserve Longwood's trees and forested areas. These models may also be used to help guide decisions about tree succession and replacement based on the identification of taxa that may benefit from the climate scenarios predicted by modeling.

For example, using the USDA Forest Service Climate Change Atlas most severe climate change scenario for 134 tree species in the eastern United States, the habitat (and growing conditions) for *Magnolia acuminata* (cucumber tree), *Tsuga canadensis* (eastern hemlock), and *Liriodendron tulipifera* (tulip tree) in the area of Longwood Gardens (southeast Pennsylvania) would be eliminated. This means that, over time and with an emerging climate change scenario that follows the most severe prediction, these trees will likely require more intense preservation measures. On the other hand, species such as *Quercus shumardii* (Shumard oak), *Catalpa speciosa* (northern catalpa), and *Ulmus crassifolia* (cedar elm) should see large increases in suitable habitat. Hence, their preservation requirements may become more benign and, if not currently part of the collection, they may be listed as potential replacements on a tree or collections succession plan.[26] See chapter 11 for more on climate change models and their use in collections research.

Now, let's go one step further in this progression of tree preservation to consider aging and declining trees. Curators and their staffs will be confronted with making decisions about the removal (and deaccession) of historic and venerated aging specimens based on both safety and aesthetic concerns. Some gardens have taken on this situation in a manner to prolong the life of these older trees, especially for trees that are highly coveted by their visitors for their age, rarity, or historical significance. Such trees may garner financial support from their adoring public for the necessary preservative technologies and materials required to keep them in place.

> In a botanical garden, park or arboretum, a collection of historic trees comes with the benefits of shade, beauty and a display of natural history but those benefits come with the challenges of safety, liability, aesthetics and a commitment to environmental protection. As a result, institutions of public horticulture are seeking innovative means of understanding, showcasing, and preserving their historic trees.[27]

European gardens and arborists have taken the lead in work to prolong the lives of ancient and historic trees, in particular the National Trust of Great Britain; the Royal Botanic Garden, Kew; and a professional group known as the Ancient Tree Forum (ATF). From a preservation standpoint, this work involves some rather unorthodox and controversial techniques: retrenchment pruning and natural fracture pruning or coronet cutting, or a combination of these. First, what is retrenchment pruning?

> Retrenchment pruning is a phased form of crown reduction, which is intended to emulate the natural process whereby the crown of a declining tree retains its overall biomechanical integrity by becoming smaller through the progressive shedding of small branches and the development of the lower crown (retrenchment). This natural loss of branches of poor vitality improves the ratio between dynamic (biologically active) and static (inactive) mass, thus helping the tree as a whole to retain good physiological function.[28]

Many very old trees will begin to show upper crown decline as they age due to a number of factors. In some species, epicormic shoots will form in a lower position at the base of the original crown to begin a process of crown reorganization, or retrenchment. Retrenchment pruning, through several phases, will reduce the older, declining crown down to the size of the emerging new, or retrenched, crown. Trees less inclined to produce epicormic shoots and a retrenched crown may be encouraged to do so through retrenchment pruning.

Regarding natural fracture pruning and coronet cutting, Neville Fay in "Natural Fracture Pruning Techniques and Coronet Cuts" uses the following definition:

> Natural fracture techniques involve pruning methods that are used to mimic the way that tears and fractured ends naturally occur on trunks and branches. A coronet cut is a type of natural fracture technique that is particularly intended to mimic jagged edges characteristically seen on broken branches following storm damage or static limb failure.[29]

Natural fracture techniques and coronet cuts may be used to encourage epicormic growth for the development of a retrenched crown. They are also thought to provide natural microsites for symbiotic and other kinds of micro-organisms as well as local wildlife typically found within damaged trees. Older trees will also need the same arbor care as other trees—aeration, mulch and other soil compaction mitigation techniques, supplemental nutrition, mycorrhizal and other helpful symbionts, and so forth—to encourage longevity. Some ancient trees that have declined past the point of further preservation as part of the living collection may be preserved as habitat for other organisms that shelter and feed on the remains of such trees. Whether accessioned in the collection, or simply a part of the garden landscape, older trees that once may have simply been removed and replaced may be preserved with special attention.

Contracting Landscape Services

Before moving on to more specialized collections preservation needs, I want to give some consideration to the use of landscape contracting services for routine landscape management, for example, turf mowing. The seasonal demands of landscape horticulture may tax a garden's ability and resources to cope with the more comprehensive demands of collections preservation. One solution to this challenge that may lower costs and increase productivity is outsourcing landscape maintenance services to professional contracting companies. Many botanical garden personnel may wrinkle their noses at this notion, but it is a trend that is taking hold at many institutions that choose to direct the bulk of their resources and energies to collections preservation and less to landscape housekeeping such as turf care. The most important aspects of this process are to find a capable contractor and provide a clear definition of the work to be done.

Landscape maintenance agreements should be based on performance and price. The contractor should perform the specified work at the lowest price or be replaced by the next bidder on the list. Use a contractor qualification questionnaire to screen candidates. In "Outsourcing: A Maintenance Alternative," Lawrence Labriola recommends the following questions:

- Type of organization and the ownership
- The company's gross sales for the last three years. Look for a positive trend with a total volume of approximately seven to ten times your project size. This will guarantee the benefits of larger staff, expertise and buying power.
- Total sales volume per type of services provided
- Number of full-time, part-time and seasonal staff. What is the seasonal nature and probability of properly trained staff?
- Scope of services provided by company staff and type of services subcontracted
- Names and addresses of subcontractors and major suppliers to help establish the reputation of their associates and the confidence the suppliers have in the contractor

- Workers' compensation experience modification factor for the last three years. This will indicate the success of safety programs, and what advantage the company may have in lower insurance costs.
- Financial references, banks, credit lines, bonding company, and bonding limit
- Names of the last three clients lost and why
- Present client list, scope of contracted work, years contracted, contact name, and phone numbers
- List of all equipment owned, type, make, and age
- Copy of safety policy and procedures
- Résumés of key personnel[30]

From these questionnaires and the results of interviews and so forth, select a few contractors to bid on your work. Job specifications for bidding should be divided into manageable units with a request for individual prices and staff hours for each unit. Consider drafting contracts that will last for three to five years with a provision to rebid the contract every five years.

The remaining portion of this section on plant collections highlights two special collections requiring practices in preservation that will serve as examples of specific preservation programs. The first of these, field gene banks, is also the focus of botanical garden research in both preservation and conservation in that the science and technology associated with them is still evolving and they require a specialized practice. Consequently, field gene banks may require staff with special skills pertinent to that type of collection. Because of the research opportunities posed by field gene banks, this subject will also be taken up in the chapter 10, "*Ex Situ* Conservation." The last subsection deals with preservation practices for historic plant collections.

Field Gene Banks

Maintaining populations of plants in protected places may be a necessary conservation strategy for some species; these are known as field gene banks. However, at botanical gardens, these may also be a part of living display collections.[31]

Field gene banks of whole plants are traditionally established for germplasm preservation, conservation, and research. Toward that end, target 8 of the Global Strategy for Plant Conservation (GSPC) calls for "at least 75% of threatened plant species in ex situ collections, preferably in the country of origin, and at least 20% available for recovery and restoration programmes."[32] The growth and importance of gene banks in general at botanical gardens is a response to this target and an individual commitment to plant conservation.

Field gene banks may also serve horticultural purposes such as plant breeding. These collections are generally not on public display and often lack public interest and appeal in the typical sense, although smaller gardens, or those with restricted budgets, may participate in field gene banking on a limited scale making special provisions for gene bank collections grown in displays or exhibits or preserved as tissues. Also, these research collections, often established at satellite areas of the botanical garden, may be accompanied by an experimental propagation facility. Field gene banks of plants are usually established in nursery blocks, screenhouses, shade houses, or greenhouses. Greenhouses are typically used to grow and preserve tender taxa and propagate plants destined for other banking locations. They are also used for controlled pollination, post-entry quarantine, pathogen testing and elimination, flower forcing for identification or seed production, and short-term research projects.[33]

Special attention needs to be given to the funding requirements for field gene banks. Begin by ensuring that the institution's mission and policy documents indicate a clear mandate for this type of programming. If not, begin a process to make those changes before moving forward with this type of resource-demanding program.

Once the institutional commitment and funding are secured, field gene bank collections should be acquired using the appropriate collecting strategies discussed in chapter 3 on building genetically diverse collections. A good starting point for decisions about what plant groups and taxa a garden might build a field gene bank program around are its core collections. These are groups for which the garden will already have a long-standing programmatic and financial commitment. The next step might be based on the results of a gap analysis of institutional collections and regional, national, and international conservation needs that revolve around your core collections. Add to that an investigation of collections-sharing opportunities within those gaps involving your core collections. For more on starting and developing field gene banks and other types of conservation collections, refer to chapters 9 and 10.

The protocols for preserving a field gene bank of living collections require documentation in a procedures manual to help ensure the integrity of the collection and how it is cared for. The manual should detail step-by-step procedures for all the necessary tasks to preserve the collections and the means by which they are organized and documented in the field. What follows are a broad set of recommendations for preserving a field gene bank collection of plants. Specific standards should be developed by each institution as part of the research conducted on their unique field gene bank collections within their unique physical context. This research will involve the study of the biological characteristics of the species, its phenology, reproductive mechanism, and population structure, among other things.[34]

The nursery block component of a field gene bank is usually established to preserve large numbers of mature species that do not readily produce seed or produce recalcitrant seed. These collections are also useful as a source of propagules, for identity verification, evaluation, and other types of research. Establishing these collections in blocks and rows allows for maximum preservation, tracking, documentation, and access.

> Most CPC [Center for Plant Conservation] field gene banks support long-lived tree species with five to 20 unrelated individual plants per conservation collection. Most require collaborations across institutions, because it is unlikely that a single institution will have enough space available to house the 100 or more individuals needed to capture the desired genetic diversity of a species.[35]

In this regard, field gene banks are established, organized, and maintained in much the same way as field nurseries. Growing sites are selected and prepared to help ensure good growing conditions for the individual plants. The area should be large enough to allow for crop rotations (herbaceous, short-lived accessions) and soil building as well as potential program expansion. As recommended previously, management of good soil health, particularly the organic matter content, is a crucial practice. Since most field gene banks are planted in blocks, this facilitates soil management without disturbing the growing plants. Following soil tests, composts and other similar sources of organic matter may be applied as mulches and incorporated as soil amendments. Cover crops may be grown between the rows, such as rye grasses and legumes, to help control weeds and improve the nutrition and organic matter content. If well mapped and documented, companion plants among the collection may be interplanted for mutual benefit, for example, coffee under macadamia nuts (see below).

Blocks of plants may be segregated according to a number of conservation, population biology, or other research criteria; genetic integrity is an important one (to minimize artificial selection and genetic drift).

> For many exceptional species, living plant collections are the only currently available ex situ conservation option, and the maintenance of these living collections introduces numerous genetic and demographic challenges associated with small, isolated populations. If not curated correctly, these small populations are subject to founder effects, genetic drift, and inbreeding, and can experience selective pressure from biotic and abiotic conditions in the ex situ environment.[36]

Chapters 9 and 10 offer more on genetic and demographic challenges for managing conservation collections as an element of collections research.

"Evaluation may require a greater number of plants or a special field layout suitable for replicated experimental designs. Gene banks mandated to propagate disease-free planting stock for farmers may have additional specific requirements related to distribution."[37] Field gene banks should have adequate space to grow a range of ecotypes or lines from different sources and to establish temporary populations for replanting in bulk. In some cases, propagation methods may enter into decisions about field organization and spacing. Those plants—woody and herbaceous—that spread via vegetative parts may need special spacing and cultivating practices; vigorous spreaders may require container cultivation. Field gene bank growing sites should be well separated from areas of the institution open to the public and from other nursery growing areas; they may require fencing.

Special accommodation will need to be made for some plants needing nurse plant partners for adequate growth in early stages or the presence of canopy trees (e.g., coffee). Vining plants and lianas will need adequate support for proper growth. Special beds and planting areas may need to be constructed for some plants strictly adapted to special soils such as halophytes, xerophytes, and various types of wetland species.

As the field blocks and screenhouse and shadehouse beds, as well as greenhouse beds and benches, are planted with field gene bank collections, they should be clearly mapped and labeled using the technologies and materials of greatest accuracy and permanence at the institution's disposal. The location of individual accessions should be clearly defined. Accession groups, as well as the individual plants, should be provided with accession labels carrying the standard information as well as the field collection numbers. Gene bank documentation must be scrutinized and audited at the highest level of documentation priority.

Plant health considerations will naturally impact how the growing sites (blocks) are arranged, although some accessions may need the special protection provided by screenhouses or greenhouses (more on those below). Screenhouses are necessary to protect accessions vulnerable to insect-vectored pathogens and also to control pollination. Tender accessions may also need the protection of a greenhouse. As outlined at the beginning of this chapter, gardens must employ sustainable practices to control the environment and biological competitors.

The value of carefully selected and preserved field gene bank collections requires serious concern for their health. Newly acquired plants and propagules must be inspected and screened in an isolation or quarantine area as described in chapter 3. This will be made easier with prior knowledge of key pests associated with the germplasm collected, for example, Emerald Ash Borer on germplasm collections of *Fraxinus* species. The Food and Agriculture Organization (FAO) and International Board for Plant Genetic Resources (IBPGR)/International Plant Genetic Resources Institute (IPGRI) crop-specific publication *Technical Guidelines for the Safe Movement of Germplasm* is useful for inspecting and handling of newly acquired plants and propagules, among other issues.

As mentioned earlier, consideration of the disease threat to collections may necessitate growing them in screenhouses, greenhouses, or perhaps as *in vitro* accessions. A major concern in this category, especially for food crops, are viruses and their vectors. Secondarily, viroids, phytoplasmas, bacteria, fungi, and nematodes are problematic and difficult to detect.[38] It is important to eliminate a virus from infected accessions, or the infected accessions themselves, not only because they threaten the entire germplasm collection but also because their presence may skew any characterization and research conducted to improve preservation and conservation practices.

Some viruses may be identified from visual signs and symptoms. However, many must be detected by special means using serological techniques, molecular probes, or virus indexing. Serological and molecular methods are more accurate but may not be available for the taxa of concern. Indexing is more broadly applicable but is not as reliable an indicator depending upon the activity level of the virus. Virus indexing may involve sap or graft inoculation of indicator plants, and the timing of these treatments may vary depending upon the indicator taxa involved.[39] Viruses may be eliminated by chemical treatments, *in vitro* heat treatments, or meristem cultures. Virology, virus control, and virus indexing require special expertise that may need to be obtained from outside practitioners. Many, if not all of the U.S. Department of Agriculture, Agricultural Research Service, National Clonal Germplasm Repository staffs are trained in the control of viruses and virus-like organisms, including virus indexing. They, and their operations manuals, may be good resources on this subject and field gene banks in general. Two other very helpful resources are the Center for Plant Conservation's *CPC Best Plant Conservation Practices to Support Species Survival in the Wild* (2019) and the FAO's *Genebank Standards for Plant Genetic Resources for Food and Agriculture* (2014).

I will offer one last item about controlling biological competition from weeds in field gene banks. Under normal conditions, this is of concern for all botanical garden plant collections, especially for young plants that are more vulnerable to competition from weed plants. For field gene banks, this concern extends to spontaneous seedlings of uncertain genetic makeup that may volunteer in the planting beds and blocks. Such volunteer seedlings must be prevented from germinating and/or rogued out when they appear.[40] Block planting makes it easier to detect and control these unwanted seedlings.

A general set of cultural standards for greenhouse gene bank collections might include, but are not limited to, the following:

- Remove flowers and fruit.
- Remove excess foliage.
- Plants on benches will be at least one pot's distance apart.
- Remove dead plants as soon as possible.
- Plants will be properly labeled.
- Pasteurize media stored outside at 160°F for one hour.
- Place organic matter in special containers and empty daily.
- Keep pots, floors, and areas adjacent to greenhouses weed-free.
- Maintain daily activity records.[41]

The screenhouse is often used for controlled pollination and to preserve healthy, pest-free, and virus-free plants. A general set of preservation standards for the screenhouse might include, but are not limited to, those standards also applied in the greenhouse plus the following:

- All plants are isolated, examined, and treated if necessary for pest problems before entering the screenhouse.
- Exclusionary screens and other barriers are routinely inspected and repaired.
- Screenhouses are outfitted with double door entries that are functional and properly used.
- Minimal access is provided to those collections most susceptible to viruses.
- Diligent inspection of all persons is required to check for insect passengers on clothing and skin.
- A strict weed control program is maintained in and around the screenhouses.
- Screenhouses will be monitored by various means for pest populations.
- Organic debris is regularly removed.
- Daily activity records will be maintained.[42]

Best Practice 5.2: USDA–NCGR Corvallis, Field Gene Bank Preservation

The National Clonal Germplasm Repository (NCGR) is a branch of the Agricultural Research Service research agency of the U.S. Department of Agriculture (USDA). The repository is a gene bank that preserves genetic resources by various means, such as field collections of plants and *in vitro* and cryopreservation of plant tissues. Seeds are preserved at the U.S. Department of Agriculture–Agriculture Research Service, Agricultural Genetic Resources Preservation Research laboratory, a USDA gene bank.

There are nine clonal repositories located throughout the United States. The Corvallis, Oregon, NCGR is the repository for temperate small fruit, pears, hazelnut, butternut, and specialty crops. Below is a *synopsis* of field gene bank preservation specifications and procedures for *Vaccinium*, one of the repository's primary collections, taken from the NCGR Corvallis Operations Manual. *Vaccinium* germplasm is preserved in greenhouses, screenhouses, and field blocks. There may also be preserved tissues (*in vitro*, cryopreserved, and DNA samples) of *Vaccinium*.

Greenhouse Preservation (for Post-Quarantine Treatment and Evaluation)
- Greenhouse is minimally heated to prevent freezing and accommodate winter chilling requirements.
- Insect-proof screens subdivide the greenhouse into sections.
- Plants entering the greenhouse must pass through the pest treatment room for inspection and, if necessary, treatment.
- All accessions are properly labeled.
- Greenhouse cultural practices for all accessions are outlined, including the removal of all flowers and fruits and dead and damaged foliage (and dead plants).

Clonal Propagation of Vaccinium
- Semi-hardwood cuttings of pencil thickness in late spring and summer under mist and 16-hour illumination.
- Hardwood cuttings in winter, stored in the cooler at 4°C until May, some treated with a rooting growth regulator, placed under mist.
- Rooted cuttings potted in artificial media and fertilized with an acidic formulation.

Screenhouse Preservation (for Pest-Protected and Dwarf or Prostrate Permanent Collections)
- Restricted entry.
- Plants entering the screenhouse must pass through the pest treatment room for inspection and, if necessary, treatment.
- No container plants are to be moved without the screenhouse manager's permission.
- All accessions must be properly labeled.
- General pest control for all taxa in the screenhouse is outlined, including maintaining the integrity of the screenhouse and double doors, and documenting all pest-monitoring data and control activities.
- General cultural practices, including container spacing and flagging plants needing special treatments.
- *Vaccinium*-specific screenhouse preservation practices:
  - All plants grown in 2 gallon containers in a 100 percent composted hemlock bark.
  - Combined nutrition program of annual slow release 19-7-10 Sierrablend, fertigation with Peter's acid special, annual micronutrient and chelated iron applications.
  - Key pests are whitefly, fungus gnats, root weevils and root rots; integrated control measures are specified.

Field Preservation (for Mature, Bush-Type Accessions of Primary Collections)
- Pest tolerance is much higher for field collections allowing for more benign and sustainable pest management strategies.
- Between row areas are maintained in low maintenance turf cover.
- Irrigation is supplied by overhead impact heads or drip irrigation.
- Weed control within the rows is primarily mechanical with limited use of contact herbicides.
- Permanent labels are applied to all plants with the genus code, plant name, accession number, and row and position number.
- *Vaccinium*-specific preservation practices:
  - Field blocks are sectioned and labeled according to species, cultivars, and breeding lines.
  - Aged fir sawdust mulch to 2 inches is applied annually within the rows.
  - *Vaccinium* is a heavy nitrogen user; 100 pounds of nitrogen/acre is applied annually as a granular 20-12-8 fertilizer with micronutrients in three applications: mid-March to April, May, and June.
  - Dormant pruning is done to lift the crowns, encourage fruiting after the first two years, and thin older canes for light penetration.

The preservation plan also outlines the NCGR sustainable, or integrated, pest management program and itemizes all the key pests of concern for the entire collection, including their natural histories, signs and symptoms, monitoring protocols, and control strategies.

The security and integrity of field gene banks requires that they be established in duplicate. As mentioned previously, the Center for Plant Conservation (CPC) collections goal is 100 replicates of woody plants per collection. This standard alone, applied to large trees, will claim a significant amount of space. Two or three duplicates of this collection may be prohibitive unless botanical gardens collaborate to hold and preserve duplicate collections—a practice that creates useful distance for minimizing the loss of genetic integrity within the shared collection. Field gene banks of herbaceous plants may contain greater duplication (6–10) over smaller tracts of land, perhaps a better programmatic choice for smaller gardens.[43] Adequate duplication may be obtained through seed and tissue collections, two other types of gene banks open to smaller gardens. The preservation methods for these will be outlined later in this chapter. In addition to providing for the preservation of genetic integrity, duplicate collections also helps ensure security from losses due to environmental anomalies and pest infestations.

Historic Collections

Historic plants and landscapes offer the public unique opportunities to explore human and natural history and gain a perspective of landscape as a vital part of our cultural heritage. To optimize this public potential, how do we identify historic plants, evaluate their historical relevance, and choose appropriate preservation treatments for them?

Historic landscapes and gardens need a design and preservation policy to help govern these activities and retain their historical integrity. In addition, the plants on the site must be evaluated to determine their significance and relationship to the entire landscape. To accomplish these tasks, gardens will rely heavily upon personal diaries or journals, agricultural records, photographs, and other types of documentation. Lauren Meier, in "The Treatment of Historic Plant Material," is more specific:

> Through the process of assembling documentary data and field survey information, the historic vegetation location, use, appearance and changes should be substantiated to the greatest extent possible. The existing vegetation should be inventoried and evaluated, including extant historic features as well as more recent additions and invasive or volunteer plant material. The condition of these vegetation features should be determined as part of a field survey in order

to assess the overall health of the plant material and any specific treatment issues or needs. Finally, the existing appearance of the vegetation should be analyzed in relation to the historic documentation. The feature's condition, relationship to historic vegetation and the overall management objectives for the property will help guide the selection of an appropriate preservation treatment.[44]

Over time, historic gardens will invariably require various treatments designed to maintain their character. These may be preservative treatments designed to stabilize vegetation and landscapes to conservation activities to restore or reconstruct these features. Meier identifies five different types of treatment:

- *Protection and stabilization* are usually temporary treatments to mitigate deterioration until a more permanent treatment can be implemented. Large and significant trees are often protected with lightning rods. Overgrown plants are often pruned in a manner to reduce their impact on each other or adjacent structures.
- *Preservation* treatments maintain the historic character of a landscape as it has evolved. A significant hedge will be maintained and managed in a way to preserve its landscape role and aesthetic. [See the subsection on tree preservation above.]
- *Rehabilitation* is any treatment that provides for contemporary and future uses of a landscape while retaining its historic character. The introduction of new plants may fall into this category of treatment. In some cases, the results of these treatments should be interpreted to assist visitors in distinguishing new features from historic ones.
- *Restoration* is self-explanatory: restoring plants, vegetation, and landscape appearance to that of an earlier, significant period. This work must be based on sufficient documentation to provide guidance in making decisions about landscape changes. Existing, well-documented historic plants will naturally be left in place. The question arises as to what to do about mature and overgrown original plantings. Meier recommends that they be judged on their appropriateness of scale relative to the original design intent. Pruning and thinning may correct problems of scale. If not, then the plantings should be replaced.
- *Reconstruction* is the process of re-creating or rebuilding a landscape that has disappeared. This type of treatment is completely dependent upon sufficient documentation. Dramatized period re-creations should be avoided for the sake of authenticity.[45]

Best Practice 5.3: English Garden Restoration,
Stan Hywet Hall and Gardens
The English Garden at Stan Hywet Hall was originally designed in 1915 by Warren Manning and then reconceived and restructured by Ellen Biddle Shipman in 1928. From 1946 until 1990, the garden had been neglected and fell into disrepair. The work to restore and preserve the English Garden began when the Stan Hywet Hall and Gardens Board of Trustees identified this part of the property as a priority and proceeded according to the following steps starting in 1989:

- Obtain qualified consulting: The board hired the firm of Doell and Doell, garden historians and landscape preservation planners, to develop a preservation and restoration process plans.
- Identify principal historic period: Three major historic periods were identified for the English Garden: 1915–1928, Manning design; 1928–1946, Shipman design; 1946–1989, decline and transformation.
- Identify design intent: Manning period intent was a walled, "hidden" garden with a formal, cross-axial symmetry, central garden pool and sculpture focal point by Walter Dryden Paddock and plant list. The concept was confirmed in the absence of a planting plan through existing office correspondence and photographs of the finished garden showing the design. The Shipman redesign intent was documented in two planting plans with specific designs for perennial plants and shrub focal points within the structure of the Manning plan. Photographs also document the implementation of the Shipman planting plan redesign.
- Identify later transformations: Maturation of surrounding trees shaded out much of the original plantings that were then changed to accommodate the new growing conditions. The original Shipman redesign intent was lost.
- Determine historic significance and integrity: This stage is critical to the development of treatment plans:

  By documenting the landscape at different periods of its history, individual features can be attributed to a particular period when they were introduced and the various periods when they were present, thus establishing the significance and integrity of a landscape. Significance refers to the meaning or value ascribed to a cultural landscape; integrity is a measure of a landscape's authenticity as evidenced by the survival of the physical characteristics that existed during the property's historic period.[46]

It was determined that the physical structure provided by Manning and Paddock, and the planting plan provided by Shipman, gave the English Garden its greatest significance and integrity. All three of these artists also have important historical legacies.

- Develop preservation concept: "The preservation concept for a historical treatment establishes the goals and policies that become the standard for all aspects of the project including the selected treatment."[47] Since the hardscape and interior plantings had become so deteriorated, and there was good documentation on the plants and design of the interior beds, it was determined that a restoration treatment was the best option. The result was the restoration of Manning's hardscape and surrounding evergreen screen, Paddock's sculpture, and Shipman's planting plan.

- Develop treatment process plan (plan of work): According to this plan, Manning's original surrounding evergreen screen was removed and replanted with smaller trees to "hide" the English Garden without shading its interior. The Shipman plantings were restored based on the original plans, and the Paddock sculpture was cleaned and restored to its original condition. "Restoration seeks to 'put back what was once there as accurately as possible' and, as such, several factors must be considered in the decision-making process including: historical significance, extant historic resources, condition of historic resources, and selection of the restoration period."[48]

Replacing or making substitutions for historic plants is often a vexing issue. Curators should always seek to make replacements by propagating the originals as clones. Regarding substitutions, Meier makes an important point: "With substitutions, care should be taken to match the visual, functional and horticultural characteristics of the historic plant as closely as possible in form, shape and texture, as well as such seasonal features as bloom time and color, fruit and fall foliage."[49]

Municipal arborists know exactly what kind of challenge this can be in their struggle to find an analog for *Ulmus americana*, the American Elm. Curators may find themselves selecting analogs for irreplaceable plants based on a triage of historically important characters.

Textbox 5.3. Recommendations for Preserving Plant Collections

Basic
- The botanical garden employs up-to-date, sustainable horticultural and botanical technologies and practices for the preservation of its collections.
- Basic preservation standards are specified in the garden's collections management policy.

Intermediate
- The botanical garden employs a preservation management system to effectively and efficiently plan and apply preservation practices to its collections.

Advanced
- The botanical garden employs a computerized preservation management system for thematic subsets of the collections to schedule, track, document, and budget preservation practices.

Gene Bank Collections

"A gene bank is a collection of propagating materials that are checked for viability and then stored under conditions that retain viability for long periods."[50]

For the purposes of this section, "gene bank" refers to collections of propagules, such as seed, pollen, modified roots and shoots (tubers, rhizomes), and tissues grown as thallus cultures or plantlets. The storage, or "banking," of these propagules involves the use of low temperatures and may require low humidity, both of which slow down metabolic processes within the living material. The propagules must be checked regularly for viability and, if necessary, re-collected or regenerated. Traditionally, most gene banks contain propagules from crop plants, but this is changing to include a greater proportion of wild species.

Gene banks are created primarily to preserve germplasm and the genetic diversity of agricultural crops and rare or threatened taxa for conservation and research purposes. Therefore, the purpose and establishment of gene banks will also be covered in chapter 10. "The key principles at the core of gene bank operations are the preservation of germplasm identity, maintenance of viability and genetic integrity, and the promotion of access."[51] Although this description comes from the large and comprehensive Millennium Seed Bank, Royal Botanic Gardens, Kew, United Kingdom, effective gene banks may be rather simple and inexpensive to establish and operate; the accessions from these banks may be shared with the Millennium Seed Bank, among others. Domestic freezer refrigerators, chemical desiccants, and various airtight containers may be all that is required. Two major liabilities are an unreliable power supply and incomplete, inaccurate documentation.

A collection of seed samples stored under special conditions to ensure their long-term survival is known as a seed bank. "Conventional seed banking is a fundamental plant [preservation] and conservation practice."[52] Seed banks have many advantages over, or in addition to, whole plant collections: the seed is easy to store, and the facility takes up little space and needs few staff for routine maintenance. Seed bank operation and maintenance is quite straightforward, although we should not lose sight of the fact that "it is critically important that all gene bank processes adhere to the standards necessary to ensure that acceptable levels of viability are maintained."[53]

Seeds are stored at a low temperature—5–10°C for short-term storage down to −18 ± 3°C for long-term storage—after they have been dried to a low moisture content: 15 ± 3 percent. Viability should be tested at regular intervals, and if it drops below a predetermined level, new collections should be made or regenerated.

Pollen is an unconventional vehicle for germplasm preservation, although it may be useful for base collections of species that produce recalcitrant seed. Pollen, like seed, may be subdivided into orthodox and recalcitrant groups based on desiccation tolerance. Orthodox pollen is freeze-dried to 5 percent water content or dried and stored using the protocol for orthodox seed.

Modified roots and stems are kept in cold storage at temperatures just above freezing for comparatively short periods of time. Tissues and plantlets are stored *in vitro* for short-term preservation or, for long-term cryopreservation, stored in liquid nitrogen.

Seed Collections

Effective seed banking begins with the acquisition of quality seed lots. This measure includes as near complete a representation of biodiversity with that particular taxon as possible. At this point, I refer you back to the field-collecting section of chapter 3 on building collections. Be reminded that the care given seed collections during fieldwork and before they are stored within the seed bank is critical to their viability. Seed collections will need to be insulated from drastic environmental changes during collection and transport to the gene bank using appropriate packaging, described in chapter 3. One other critical item to double-check when moving from the seed acquisition stage to the seed storage stage of gene banking is the possession of the correct and accurate passport data (documentation) associated with each seed lot. In addition, make sure that all the legal documents associated with field-collected seed as well as Convention on Biological Diversity (CBD) documentation necessary for access to genetic resources of the country where the collecting took place and phytosanitary inspection is included and in order.[54]

Before submitting seeds for seed banking, it must be determined if they are orthodox or recalcitrant. Orthodox seeds are those that will store well for longer periods of time under conditions of reduced temperature and humidity, or, more specifically, from the CPC's *Best Plant Conservation Practices* manual:

> A seed's physical and physiological states determine whether it can be stored by conventional means or not. *Seed water content*, relative humidity and temperature are key interacting factors contributing to *longevity*. In order for a seed to be stored and survive freezing temperatures, a prerequisite is that it is capable of surviving *desiccation* or removal of most of the water in its cells.[55]

Many crop plants and temperate wild species fall into this category.

Recalcitrant seeds do not generally store well under any conditions. The following is an excerpt from the CPC manual:

> While orthodox seeds can be stored conventionally, *recalcitrant* seeds cannot be stored conventionally. The term "recalcitrant" anthropomorphizes seed responses to water loss. A recalcitrant seed tolerates some water loss, but not the extreme level survived by orthodox seed. Water remaining in recalcitrant seeds forms lethal ice crystal during conventional storage. "Recalcitrant" is also used to describe a seed that is difficult to germinate, which happens when seeds lack embryos (that is, "empty" seeds), have fastidious germination requirements (that is, *dormant* seeds or those with rudimentary embryos), or age quickly (possibly *intermediate*-type seeds).[56]

Included in this category are many tropical species and some temperate trees in the *Fagaceae*, such as *Quercus* and *Castanea*. Long-term storage of recalcitrant seeds is achieved with excised embryos stored in liquid nitrogen (LN). In the CPC definition of recalcitrant seeds, there is reference to "intermediate-type seeds." These are seeds that cannot be clearly classified as orthodox or recalcitrant (sometimes called "exceptional"). As the name implies, they may be stored using a modified, or intermediate, approach to conventional storage. Unlike recalcitrant seeds, intermediate seeds may be stored whole in LN without the need to excise their embryos, which is a much easier process. The CPC manual characterizes intermediate seeds by the following storage-related symptoms:

- Longevity is highest if seeds are dried to between 45 and 65 percent RH compared to the 15 to 20 percent RH optimum observed for orthodox seeds.
- Seeds age faster when stored at conventional freezer temperatures compared to refrigerated temperatures. Faster aging might be detected within days, months, or years, making it difficult to identify which species' seeds are intermediate.
- Longevity of seeds increases with drying and cooling (as with orthodox seeds), but seeds still age rapidly during conventional storage and will die within about five years.[57]

"To retain viability, recalcitrant seeds are stored at as low a temperature as possible under conditions that retain relatively high seed moisture levels and ensure a supply of oxygen for respiration."[58] Other alternatives for the long-term preservation of taxa with recalcitrant seed is their growth in a field gene bank and storage of tissues (embryos) *in vitro* or in cryopreservation. It has recently been determined that tropical grain legumes, *Citrus* spp., tropical grasses, and many taxa with small seeds are orthodox.[59] An outline of both *in vitro* and cryopreservation is presented later in this chapter and in chapter 10 on *ex situ* collections.

Another important determination that must be made before preparing seed accessions for storage is their maturity. Small amounts of immature seed (<10 percent) may be culled from the collection. However, larger amounts, especially for rare and endangered taxa, should be separated for postharvest ripening. Well-ventilated conditions with ambient temperatures protected from rain may be suitable. Some seed accessions may need a ripening catalyst, such as exposure to ethylene gas in a closed container. These collections should be carefully monitored and subjected to long-term storage conditions as soon as they are mature.[60]

There are two basic types of seed bank collections. A *base collection* is one that is stored under optimal, long-term conditions and is not altered or disturbed unless indicated by viability testing. An *active collection* is one that is stored under short-term conditions for the purposes of access and distribution—an *Index Seminum* would be a good example.[61] Refer to the section "Risk Assessment, Emergency Preparedness, and Insurance" at the end of this chapter for information on safety duplication of seed bank collections.

The following are a set of combined recommendations from J. G. Hawkes in *Botanic Gardens and the World Conservation Strategy* and the FAO's *Genebank Standards for Plant Genetic Resources for Food and Agriculture* for preserving base collections of seeds:

- Conduct a prestorage germination and/or tetrazolium test to help determine viability and dormancy (results should exceed 85 percent[62])
- Dry seeds slowly to a humidity level of 8 percent (15 ± 3 percent[63])
- Store seeds at –18°C (–18 ± 3°C [64]). Active collections and seeds of unknown cold capacity may be stored at 4°C (5–10°C[65])
- Conduct a sequential germination test every three years or, where experimentation or the literature dictates, at longer intervals. Consider possible dormancy requirements when attempting to germinate seeds. Active collections need not be tested after the initial, prestorage tests. Viability should not fall below 85 to 90 percent of standard germinability. Table 5.1 is a plan for sequential germination tests from Hawkes.[66]
- Storage receptacles may be (1) aluminum cans; (2) glass tubes, phials, or bottles with a sealing aluminum screw top; (3) Pyrex glass ampules heat sealed for very long-term storage; (4) plastic-aluminum foil bags or seed envelopes. These containers should be vacuum sealed. Active collections do not need to be sealed.[67]

Table 5.1.   Sequential Germination Tests

Number of seeds tested	Regenerate if number of seeds germinated is ≤	Continue test if number of seeds germinated is between	Maintain status quo if number of seeds germinated is ≥
40	29	30–40	
80	64	65–75	76
120	100	101–110	111
160	135	136–145	146
200	170	171–180	181
240	205	206–215	216
280	240	241–250	251
320	275	276–285	286
360	310	311–320	321
400	345	346–355	356
440	380	381–390	391
480	415	416–425	426
520	450	451–460	461
560	485	486–495	496
600	520	521–531	532

Source: J. G. Hawkes, "A Strategy for Seed Banking in Botanical Gardens," in *Botanic Gardens and the World Conservation Strategy*, ed. D. Bramwell et al. (London: Academic Press, 1987), 140.

Seeds should be dried to the appropriate moisture level prior to storage using a desiccant or a dehumidified chamber. Envelopes of silica gel are a desiccant that is often used to dry seeds. They may be placed in sealed jars for varying lengths of time to effectively lower their moisture content. Avoid using various heating elements for drying seeds as higher temperatures may damage the embryos. The oil content of seeds also impacts their moisture content. Therefore, equilibrium relative humidity (eRH) is the best measure of equilibrium moisture content. However, the eRH in the seed storage containers will change as the storage temperature varies from the drying temperature.[68] Water sorption isotherms are used to determine the critical moisture level at a given storage temperature expressed as a percentage of the total seed weight.[69]

Well-prepared seed stored under optimum conditions may remain viable for a matter of decades and longer. However, the effective length of storage for prepared seed is ultimately variable by species. Refer to the CPC's *Best Plant Conservation Practices to Support Species Survival in the Wild*[70] (part 1, section D) on cleaning and storing seeds; the FAO's *Genebank Standards for Plant Genetic Resources for Food and Agriculture*[71] (section 4.2); and Botanic Gardens Conservation International's "Seed Conservation Hub—Training and Resources" PDF.[72]

Needless to say, seed storage temperatures must be monitored continuously. Storage equipment and/or rooms should be alarmed to indicate when storage temperatures have changed. Include incident reports in collection documentation that include temperature differentials during those incidents. Be sure that all information on service providers is located in plain sight on the equipment and other facilities.

When sequential germination tests on base collections indicate an unacceptable loss of viability, or distribution and use of stocks from active collections reduce holdings, regeneration must take place by growing out a number of seeds and harvesting a new seed lot to be placed back in storage. As indicated by the term *germination test*, viability testing is best done using actual germination tests, not tetrazolium tests. However, where there is significant seed dormancy, other tests may be used. Samples may need to be selected and used for germination research in order to identify the most effective germination protocol before settling on a particular germination test. Germination tests should follow standard guidelines formulated by a recognized authority such as the Association of Seed Analysts, the International Seed Testing Association, or the International Board for Plant Genetic Resources. Regarding small seed collections, it may be necessary to grow plants and collect next-generation seeds for testing and storage and/or make subsequent seed collections from the wild.[73]

The Food and Agriculture Organization (FAO) of the United Nations recommends the following protocol for sequential germination tests:

> Viability monitoring test intervals should be set at one-third of the time predicted for viability to fall to 85 percent of initial viability or lower depending on the species or specific accessions, but no longer than 40 years. If this deterioration period cannot be estimated and accessions are being held in long-term storage at −18°C in hermetically closed containers, the interval should be ten years for species expected to be long-lived and five years or less for species expected to be short-lived.[74]

Generally speaking, if the viability test shows a drop below 85 percent of the test lot, regeneration of the accession should be initiated. For some species, this is an unreasonable standard. In those cases, a lower threshold should be determined based on empirical research and testing by the seed bank curatorial staff or adoption of a threshold based on testing by other seed banks. Species with lower viability percentages will likely need to be tested more often. Testing sample size should be as representative as possible without wasting too much of the overall inventory. Hence, the sample size will vary based on the total number of accessions. The FAO recommends a general sample size of 200 seeds.[75]

Regeneration is a critical programming element for seed banks as a means of preserving and perpetuating the collection.

> An accession will be regenerated when it does not have sufficient seeds for long-term storage (e.g., 1,500 seeds for a self-pollinating species and 3,000 for an out-crossing species) or when its viability has dropped below an established minimum threshold (i.e., below 85 percent of initial viability of the stored seeds).[76]

In addition to the above, regeneration is undertaken when the accession inventory is depleted due to repeated use and/or distribution. The ultimate goal is to pinpoint regeneration needs in order to preserve seed accessions while also preserving their genetic integrity, which may be compromised by needless or excessive regeneration. Regeneration of out-crossing taxa, in particular, should be minimized, especially if the inventory is stable, to reduce the likelihood of deleterious changes to the gene pool.

Ideally, regeneration should be done under the same conditions where the species sample was originally collected. More practically, reduce competition among regenerates as much as possible so that genotypes less well adapted to the prevailing conditions will have a chance to set sufficient quantities of seed.

To minimize the potential for inbreeding depression and genetic drift, grow out as many as 30 plants or more for regeneration from the most original sample. Isolate these lots in insect-proof glasshouses, screenhouses, or gauze chambers. Cross-pollinate them with select insect pollinators. For self-pollinating species, separate sample plots from each other and simply collect similar amounts of seed from each plant and bulk these together into one sample.

The documentation requirement for all of the above operations is imperative, and these records must be precise. The documentation should contain the original passport data for each accession, an activities record, a distribution record, and a catalog record for evaluation and characterization data and information. The software programs described in chapter 4 will accommodate gene bank documentation. The software program BRAHMS was originally developed for just such an application. There are also gene bank–specific information management systems, "such as GRIN-Global, GENESYS, Mansfield Database (IPK) and SESTO (NordGen), which have been specifically developed for gene banks and their documentation and information management needs."[77]

High-value seed bank accessions should be shared in duplicate to help ensure their security. This subset of the original collection is considered the secondary most original sample. Of course, with the duplicate security sample comes all the duplicate documentation. Partnerships for the exchange of security duplicate collections are usually of two types: "black box" arrangements involving duplicates that are simply stored and secured by the recipient seed bank without any programmatic access, as well as partnership sharing where the recipient has curatorial access to the duplicate collection as though it is their own. In the black box arrangement, the depositing institution must ensure that the duplicate collection is of an acceptable "most original" quality. In either case, there must be a written legal agreement between the depositor and the recipient institutions. Finally, depositors should be aware of the differing environmental conditions, no matter how slight, between the depositor location and facilities and that of the recipient institution as it impacts sustaining viable accessions and conducting germination tests and regeneration.

Knowledge of and sensitivity to the natural history, breeding biology, and cultivation of seed bank accessions is paramount, underscoring the curatorial expertise necessary for a well-managed seed bank collection.

Tissue Banks

Tissue banking is an alternative preservation and conservation strategy for plants that produce recalcitrant or very few seeds, are not responsive to vegetative propagation, and may be short-lived or difficult to grow as *ex situ* whole plants in field gene banks. In response, more than 100 botanical gardens around the world have tissue culture and micropropagation facilities.[78] Tissue bank plant preservation may be the most challenging of all gene banking programs due to the tremendous variation encountered when working with a range of tissue types.

> The Standards for *in vitro* culture and cryopreservation are broad and generic in nature due to the marked variation among non-orthodox seeds and vegetatively propagated plants. This variability is a function of the inherent biology and metabolic status of the plants concerned, which influences their differing responses to various manipulations and often requires modifications of basic approaches to be made on a species-specific basis.[79]

This type of preservation program is often developed and managed by persons on the curatorial staff hired especially for this work often with training in microbiology and chemistry. Be that as it may, the curator must have sufficient orientation to and a basic understanding of tissue bank programming and procedures to effectively oversee this work. In addition, the curator brings both ecological and horticultural knowledge of the subject taxa to bear on developing tissue culture protocols. What follows is a general accounting of the preservation and management requirements for the day-to-day care of *in vitro* and cryopreservation of accessions. Because the treatments and preparations for tissues preserved by these approaches are, to a large extent, experimental and require ongoing research, that information will be presented in the next section on collections research, found in chapter 10, "*Ex Situ* Collections."

Much *in vitro* and cryopreservation botanical programming is devoted to preserving recalcitrant seeds of endangered plants or their excised embryos or plantlets. Among these tissues may also be samples for DNA extraction. These seeds, and their tissues, cannot be held in a dormant, and therefore less demanding, state like that of orthodox seeds. The integrity of the recalcitrant seeds, tissues, and organs involved must be preserved in a slowed metabolic state that keeps them alive and viable over as long a period as possible. The technologies and practices involved in this work revolve to a great extent around maintaining hydration at the lowest possible temperatures. Second is the need to monitor for and control contamination by various types of organisms, beginning with an initial disinfection before the seeds or tissues are stored using any of several sustainable products for this purpose.

Recalcitrant seeds are stored for short- to medium-length terms and may be viable for a period of weeks to months. The seeds must be stored in containers that will retain seed moisture and provide an atmosphere of saturated relative humidity. The "bag within a bag" approach using seeds in paper bags sealed inside polyethylene bags is commonly used. Some seed banks use plastic buckets with sealing lids. These containers are then held at the lowest possible temperatures that each taxon will tolerate without losing viability. These temperatures will also reduce and slow fungal and other types of infection. Storage temperatures must be held constant to reduce the likelihood of condensation developing inside the storage containers. "For recalcitrant seeds of temperate origin, temperatures of 6 ± 2°C are generally suitable for storage, while for the majority of seeds of tropical/sub-tropical origin, 16 ± 2°C is normal range."[80]

Accessions should be ventilated and examined periodically—intervals that must be determined by each facility based on empirical trials and observations. Contaminated seeds must be culled out and eliminated, and the remaining inventory must then be disinfected and restored to a sterile container. Similarly, a check of viability may be done during the ventilation intervals or at times specifically scheduled for this purpose. Any one of several possible conditions may be revealed during these viability checks: (1) no change in condition or viability; (2) viability is good, but hydration levels have declined, calling for an examination of the storage containers used; (3) hydration levels are good, but viability has declined, indicating the possible end of the useful storage period at that temperature; (4) seeds show signs of germination indicating that the useful storage period may have been reached at that temperature. All these signs may give rise to manipulations and experimentation with the storage conditions and practices used in an effort to improve the storage environment and extend the possible storage time. It is possible that contamination is so chronic and intractable that another means of species preservation must be found, such as *in vitro* culture.

In vitro culture, like recalcitrant seed storage, is a rather short-term preservation strategy, in this case lasting from several months to a matter of a few years. This method, otherwise known generically as tissue culture and micropropagation, has most often been used for clonal selections of crop plants. *In vitro* cultures usually contain plantlets or shoots often derived from excised embryos and shoot tips. These cultures primarily serve as sources for disease-free taxa for distribution, multiplication, and cryopreservation. *In vitro* cultures are established in a variety of sterile containers ranging from test tubes and petri dishes, to jars and other types of containers with a prepared media formulated for the specific needs of the living explants. The cultures are held in environmentally controlled and artificially illuminated spaces. A major concern for the preservation of *in vitro* cultures is the control of infectious organisms, a serious threat when working with vulnerable tissues and nutritive media.

Unlike commercial tissue culture/micropropagation operations that seek to maximize growth, *in vitro* culture for preservation and conservation purposes may begin with a stage of rapid multiplication but then takes the opposite tack of seeking slow growth to extend the preservation period of the cultured shoots or plantlets. This requires high levels of manipulation and experimentation to find the correct variables to meet the goal of slow, viable growth. This means that a great deal of time and work is invested in developing the correct media and growing conditions for each taxon, parameters that are fiercely guarded among commercial operations.

This begins with the use of stock, or mother, plants as the source of *in vitro* tissues that are in good physiological condition. Since "experimentation with a range of permutations and combinations of the means to achieve satisfactory slow growth are imperative when first working with explants of a species,"[81] this work usually requires the skills of someone trained in *in vitro* lab and micropropagation techniques, often a microbiologist. Following the establishment of stable and disease-free cultures, the daily care of *in vitro* accessions may be placed in the charge of specially trained horticulture staff.

The preservation of *in vitro* conservation collections is highly dependent upon aseptic conditions. The high risk of contamination and pest infestation is first mitigated by keeping the facilities and equipment clean, if not aseptic. Before entering the tissue culture lab, staff should change clothes and shoes or wear Tyvek suits. Fresh plant material must be disinfested before being brought into the laboratory. Laboratory floors must be routinely wet mopped and all work surfaces wiped down with a sustainable disinfectant. Laminar flow hoods must be wiped down with a disinfectant and their filters regularly changed. The same goes for the HEPA filters for the laboratory's ventilation system.

Regarding the general growing conditions for the preservation of *in vitro* accessions, "optimal storage temperatures for cold-tolerant species may be from 0°C to 5°C or somewhat higher; for material of tropical provenance the lowest temperatures tolerated may be in the range from 15°C to 20°C, depending on the species."[82] The specifications for the proper illumination of *in vitro* cultures—spectral quality, intensity, and duration—will have to be determined through research and experimentation by the lab manager and shared with the horticulture staff. However, spectral quality for most plants is a known specification while intensity and duration may vary by taxon. Staff will also need to be trained to monitor *in vitro* cultures for various types of explant stress indicators, genetic instabilities, and the signs and symptoms of *in vitro* contamination.

At the end of their viable storage period and/or when *in vitro* cultures are called upon for use in replanting, they must be transferred to a suitable *ex vitro* environment and established as young plants. This involves a staged hardening-off period with plantlets gradually exposed to conditions that assist them in becoming more autotrophic and resilient to the conditions of their final planting location. A gradual exposure to changing conditions of relative humidity, growing media, fertility rates, irrigation, and illumination must be carefully managed to successfully acclimate and grow young plants into useful transplants. Well-documented standard tissue culture and micro-propagation techniques will serve as a useful guide for this transition period.

Cryopreservation—a complex process of deep freezing—is another means of preserving plants for conservation purposes that may not be more eas-ily preserved by other means. "Cryopreservation permits cells or tissue to be stored for an indefinite period in [liquid nitrogen] LN (–196°C) where metabolic activities are suspended."[83] At the time of this edition, there are only three botanical cryopreservation facilities in the United States. Two of these are operated by botanical gardens: the Cincinnati Zoo and Botanical Garden and the Huntington Library, Art Collections and Botanical Garden. Cryopreservation is applied to select types of seeds and tissues of a type used for *in vitro* culture and may, in fact, be derived from *in vitro* cultures.

Although there is a basic understanding of and protocol by which species may be preserved through cryopreservation, the specific details and proce-dures for individual species are the subject of research and experimentation. Once these procedures are identified and used for the cryopreservation of various taxa, the care of these specimens involves the operation of the LN vessels containing the specimens and the facility in which they are housed. The maintenance of reliable power supplies are critical and backup genera-tors may be necessary. Refer to chapter 10 on *ex situ* collections for more details on establishing and working with cryopreserved specimens.

Before leaving the subject of gene banks, and tissue banks in particular, I want to reiterate the critical importance of documentation. Test tubes, vials, and cryopreservation vessels must be well organized and labeled for collection control and tracking. Curators, and their staffs, must remember that these are another manifestation of the institutional collection, each lot of seed and each tissue an accession with an identifying number and registration data.

All of the various containers used in the above preservation methods must be clearly and accurately labeled with printed (sticky) labels containing bar codes or numbers and text. Looking at the documentation records of a select few botanical gardens, I would expect to find locator data to indicate, first, that a particular accession may be found on the grounds, in the nursery area, and in the gene bank as seeds and tissues, either *in vitro*, cryopreserved, or both; and, second, that they may all be easily located by maps, organization charts, and their individual labels.

Textbox 5.4. Recommendations for Preserving Gene Bank Collections

Basic
- Gene bank collections are consistent with and supported by the garden's mission and collections management policy.
- *Index Seminum* collections of orthodox seeds are thoroughly cleaned, germination tested, air dried, and stored in containers at 4°C until disseminated.
- *Index Seminum* collections of recalcitrant seeds are collected when ripe, thoroughly cleaned, and disseminated immediately.

Intermediate
- Base collections of seeds are slowly dried to 8 (15 + 3) percent moisture content and stored in sealed metal, glass, or foil containers at –18°C (–18 ± 3°C).
- Base collections of seeds are subjected to a sequential germination test every three years or, where experimentation or the literature dictates, at longer intervals.
- Base collections of seeds are recollected or regenerated by growing out a number of seeds and harvesting a new seed lot to be placed back in storage when viability drops.
- Long-term gene bank collections are monitored and alarmed for the event of cold storage or dehumidification failures.

Advanced
- Pollen collections are stored following the protocol for base collections of orthodox seeds.
- Tissues are established *in vitro* and placed in cold storage for short-term preservation or, for long-term cryopreservation, stored in liquid nitrogen.

Plant Propagation and Production Programs

As was first stated in chapter 3, a program of plant propagation is critical to the renewal and augmentation of the plant collections. Planned, often cyclical programs to repropagate existing plants in the collections should be standard practice for botanical gardens. This process, including a method for establishing propagation priorities, should be articulated as an integral part of the collections management plan.

Monitor and evaluate plant collections regularly in order to assess which ones will require repropagation. This may be implemented as a regular part of preservation monitoring and evaluation. The curator and propagator must establish standards for these evaluations based on the collections management policy and plan. Unique, historical, and conservation collections may require special attention and will likely become priorities in the propagation cycle. Guerrant and McMahan, in "Practical Pointers for Conserving Genetic Diversity in Botanic Gardens," recommend that these programs, to be most useful, be sensitive to gene pool maintenance, particularly suspected gene pool size, artificial founder effects, and hybridization.[84]

For the most part, renewal propagation will be by vegetative means to retain clones. Renewal of seed bank material, as was discussed earlier, will require the propagation of existing seeds and controlled pollination. To accomplish this, gardens must be staffed with experienced propagation staff equipped with the right facilities and resources to ensure success. (I will offer more on preserving the genetic integrity of conservation collections during propagation in the next section on collections research.) Many taxa in the collections of botanical gardens are difficult to propagate and are often not covered in the propagation literature. In these cases, propagators will likely implement a triage-based protocol based on taxonomic affiliations, ecological clues, and standard practices. Propagators must be prepared to research and explore layering, grafting, tissue culture, and other specialized techniques.

Propagation activities should be comprehensively documented. The propagation file may be part of a larger activity database that also contains other collections preservation activities. Tracking propagules through the propagation and production process is of critical importance—transplanting is a particularly troublesome juncture. Be sure that all propagules and resulting plantlets are adequately labeled; refer to chapter 4. From the standpoint of scheduling and tracking, keep in mind that it may take months or years to successfully propagate an accession or a taxon. During this period, preserve old accessions until a replacement is propagated if you want to preserve specific plant lines.

Textbox 5.5. Recommendation for Propagating Collections

Basic
- The botanical garden conducts a program of planned, cyclical propagation renewal of its collections, choosing the means most appropriate to preserve genetic diversity.

Preserving Exhibits, Displays, and Landscapes

Preserving and caring for exhibits, displays, and interpretive landscapes is more than simply an extension of preserving the collections. These botanical garden elements are created through a process of art, design, interpretive practices, educational techniques, and horticultural science. The tenets of many of these disciplines are not familiar to collections preservation staff, who must rely on other garden staff for guidance. The development and preservation of exhibits, displays, and landscapes is often an interdisciplinary process that may involve several staff members with varying specialties appropriate to the task.

Preserving exhibits, displays, and interpretive landscapes is facilitated by a written description of their programmatic goals, objectives, and concepts along with a schematic design illustrating how they will be implemented. These documents may then be used by the curatorial staff as a benchmark or principal reference point in establishing and evaluating preservation requirements. To start with, this information may be used by the curatorial staff to develop a preservation assessment of any new exhibits, displays, or landscapes for project feasibility and budget analysis. Particular consideration should be given to public access and circulation—one of the greatest challenges to landscape preservation, particularly for historic landscapes.

The curatorial staff should then specify the preservation requirements and practices pertinent to any given exhibit, display, or landscape in a preservation manual that serves as a staff guide. Distribute these guidelines for review by other garden professionals and/or consultants involved in the development of a particular garden feature before implementing any special preservation practices. Finally, the curatorial staff and other "exhibit team" members should establish evaluation guidelines and criteria, perhaps as a checklist, to monitor and evaluate the programmatic effectiveness and preservation of an exhibit, a display, or a landscape.

The preservation of natural areas and their dramatization in gardens is of growing interest to botanical gardens and their visitors. Many gardens are fortunate to have large parcels of undeveloped and natural area landscapes. These may serve simply as buffer areas, or, at the other end of the spectrum, they may be integral components of botanical garden programs. Either way, they will require some measure of preservation ranging from protection to active management. Some important issues will be inventorying the site and its resources: the character of the vegetation, the presence of invasive and disruptive plants, the presence of threatened or endangered species, the degree of and tolerance to human impact, and the stability of the natural area. At this point, I want to reiterate the information in the subsection on tree preservation earlier in this chapter on forest mycorrhizal networks and their growing importance in understanding forest ecology. Invariably, some type of management strategy will have to be developed for the site. John Ambrose, in "Conservation Strategies for Natural Areas," offers a word of caution: "Caution should be exercised in tinkering with natural ecosystem dynamics beyond responding to clear external threats. One should not be lured into thinking that the responsible management of natural areas must include active intervention."[85]

There are also many smaller gardens that have discovered creative ways to preserve native plants and educate their visitors with smaller, dramatized, or simulated natural area garden displays. Woodland gardens fall into this category, and perhaps there is no more definitive one than the Native Plant Trust's Garden in the Woods. Woodland gardens, though they often draw inspiration from natural areas, must be managed and preserved differently. Garden personnel must be aware of the needs of the tree overstory, for example, composition, density, potential hazards, and its continuity over many years. Cultivation and preservation practices must take into consideration the integrity of the display as to its naturalistic appearance and interpretive value.

Finally, historic landscapes and gardens present a unique set of special challenges for the preservation staff of any botanical garden. "Beyond the horticultural skills necessary to maintain a landscape, those responsible for a designed historic landscape have the additional burden of preserving a work of art created with dynamic materials whose spatial relationships change constantly."[86]

The challenges may be manyfold depending upon the historical nature of the landscape. The preservation requirements of a historic landscape—the significance of which is derived from its design, such as Fletcher Steele's Naumkeag—are quite different from those of Washington's Mount Vernon. Historic gardens must develop design and restoration criteria to help determine the significance of garden elements. These gardens must also make diligent use of any documentation and data that help guide preservation, minimize the implementation of stereotypes, and determine levels of visitor access.

Textbox 5.6. Recommendations for Preserving Displays, Exhibits, and Landscapes

Basic
- Botanical gardens have written descriptions of the programmatic goals, objectives, and concepts along with schematic designs to guide the preservation of their exhibits, displays, and landscapes.

Intermediate
- Botanical gardens have a design policy to govern the development and preservation of displays, exhibits, and landscapes.
- Botanical gardens have exhibit, display, or landscape preservation manuals that serve as a staff guide.

Risk Assessment, Emergency Preparedness, and Insurance

Botanical gardens with comprehensively planned, developed, and preserved collections should be, by their very nature, well prepared for and resilient to any emergencies. Nevertheless, a pointed concern for emergency preparedness and an accompanying audit may reveal any gaps and vulnerabilities, and it is a worthwhile process. Risk assessment—as a part of emergency preparedness—should naturally be conceived as a proactive endeavor.

A risk assessment for emergency preparedness begins with a thorough knowledge of your plant collection, the curator's raison d'être. If, as part of your curatorial program, you have maintained an up-to-date inventory of collections that includes evaluations and a growing catalog of characterizing data, you are in a good position to prioritize your collection for risk assessment and other emergency preparedness measures. If, however, you are behind in these documentation efforts, risk assessment and emergency preparedness is an unassailable reason for making this work a priority. Abby Meyer, in her article, "What's Our Backup Plan? A Look at Living Collections Security" for *Public Garden*, prescribes several useful steps for gaining knowledge toward securing your collection:

Your institution can take a few deliberate actions to minimize threats to collections security—while at the same time making a big conservation impact. And while they may seem basic, they are worth reviewing and engaging with at your institution:

- *Complete regular collection inventories.* Knowing what you have is a good starting point (Dosmann, 2012). Whether you set a three- or thirteen-year inventory goal, starting is the key. Incorporation of activities into regular workflows can help to identify manageable tasks to tackle through time.
- *Share collections data with the broader community.* Your collection can only be useful if potential users know about it. BGCI's PlantSearch is one of the only options to connect your collections to the global botanical community; contributing is free and simple.
- *Assess your collection.* Collection assessments can uncover previously overlooked or unknown information, and engage your staff in collection management activities. PlantSearch can help you identify the taxa in your collection that are threatened and underrepresented in collections. Comparing with other collections can also help you identify strengths and gaps for your collection.
- *Duplicate and distribute plants.* Duplicate priority specimens that are most vulnerable and establish backups within and outside of your institution. Propagation of priority plants may be complicated by taxonomic uncertainty, difficulty in producing viable propagules, or lack of available germplasm to use. It also often requires input from several individuals, which involves an additional level of coordination.
- *Support legal and ethical plant exchange.* Inform yourself and your staff about sharing plant material. Review or establish your institution's policies and practices surrounding plant exchange. This includes obtaining or renewing phytosanitary certificates and documentation associated with relevant legal or international frameworks such as the Convention on Biological Diversity.
- *Collaborate.* Find ways to work with other institutions in collections development. Seek out institutions that hold globally unique species your institution could accommodate. Other collaborative activities could include taxonomic verification, exchange of propagation knowledge, in situ versus ex situ species gap analyses, and expeditions aimed at diversifying ex situ collections.[87]

Bullet points 1 and 3 will assist you in assigning risk priorities to your collection. Ideally, curators would like to be able to protect and preserve all collections, but in the face of an emergency this may not be possible. Setting priority levels will help you decide which accessions need special protection and/or duplication and replacement in case of an emergency—the old adage "don't put all of your eggs in one basket" applies, particularly for collections of seeds and tissues. It's quite possible that for preservation purposes you may have already assigned various individual collections a standard of care priority that may be a useful equivalent for emergency preparedness. Kristine Aguilar, plant records manager at Longwood Gardens, suggests the following priority levels:

> *High Priority* will be assigned to accessions that are impossible to replace. These accessions may be rare in the wild and or in cultivation. They may be rare based on their age, and/or may be historically or scientifically significant to your garden or the horticulture community.
> *Medium Priority* will be assigned to accessions that are difficult or expensive to replace based on their monetary worth, size or availability in commerce.
> *Low Priority* will be assigned to accessions that are easy and inexpensive to replace.
> *No Priority* will be assigned to accessions that would not be replaced.[88]

Once the above priority levels have been assigned, identify the principal threats to a collection beginning with those considered a high priority (e.g., accessions extinct in the wild). The threats to collections fall into several categories: natural disasters, pest and disease outbreak, theft and vandalism, and controlled environment breakdown (e.g., greenhouse). Abby Meyer's last three bulleted items represent one of the most far-reaching and efficacious set of measures for securing collections against loss: duplication, distribution, and collaboration—an important curatorial practice and preservation measure that has been underscored many times in previous chapters. Collections distribution and sharing may be undertaken as part of a large multi-institutional program (e.g., CPC's National Collection) or among a small consortium of regional collaborators sharing plants, propagules, and tissues as suggested by Meyer. At the very least, gardens should ensure that they have backup germplasm of high priority collections in the form of multiple accessions of living plants in the collection and/or stored seeds, and such. Refer to figure 1.4, "Steps for Splitting Accessions by Maternal Line for Duplicate Storage and Testing," in the CPC's manual, *CPC Best Plant Conservation Practices to Support Species Survival in the Wild*.

Within gardens located in areas prone to repeat and damaging weather events, such as hurricanes, curators have taken steps to identify the most benign and protective microclimates for the preservation of high-priority collections and, if necessary, begun a process of establishing duplicate accessions and/or transplanting those collections to those areas. For the purpose of insurance claims, high-priority collections may also be evaluated using *The Guide for Plant Appraisal* (10th ed.) by the Council for Tree and Landscape Appraisal.[89]

Safety duplication for collections and their documentation is an important means of emergency preparedness, and it is certainly a necessity for institutions located in high-risk areas. Both of these subjects have been addressed in earlier portions of this chapter and in chapter 4 on documenting collections.

The risk assessment described above will better position the garden to select and negotiate on insurance contracts with an insurance provider.[90] When purchasing insurance coverage, these points warrant special attention:

1. Does the coverage clearly define the property the garden wishes to cover? It should be clear which collections and collections infrastructure are covered.
2. Are the territorial limits of coverage adequate to include satellite locations and so forth?
3. Are the perils insured against and the exclusions realistic? Many policies include all risks with some named exclusions.
4. Are the procedures for establishing valuations realistic? Make sure that these provisions keep pace with updated valuation methods.
5. Is it prudent to include deductibles in order to reduce premiums? The garden will have to consider its own ability to cover the expense of losses falling below this limitation.[91]

I want to emphasize that botanical gardens (and other museums) are unique in holding objects in trust for current, and perhaps more important, for future use. Fulfilling this trust means that botanical gardens must make a commitment to do everything that is possible to protect and preserve their collections for as long as possible. "A governing body that believes in collections preservation and that is willing to take a step or two each year towards improving the care that its collection receives is, in the long run, one that is true to its purpose in the fullest sense."[92] I recommend that you consult the American Alliance of Museums document, *Developing a Disaster Preparedness/ Emergency Response Plan* and perhaps more importantly, take the American Public Gardens Association Disaster Readiness Training.[93] From Pam Allenstein, Plant Collections Network Manager at APGA, "part of our Disaster Readiness Initiative launched this year, this on demand training program uses recommended FEMA process and offers specialized guidance for public gardens. Safeguarding plant collections and their associated documentation, as well as biological invasions, are addressed. Training is offered as a member benefit so APGA login credentials are needed to access the program."[94]

Notes

1. G. E. Burcaw, *Introduction to Museum Work*, 3rd ed. (Lanham, MD: AltaMira Press, 1997), 102.

2. American Association of Museums, "Stewards of a Common Wealth," in *Museums for a New Century* (Washington, DC: AAM, 1984), 42.

3. Clive R. Lundquist, R. Sukri, F. Metali, "How Not to Overwater a Rheophyte: Successful Cultivation of 'Difficult' Tropical Rainforest Plants Using Inorganic Compost Media," *Sibbaldia* 15 (2017).

4. Peter Wyse Jackson, "Developing Botanic Garden Policies and Practices for Environmental Sustainability," *BG Journal* 6, no. 2 (2009): 3.

5. American Public Gardens Association, "Sustainability Index," https://www.publicgardens.org/sustainability-index.

6. American Public Gardens Association, "Sustainability Index."

7. Climate Toolkit, "The Climate Toolkit for Museums, Gardens and Zoos," https://climatetoolkit.org.

8. R. L. Metcalf and W. H. Luckmann, eds., *Introduction to Insect Pest Management* (New York: Wiley, 1975).

9. American Public Gardens Association, "Sentinel Plant Network," https://www.publicgardens.org/programs/sentinel-plant-network/about-spn.

10. American Public Gardens Association, "Sentinel Plant Network."

11. Arnold Arboretum, *Landscape Management Plan* (Jamaica Plain, MA: Arnold Arboretum, 2011), v.

12. Arnold Arboretum, *Landscape Management Plan*, iii.

13. Andreas Ensslin and Sandrine Godefroid, "How the Cultivation of Wild Plants in Botanic Gardens Can Change Their Genetic and Phenotypic Status and What This Means for Their Conservation Value," *Sibbaldia* 17 (2019): 52.

14. Ensslin and Godefroid, 53.

15. Ensslin and Godefroid, 62–63.

16. R. W. Harris, "A Management Approach to Maintenance" (presented at the Golf Course Superintendent's Institute, Minneapolis, MN, March 1976).

17. *Wikipedia*, "DMAIC," last updated June 28, 2021, https://en.wikipedia.org/wiki/DMAIC.

18. Monika A. Gorzelak, Amanda K. Asay, Brian J. Pickles, and Suzanne W. Simard, "Inter-plant Communication through Mycorrhizal Networks Mediates Complex Adaptive Behaviour in Plant Communities," *AoB Plants* 7 (2015): plv050.

19. M. Liang, D. Johnson, D. F. R. P. Burslem et al., "Soil Fungal Networks Maintain Local Dominance of Ectomycorrhizal Trees," *Nature Communications* 11, no. 2636 (2020), https://doi.org/10.1038/s41467-020-16507-y.

20. University of Maryland–College Park, "Tree Management Plan 2017," October 26, 2017, https://arboretum.umd.edu/sites/default/files/2017TreeManagementPlan.pdf.

21. Madison Park Conservancy, *Madison Square Park Tree Conservation Plan* (New York: MPC, 2017), 15.

22. Madison Park, *Madison Square*, 16.

23. Madison Park, *Madison Square*, 25–26.

24. Madison Park, *Madison Square*, 30.

25. Shawn Kister, "Tree Management and Climate Change," *Public Garden* 31, no. 2 (2016): 28.

26. Kister, "Tree Management," 29.

27. Laurie Metzger, "Historic Tree Collection Management: A New Vision for Old Trees" (master's thesis, University of Delaware, 2014), xi.

28. Guy Meilleur, "Retrenching Hollow Trees: An International Practice," http://www.historictreecare.com/wp-content/uploads/2012/05/RETRENCHING-HOLLOW-TREES-FOR-LIFE-131126.pdf.

29. Neville Fay, "Natural Fracture Pruning Techniques and Coronet Cuts," April 2003, https://www.semanticscholar.org/paper/Natural-Fracture-Pruning-Techniques-and-Coronet-Fay/6da7d4f3c302240af3cacc96e25a433406f31f88.

30. L. Labriola, "Outsourcing: A Maintenance Alternative," *Public Garden* 10, no. 2 (1995): 25.

31. Center for Plant Conservation, *CPC Best Plant Conservation Practices to Support Species Survival in the Wild* (Escondido, CA: CPC, 2019), 2-22.

32. Susan Sharrock, *A Guide To The GSPC* (Richmond, Surrey: BGCI, 2012), 18.

33. Kim E. Hummer, ed., *Operations Manual* (Corvallis, OR: National Clonal Germplasm Repository, 1995), 6-1.

34. Food and Agriculture Organization, *Genebank Standards for Plant Genetic Resources for Food and Agriculture*, rev. ed. (Rome, Italy: FAO, 2014), 76.

35. Center for Plant Conservation, *CPC Best Plant Conservation Practices*, 2-22.

36. Jeremie B. Fant et al., "What to Do When We Can't Bank on Seeds," *American Journal of Botany* 103, no. 9 (2016): 1541–43.

37. B. M. Reed, F. Engelmann, M. E. Dulloo, and J. M. M. Engels, *Technical Guidelines for the Management of Field and In Vitro Germplasm Collections*, IPGRI Handbooks for Genebanks 7 (Rome, Italy: International Plant Genetic Resources Institute, 2004), 27.

38. Reed et al., *Technical Guidelines*, 3.

39. Reed et al., *Technical Guidelines*, 9.

40. Food and Agriculture Organization, *Genebank Standards*, 85.

41. Hummer, *Operations Manual*, 6-3.

42. Hummer, *Operations Manual*, 6-9.

43. Reed et al., *Technical Guidelines*, 12.

44. L. Meier, "The Treatment of Historic Plant Material," *Public Garden* 7, no. 2 (1992), 25.

45. Meier, "Historic Plant Material," 25.

46. Mary C. Halbrooks, "Decision Making in the Restoration of a Historic Landscape," *Public Garden* 20, no. 1 (2005): 18.

47. Halbrooks, "Decision Making," 18–19.

48. Halbrooks, "Decision Making," 19.

49. Meier, "Historic Plant Material," 27.

50. David R. Given, *Principles and Practice of Plant Conservation* (Portland, OR: Timber Press, 1995), 130.

51. Udayangani Liu, Elinor Breman, Tiziana Antonella Cossu, and Siobhan Kenney, "The Conservation Value of Germplasm Stored at the Millennium Seed Bank, Royal Botanic Gardens, Kew, UK." *Biodiversity and Conservation* 27 (2018), https://doi.org/10.1007/s10531-018-1497-y.

52. Center for Plant Conservation, *CPC Best Plant Conservation Practices*, 1-3.

53. Food and Agriculture Organization, *Genebank Standards*, 9.

54. Food and Agriculture Organization, *Genebank Standards*, 19.

55. Center for Plant Conservation, *CPC Best Plant Conservation Practices*, 1-4.

56. Center for Plant Conservation, *CPC Best Plant Conservation Practices*, 1-5.

57. Center for Plant Conservation, *CPC Best Plant Conservation Practices*, 1-6.

58. S. A. Eberhart et al., "Strategies for Long-Term Management of Germplasm Collections," in *Genetics and Conservation of Rare Plants*, ed. Donald Falk and T. Holzinger (New York: Oxford University Press, 1991), 139.

59. J. G. Hawkes, "A Strategy for Seed Banking in Botanic Gardens," in *Botanic Gardens and the World Conservation Strategy*, ed. D. Bramwell et al. (London: Academic Press, 1987), 140.

60. Food and Agriculture Organization, *Genebank Standards*, 22.

61. Eberhart et al., "Strategies," 137.

62. Food and Agriculture Organization, *Genebank Standards*, 30.

63. Food and Agriculture Organization, *Genebank Standards*, 22.

64. Food and Agriculture Organization, *Genebank Standards*, 22.

65. Food and Agriculture Organization, *Genebank Standards*, 22.

66. Hawkes, "Seed Banking," 144.

67. Hawkes, "Seed Banking," 141.

68. Food and Agriculture Organization, *Genebank Standards*, 25.

69. Food and Agriculture Organization, *Genebank Standards*, 25.

70. Center for Plant Conservation, *CPC Best Plant Conservation Practices*, 1–31.

71. Food and Agriculture Organization, *Genebank Standards*, 24.

72. Botanic Gardens Conservation International, "Seed Conservation—Training and Resources, https://www.bgci.org/resources/bgci-tools-and-resources/seed-conservation-hub-and-training-resources/

73. Center for Plant Conservation, *CPC Best Plant Conservation Practices*, 1-6.

74. Food and Agriculture Organization, *Genebank Standards*, 30.

75. Food and Agriculture Organization, *Genebank Standards*, 30.

76. Food and Agriculture Organization, *Genebank Standards*, 36–37.

77. Food and Agriculture Organization, *Genebank Standards*, 51.

78. Botanic Gardens Conservation International, GardenSearch (database), https://www.bgci.org/resources/bgci-databases/gardensearch/.

79. Food and Agriculture Organization, *Genebank Standards*, 116.

80. Food and Agriculture Organization, *Genebank Standards*, 131.

81. Food and Agriculture Organization, *Genebank Standards*, 135.

82. Food and Agriculture Organization, *Genebank Standards*, 135.

83. Food and Agriculture Organization, *Genebank Standards*, 139.

84. E. Guerrant and Linda McMahan, "Practical Pointers for Conserving Genetic Diversity in Botanic Gardens," *Public Garden* 6, no. 3 (1991): 20.

85. J. Ambrose, "Conservation Strategies for Natural Areas," *Public Garden* 3, no. 2 (1988): 18.

86. Susan Maney-O'Leary, "Preserving and Managing Design Intent in Historic Landscapes," *Public Garden* 7, no. 2 (1992): 15.

87. Abby Meyer, "What's Our Backup Plan? A Look at Collections Security," *Public Garden* 33, no. 4 (2018): 9.

88. Kristina Aguilar, *Assigning Disaster Priorities to Your Collection* (Kennett Square, PA: Longwood Gardens, 2010).

89. International Society of Arboriculture, *The Guide for Plant Appraisal* (10th ed.) Council for Tree and Landscape Appraisal. Atlanta, GA: ISA, 2018.

90. M. Malaro, "Collection Management Policies," in *Collections Management*, ed. A. Fahy (London: Routledge, 1995), 18–19.

91. Malaro, "Collection Management Policies," 20.

92. David McInnes, "Commitment to Care: A Basic Conservation Policy for Community Museums," *Dawson & Hind* 14, no. 1 (1987): 18.

93. American Public Gardens Association, Disaster Readiness Training, https://www.publicgardens.org/program/disaster-readiness/training.

94. Pamela Allenstein, personal communication, October 2021.

CHAPTER SIX

~

Collections and Public Programs

"Museums' collections and knowledge underpin their role as educational institutions, a role that places them among social agencies concerned with the life-long process of learning. At the heart of this role is the public exhibition of collections, which allows museum visitors to experience a direct, personal relationship with works of art, artifacts, specimens . . . publishing, outreach activities, school programs and so forth all enrich this relationship. A unique aspect of museums' educational role is their ability to appeal to all ages. Museums have the flexibility of being available to all and not being tied to a curriculum."[1]

It is a basic tenet of museology that all museums strive to make themselves and their collections accessible—physically, intellectually, and emotionally. Consistent with this tenet is the traditional view that botanical gardens have an important educational role. Public education in botanical gardens involves "thoughtful application of audience analysis and principles of teaching and learning to the processes of interpretation, exhibition, and—where appropriate—to collecting and research."[2]

Botanical gardens fulfill their educational obligation and role with a range of interpretive and educational programs. Within this context, it is the role of the curator, or curatorial staff, to support and create collections-based interpretive and educational policies and programs as part of an interdisciplinary, public programming team. This is the focus for this chapter as opposed to an exploration of the planning, design, and implementation of educational programs and operations themselves. Before going any further, I should point out that, in a public programming context, the curator and curatorial staff will likely defer to the professional judgment of a garden educator who has special expertise and, hopefully, the time to carefully consider educational needs throughout the garden. The curatorial staff should appreciate that garden educators act as audience advocates and work to provide meaningful and lasting learning experiences for a diverse public.

The relationship between curatorial practice, collections management, and public programming should be a symbiotic one. After all, the collections are the center and foundation of the botanical garden experience. "Public programming enables access to museum collections and the knowledge and values they embody."[3]

Therefore, a consideration of public programming needs plays into policies affecting collections development and management, not to mention every aspect of botanical garden endeavor. Concomitant to this, public programs should be based upon the collections. This symbiosis is best articulated in a public programming policy closely aligned to the museum's mission statement and collections management policy. Garden administrators, curators, and educators must be cautious and clearly oriented to this need because it is with public programs that garden stakeholders, donors, and others often seek to wield their influence—often in digressive directions.

Public Programming Policy

The garden's public programming policy outlines how it will interpret its collections to the public. The curatorial staff play an important role in the formulation of this policy by helping to ensure that it relates appropriately to the collections and the collections management policy. This policy must also serve to ensure that the garden's collections adequately support interpretive themes, including the use of collections for interpretive exhibits.[4] This can be a sticking point between garden curators and educators over the apparent dichotomy that exists between collections preservation/conservation and various types of collections access required for education. Use this policy to stipulate access issues and their resolution. Also, curators should recognize that professional garden educators are not only aware of the prime importance of collections preservation but also are skilled in developing educational avenues that respect those needs. Finally, collections staff should ensure that they are part of any stipulated evaluation process.

Textbox 6.1. Recommendations for Public Programming Policy

Basic
- The botanical garden has a written public programming policy that is relevant to its collections.

Intermediate
- The botanical garden conducts periodic reviews and evaluations of its public programming utilizing an interdisciplinary team composed of collections management staff and others.

Interpretation

The curatorial staff should play a prominent role in overseeing and guiding the use of collections in interpretive and museum education programs. The intrinsic involvement of curatorial staff in interpretive activities is expressed in the following definition: "Interpretation may be defined as activities that responsibly explain, and/or [exhibit] the collections in such a personalized manner as to make its background, significance, meaning and qualities appealing and relevant to the various museum publics."[5]

It is necessary for the curatorial staff to work closely with garden interpreters to create interpretive strategies that foster an understanding of what is known in museology as "object language"[6] among visitors, leading to a visual literacy about plants. Visual literacy may be defined as a shared understanding of the assigned meaning of a common body of visual information—that is, the attributes of the individual plants, the relationships between exhibited plants, and the cumulative effect of the arrangement of these components in an exhibit.[7] The curatorial staff will also be pivotal in the research, editing, promulgation, and review of abstract interpretive concepts centered on the collections.

To function effectively with interpretive staff, it is useful for curators and collections management staff to be familiar with the basic principles of interpretation:

- Interpretation is revelation and provocation based on information.
- Interpretation relates what is being interpreted to something within the experience of the members of the audience: the past is made meaningful in relation to the present.

- Interpretation presents a complete story and relates to the whole person (e.g., intellect and emotions).
- Visitors are diverse; therefore, a variety of approaches to interpretation are required.
- Visitors anticipate a relaxed, informal atmosphere.
- The social group is an important vehicle for the interpretive message.
- Interpretation is an art that can be taught and successfully learned; the art of the interpreter is the catalyst of a successful interpretive experience.
- Feedback to the interpreter is essential.[8]

Naturally, collections documentation comes to mind first in a consideration of crafting handy interpretive tools for visitors. Plant labels are one of the most rudimentary interpretive devices and often serve a dual purpose since they are necessary to the documentation program as well. Another very simple documentation-derived interpretive tool is a printed inventory of the collections. With some creative formatting, this can serve very nicely as a comprehensive reference publication to the collections if it also includes some natural history information such as nativity and a locating key. More elaborate catalogs on segments of the collections or collections exhibits may be created from catalog records and other miscellaneous background information typical of the publications offered in other types of museums. Indexes based on particular collection features, characters, or details may also be compiled as a specialized interpretive directory. An index is usually produced for internal use but may also be used to guide public access to particular aspects of the plant collection. The cartographic records may be used to create orientational and interpretive maps for visitors. Some electronic mapping programs will be able to produce a wealth of themed maps suitable for self-guided tours.

Textbox 6.2. Recommendations for Interpretation

Basic
- The botanical garden presents and interprets its collections to the public.
- The collections management/curatorial staff play a pivotal role in the research, editing, and review of collections concepts for interpretation.

Educational Programs

The curatorial staff may be expected to have the same level of involvement in educational programs as for interpretive programs. For the purposes of this text, educational programs are defined as "systematic instruction, within a specified time period, in subject areas related to the collection, the results of which are capable of being measured."[9]

In addition, the curatorial staff are encouraged to participate in the delivery of these programs. This recommendation is based on the premise that no one learns as much from the educational process as does the teacher—a condition that serves to charge the investigation and familiarity of the staff with the collections. Garden educators may rightfully insist that curatorial staff undergo some training in delivering educational programs and pedagogy before taking on the task of teaching.

Citizen Science Programs

The use of volunteers in botanical garden programming is nothing new; gardens have been engaging their members and the public as volunteers since the beginning. Botanical gardens have also benefited from the scientific interests of their members and the public in a somewhat different way through sharing information and data. One area of shared interest that has been ongoing for decades is meteorology. Amateur meteorologists have shared their weather and climate data with gardens to satisfy the garden's need to corroborate this information with plant phenology, performance, and hardiness.

A long-running example of a citizen science program is the National Audubon Society's Christmas Bird Count. Begun in 1900, the society's bird count runs from December 14 through January 5 each year. Groups of volunteers, called a "circle," are led by an experienced birder to collect information about local bird populations. This wildlife census, conducted by more than 2,000 birding circles across the United States and Canada, informs the bird conservation efforts of the Audubon Society and other ornithological organizations.[10]

Lately, there have been several large and popular scientific partnerships in the area of plant phenology observations by "citizen scientists." These collaborations serve to broaden the host garden's research in this area of plant development and climate change. A good example is Project BudBurst begun in 2007 to engage volunteers in collecting data on seasonal changes in plants. Sponsored by the National Science Foundation and run by the National Ecological Observatory Network in partnership with the Chicago Botanic Garden, Project BudBurst has grown into a large network of volunteers across the United States who monitor plants for seasonal changes.[11]

Citizen science programs and projects have become so popular and preva-lent that the U.S. federal government passed the Crowdsourcing and Citizen Science Act of 2016 (15 USC 3724) authorizing these partnerships in a range of government research and science programs. The following defini-tion comes from the CitizenScience.gov (U.S.) website:

> In citizen science, the public participates voluntarily in the scientific process, addressing real-world problems in ways that may include formulating research questions, conducting scientific experiments, collecting and analyzing data, interpreting results, making new discoveries, developing technologies and applications, and solving complex problems. In crowdsourcing, organizations submit an open call for voluntary assistance from a large group of individuals for online, distributed problem solving.[12]

Citizen science is a natural for botanical gardens. Anyone who has ever worked at a botanical garden has surely connected with dedicated gardeners and amateur horticulturists who have a thirst to know more about what the garden is doing and its collections than labels, brochures, and lectures pro-vide. Many such regular and curious garden visitors will clamor for volunteer opportunities and, in those positions, will seek to involve themselves in real scientific work. Citizen science programs provide the perfect avenue for the involvement of these valuable visitors and supporters while furthering the curatorial and other programmatic goals of the institution.

In addition, citizen science opportunities provided by botanical gardens for young people can be tremendously engaging and impactful, helping them make real-world connections to the environment and the plant world as well as stronger ties to their communities.[13] Citizen science projects at botanical gardens on conservation include studies on plant demographics, reproduction, and habitat fragmentation, just to name a few. Clearly, cooperation between scientific researchers and volunteers from local communities have the poten-tial to deepen the scope of research and increase the ability to collect scien-tific data important to botanical garden conservation and research programs.

The benefits of citizen science programs to the missions of botanical gar-dens include the following:

- Connect people with nature
- Improve scientific and environmental literacy
- Make local connections to global scientific and environmental issues
- Empower individuals and communities to enact positive environmental change
- Answer scientifically meaningful research questions[14]

There is a burgeoning literature on citizen science documenting growing numbers of successful programs. Two helpful online references that will provide a general orientation to citizen science are CitizenScience.org and CitizenScience.gov. For more context-appropriate references, I suggest familiar professional organizations such as the American Public Gardens Association (APGA), Botanic Gardens Conservation International (BGCI), and the Center for Plant Conservation (CPC), as well as the American Alliance of Museums (AAM) and their publications. Don't forget local, allied natural history institutions such as zoos, nature centers, and natural history museums, all of which are engaged in citizen science programming as well.

Best Practice 6.1: Citizen Science

Denver Botanic Gardens: Denver EcoFlora Project

> "The Denver EcoFlora Project is a collaboration between Denver Botanic Gardens, New York Botanical Garden, and the Institute of Museum and Library Services to engage citizen scientists with biodiversity in the Denver–Boulder metro area."[15]

Anyone in the Denver metropolitan area can participate in this project to document the native flora of the area through images and data using iNaturalist. The project provides citizens access to novel and interesting observations and records of plants across the region with the goal to engage them in observing, protecting, and preserving the native flora. The project has its own site on iNaturalist with training in using the platform for first-time visitors. There are monthly challenges, called EcoQuests, that motivate participants to uncover biodiversity in different locations and/or by different attributes.

Textbox 6.3. Recommendations for Educational Programs

Basic
- Botanical garden collections management staff support and participate in educational programs.
- Curatorial staff involved in teaching classes receive basic training in teaching methods and pedagogies.
- The garden involves visitors, members, and volunteers in citizen science opportunities.

Displays and Exhibits

The basic interpretive medium in any museum is the exhibit. Often known simply as "gardens," exhibits are equally important for botanical gardens. "Exhibitions have the unique dimension of physical reality: they can stimulate the senses, challenge the imagination and excite new perceptions, and effectively relate information. They are an invaluable tool in educational and interpretive programming."[16] What is the difference between a display and an exhibit? "An exhibit is a display plus interpretation; or, a display is showing, an exhibit is showing and telling."[17]

The context for experiencing the collections of a botanical garden is implicit in its name: a garden. For the most part, the collections of botanical gardens are preserved and interpreted in the same space, a garden space. For most other museums, this is not the case; their collections are often best preserved in off-exhibit areas away from public spaces where they can be carefully scrutinized and environmental conditions can be more carefully controlled and monitored. This situation does make botanical garden exhibition more difficult to define and segregate from the traditional practice of growing and preserving all collections in garden settings. A common exception to this practice usually involves collections of plants requiring greenhouse protection.

Exactly what is the purpose of museum and botanical garden exhibitions? I find Barry Lord's description in "The Purpose of Museum Exhibitions" to be instructive: "The purpose of a museum exhibition is to transform some aspect of the visitor's interests, attitudes or values affectively, due to the visitor's discovery of some level of meaning in the objects on display—a discovery that is stimulated and sustained by the visitor's confidence in the perceived authenticity of those objects."[18] Because their purpose is so integral to museum education and because garden and conservatory exhibits are such a major force in public programs affecting curatorial practices, I will spend more time on the role of curators in association with these than with other public programs.

A museum's intentions and capabilities regarding an exhibition program are often articulated in an exhibitions policy. The curatorial staff play an important role in the formulation of this policy, particularly as it pertains to the types of exhibits, assessment of collections for incorporation into exhibits, the exhibition team, and the evaluation process. Exhibit policies should contain the following collections management items:

- The type of exhibits, their respective roles, and the space required

- Exhibit priorities
- Collections assessment for exhibition: strengths, weaknesses, omissions; preservation, access, security concerns
- Identification of the exhibition team and the operational authority and constraint
- Exhibit resource assessment
- Exhibit evaluation

The process of creating an exhibit is multidimensional, often requiring the concurrent development of each dimension, for example, development of the story line or interpretive theme, selection of plants, overall design, preservation and security assessment, interpretive planning, and so forth. Since exhibits are developed around and based upon the garden's collections, there will naturally be the need for a strong curatorial role in their creation. In the development process, the curator will provide critical input in the following areas:

1. *Formulating the exhibit concept.* The exhibit concept is a basic part of the exhibition planning process and is often drawn from various curatorial themes about the collections.
2. *Curatorial research.* This work may revolve around thematic research to provide contextual information and collections research necessary to make informed choices about authentic taxa to be used. Oftentimes, through the process of normal events, this work has already been done and naturally leads to an exhibit proposal.
3. *Collection evaluation, selection, and preparation.* The curator must develop a desiderata of taxa from the collections suitable for the exhibition. The curator's judgment is critical in determining the extent to which the garden's collections meet the needs of the exhibition and by what means other taxa might be acquired for this purpose.
4. *Documentation.* The curator will handle any documentation changes associated with the taxa on exhibit. The research associated with exhibits often generates precious registration and catalog data along with any other actions that must be taken to prepare taxa for exhibition, for example, propagation, specific training or grooming, and so forth. The relocation of taxa for exhibition must also be documented.
5. *Preservation.* The curator must establish and see to the preservation of taxa on exhibit as well as the entire exhibit context. Hopefully, these needs will not work against each other if the exhibit has been carefully planned and the concept is appropriate. The exhibit may warrant a

special condition or preservation profile that specifies particular preservation and exhibit care needs. There may be a need to reexamine the garden's insurance contracts as they may apply to the new exhibit and its contents. Compromise may be the order of the day, but it should always be aimed at reducing the risk of damage or loss of the taxa while preserving the objectives of the exhibit.

6. *Preparation of the exhibition brief, prospectus, or narrative.* This document assembles all the information that defines the content and purposes of the exhibit and goes by many different names. It usually serves as the cornerstone narrative to define the exhibit along with the graphic design plans. This document usually contains the core idea for the exhibit expressed as a thesis or question that the exhibit explores, the thematic framework, the thematic structure articulating how the themes relate to one another, the story line expressing the key messages of the exhibit, and the resource plan identifying all the collections available for use in the exhibit.[19]

At their most basic, exhibits and displays need to pique a visitor's curiosity. They should also focus on or contain elements that relate to visitors' current interests but will serve to inspire them anew. Ensure that the exhibit, or portions of it, are done on a human scale and give a sense of unity and place.

Textbox 6.4. Recommendations for Displays and Exhibits

Basic
- The botanical garden has a written exhibition policy that includes specifications for assessment of collections suitability and preservation requirements.
- The collections management/curatorial staff play a pivotal role in the research, editing, and review of collections concepts for exhibition.

Intermediate
- The botanical garden conducts periodic reviews and evaluations of its exhibits utilizing an interdisciplinary team composed of collections management staff and others.

It may be advantageous for curatorial staff to develop a manual for the proper care and preservation of permanent botanical exhibits or to include a special section on this topic in the collections preservation manual. Plan to interpret the contents and practices outlined in the manual through staff training programs.

Refer to figure 6.1, a flowchart for the exhibit planning and building process.

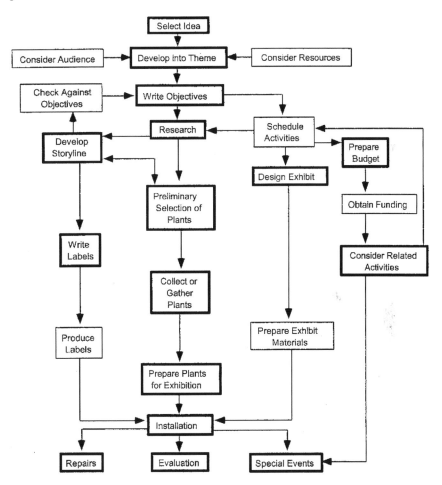

Figure 6.1. **Flowchart of Exhibit Process**

Notes

1. Communications Canada, *Challenges and Choices: Federal Policy and Program Proposals for Canadian Museums* (Ottawa: Minister of Supply and Services Canada, 1988), 27.

2. American Association of Museums, Standing Professional Committee on Education, "Standards: A Hallmark in the Evolution of Museum Education," *Museum News* 69, no. 1 (1990): 78.

3. Alberta Museums Association, *Standard Practices Handbook for Museums* (Edmonton, AB: AMA, 1990), 218.

4. C. Zimmerman, "Preparing an Education and Interpretation Policy," *Dawson & Hind* 14, no. 1 (1987/1988): 15–17.

5. John R. Dunn, "Museum Interpretation/Education: The Need for Definition," *Gazette* 10, no. 1 (1977): 15.

6. Alberta Museums Association, *Standard Practices*, 226.

7. Alberta Museums Association, *Standard Practices*, 226.

8. Alberta Museums Association, *Standard Practices*, 224.

9. Zimmerman, "Preparing an Education and Interpretation Policy," 14.

10. Audubon, "Audubon Christmas Bird Count," https://www.audubon.org/conservation/science/christmas-bird-count.

11. CitizenScience.gov, "Project BudBurst," https://www.citizenscience.gov/project-budburst/#.

12. CitizenScience.gov, "About CitizenScience.gov," https://www.citizenscience.gov/about/#.

13. CitizenScience.org, "Citizen Science: Activating STEM Learning out of School," https://www.citizenscience.org/wpcontent/uploads/2018/08/AfterSchool STEM-170510.pdf.

14. Jennifer Schwarz Ballard, Amy Padolf, Jessica A. Schuler, and Karen Oberhauser, "Engaging Diverse Audiences in Conservation and Research through Citizen Science" (presented to the American Public Garden Association Conference, Washington DC, June 17–21, 2019).

15. Denver Botanic Gardens, "Citizen Science," https://www.botanicgardens.org/citizen-science.

16. Alberta Museums Association, *Standard Practices*, 251.

17. G. E. Burcaw, *Introduction to Museum Work* (Nashville, TN: American Association for State and Local History, 1975), 115.

18. B. Lord, "The Purpose of Museum Exhibitions," in *The Manual of Museum Exhibitions*, ed. B. Lord and Gail Lord (Walnut Creek, CA: AltaMira Press, 2002), 18.

19. Lord, "Purpose of Museum Exhibits," 345–56.

PART II

~

COLLECTIONS RESEARCH AND APPLICATIONS

We still have so much to learn about plants—a huge, varied, and life-giving kingdom of fellow organisms critical to all life on this garden planet we call Earth. Plants of every kind, from monumental pillars reaching to the sky to diminutive prothalli bridging particles of soil on the ground, that are so different from us, and yet, so absolutely necessary to our existence. The botanical garden curator is a pivotal position in regard to its research programs. Research on and with the collection is the life blood for the growth and development of effective curators, curatorial programs, and botanical gardens.

The following is a synopsis of the contents of this second part of *Curatorial Practices for Botanical Gardens*:

Chapter 7 makes a case for the centrality of research at museums and, in particular, botanical gardens, including a recommendation on research policy.

Chapter 8 details the investigations, trials, and applied research that move collections management forward as improved practice, including monitoring and evaluation.

Chapter 9 deals with conservation, a major focus of botanical garden research and practice in response to climate change and species extinction, including *in situ* conservation programs.

Chapter 10 details *ex situ* collections, an increasingly important part of conservation programming in service to *in situ* conservation and species preservation, including seed/tissue/DNA/field gene bank collections and collections of conservation value.

Chapter 11 deals with species recovery programs and ecological restoration, a wide-ranging endeavor encompassing small efforts to bolster a single population of a threatened species to the complete restoration of a damaged or destroyed ecosystem. In most cases, this critically important work draws on the resources of *ex situ* collections at botanical gardens for species recovery and restoration.

Chapter 12 examines plant introduction programs, a traditional role for botanical gardens that, unlike species recovery involving plant reintroduction to the wild, is geared toward the introduction of plants to commerce for use in agriculture and horticulture. The epilogue for part II is a brief reiteration of several important aspects of collections research and some final thoughts on the subject.

CHAPTER SEVEN

~

Research in Museums and Botanical Gardens

"Museums, by collecting objects, hold the information held by real things. That information may be about man or nature or both, and may relate to art, history, or science."[1]

Research is a fundamental part of museum and botanical garden work. It is one of the principal justifications for our collections. Botanical garden research is focused on the search for truths hidden in the plant collections. The diverse and permanent collections of living plants in botanical gardens gives them a comparative advantage in research. Individual plants may be regularly and repeatedly monitored and sampled for various characteristics.

Without research, much of the potential information associated with plants in our collections remains unknown. It is the mandate of all good botanical gardens to produce information and knowledge about their collections and present it to the public. Research is the means by which this information is produced. Research enables the museum to accurately interpret their collections in a meaningful way. Research also provides botanical garden personnel with insights into the efficient and effective functioning of the institution as a whole. "The life of collections and their inestimable and highly defensible values lie in what we do with them, what new knowledge can be extracted from them, and what of past knowledge is documented as groundwork for improved insight and new interpretation."[2] Without research, our collections are at risk of becoming obsolete.

One may choose to think of research as that part of the botanical garden operation that generates the program, as opposed to those parts that disseminate or publicize it.[3] One of the principal goals of the curatorial and collections management program is to facilitate and carry out research. Museum collections are a unique asset, as well as a core strength, that provide museums with a competitive advantage over all other research institutions.[4]

Research on plants in the collections involves both considerations of the plant for what it alone can reveal, as well as what can be discovered about its context from additional sources of information. In other words, collections comprise "objects [that] serve as a body of data against which a myriad of . . . assumptions [can] be tested."[5] So, they can be "regarded as 'scientific data in waiting.'"[6] Research is the endeavor to *discover* new knowledge, *compile* facts, *interpret* information, and *establish* or *revise* accepted conclusions, theories, and laws.[7] Documenting research, the product as well as the process, is "the accumulation, classification and dissemination of information or evidence"[8] about the collections.

For the botanical garden profession to grow in strength and excellence, it must, in addition to conducting collections research, undertake research in botanical garden practice—a branch of museology. This involves every aspect of operations: garden administration, collections management, public programming, and exhibition. We must examine our procedures and practices and question why things are done in a particular way. Institutional audits are a valuable form of research.

Many botanical gardens have difficulty identifying their role in research and their research endeavors. The fact is, to successfully implement and manage the curatorial program described here, curators, collections managers, and their staffs, whether they recognize it or not, will engage in research. Collections management for research is purely good collections management. Part of the challenge for botanical gardens in determining what their research programs will be is simply to recognize the research component of their basic programs—particularly if they are comprehensively planned and implemented.

Still, we may categorize specific kinds of research to recognize more easily their nature and requirements. According to the Alberta Museums Association's *Standard Practices Handbook*, museums can contribute to the body of research in three areas:

- Summative research: compilation and synthesis of previously documented information on a subject with no attempt to formulate new insights [e.g., compiling information on a plant from various published sources].

- Applied research: in museum activities such as collecting, exhibit design, preservation or conservation practices, education and evaluation methods, etc.
- Basic research: original scholarship adding to the base of knowledge in a specific subject area [e.g., biodiversity, taxonomy, conservation biology].[9]

Gardens often do not fully recognize, or give credence to, the myriad types of summative and applied research they could and, in fact, do conduct. "What we heard most clearly during the course of our [the Commission's] work was an assertion of the significance of scholarly endeavor in the context of the museum, and serious concern that the research function is misunderstood and inadequately funded."[10]

Gardens do not undertake research in a vacuum; similar research occurs in universities and colleges and at other botanical gardens. Gardens, particularly smaller gardens and those with restrictive research budgets, should cultivate ties with these institutions.

> Since research helps to infuse collections with meaning, museums need to foster ties with institutions such as universities in order to benefit from existing pools of knowledge. In this way, museums may also be able to encourage research related to their collections and public programming objectives. These connections include sponsorship of research projects or artists-in-residence, cross-appointments, exchanges of personnel, and technological linkages through electronic media.[11]

As this quote suggests, gardens without research staff can bring in off-site researchers to work on their collections using various incentives such as grants and fellowships, in-kind support through access to facilities and housing, and professional recognition. One way or another, potential researchers should be welcomed and given access to garden resources for their work.[12] It is worth reiterating here that any institution, regardless of size or focus, should be encouraged to engage its collection in research. Michael Dosmann makes a good point:

> In fact, similar to local and small herbaria, regional or specialized gardens can contain unique collections of germplasm, have exceptional cultural histories, or grow plants under unique environmental conditions. Whether their goal is the development of collections or research agendas, the most logical route for gardens to follow is to tell their own story, to build upon their unique collection strengths and assets.[13]

The role of the curator in botanical garden research is paramount and inseparable. Again, this is particularly true at smaller gardens, where the curatorial program likely constitutes the bulk of the institution's research. What is more important, to fully understand something and to effectively teach it, one must engage in the exploration and testing of new knowledge. Curators have a tremendous opportunity and obligation to lead the way in research and to justify by research the collections they develop and guard. This is not to imply that curators be qualified for and capable of collections-based research in all of its possible forms. The research interests and goals of some gardens may be of such a specialized and exacting nature that specially trained experts in that area are required. Many gardens have scientific staffs with specializations in microbiology, forest ecology, plant breeding biology, restoration ecology, and others. Curators are important liaisons, advocates, and facilitators for these specialists in their need for access and use of the collections. Also, the botanical, horticultural, and ecological knowledge of curators regarding the collections is often useful for researchers. And, in turn, much of the knowledge generated by these specialists will also serve to inform the curator and their staffs about the nature of the collection. This will assist the curator in refining and advancing curatorial practice and collections management to the benefit of everyone. It is the curator's job to help establish and maintain these internal networks and services.

"Research leadership will inspire others and, by thus leading, draw more investigators."[14] Before leaving the subject of a justification for research and collections in general, I should note that botanical gardens can also serve as the source of research inspiration. Dosmann points out that *Science in the Pleasure Ground*, a history of the Arnold Arboretum, speaks eloquently of the often inspired relationship among gardens and research, landscape, and collections.[15] Ricklefs echoes this notion: "Beyond the value of museum collections as a resource for ecologists, specimens remain a source of insight and inspiration. . . . [T]he examples of taxonomic diversity and geographical variation . . . suggest . . . novel research problems or add to the context within which one thinks about ecological and evolutionary issues."[16] A molecular biologist may find, as Mark Chase puts it, "reciprocal illumination" in the benefits of observing both molecules and living plants.[17]

Needless to say, botanical gardens are good places to assemble and grow large collections of plants for research since it is usually part of their mission and they should be good stewards for such collections. Too often, though, institutions dwell on the historical contributions of their collections to science and overlook their current and future value. Some of this may be attributed to a limited view of the scientific disciplines best suited to use living plant collections, as well as the commonly held perspective that the research community has left gardens and collections behind with foci on molecular biology and so forth. Hence, some gardens have been reluctant to engage contemporary research users, particularly if they are from other institutions.[18] Nevertheless, our collections are of value for a number of research endeavors that include traditional pursuits such as taxonomy but extend beyond that subject. As Rae puts it, "taxonomic research . . . should not be regarded as the only worthwhile reason for botanic gardens to exist—there is far more to a botanic garden than this. The diversity of functions should not only be tolerated but encouraged."[19]

Peter Raven suggests some pertinent research programs utilizing collections:

- For long-lived plants such as trees and shrubs, hybridizing them in botanical gardens, where space is available to grow and analyze the hybrids, has great value in understanding the nature of species in many groups of plants. Such experiments require the long-term maintenance of hybrid individuals, which may require many years to attain their reproductive stage.
- Coming into more prominence in recent years is the use of living collections of plants and plants assembled for that purpose for phenological or other comparative observations that can be projected over wide areas.
- Particularly noteworthy is the evaluation of the diverse collections of plants in botanical gardens as bellwethers of the forces that will come to bear on populations of plants in the future. The plants can be used to monitor climate and other environmental changes.
- Research into the development of new horticultural varieties or into the survival of species and individual strains is very suitable for the diverse collections of plants found in botanical gardens. Such research has historically been subsidized by the Department of Agriculture and has often been well developed in land grant universities.[20]

Before embarking on a discussion of particular research programs appropriate for botanical gardens in the next chapter, it's worth extending Dr. Raven's points about what is possible and needed. Michael Dosmann has a succinct and poignant recommendation: "To maximize broad collection use, gardens will have to avail themselves to users in fields including and beyond the traditional, from ecology and natural resources to molecular biology, even including non-biological sciences."[21]

Taking a look at traditional research users of living collections, taxonomists, their comparative methods and approaches are now being extended to the use of molecular, developmental, and chemical characters as well as the conventional morphological ones. As taxonomists seek to apply these new approaches to the diversity of the plant kingdom, botanical gardens should be there to offer their wealth of genetic and molecular diversity (provided that they have planned for such an eventuality in their collecting efforts). Michael Dosmann conjures an apt description of what's available in one garden venue: "Just imagine the typical ornamental border: a polyploid circus of cuticular waxes, pigment combinations, bizarre leaf and floral morphologies, contorted habits, atypical growth rates, and unusual tolerances to environmental stresses."[22]

In some cases, gardens find themselves in a position to offer viable research material by coincidence or serendipity: they just happen to have—usually for other reasons—what a researcher needs. Perhaps many of the unusual clones of ornamental plants within the collections of botanical gardens will be exploited for genomics. They should be, and more thought should be given to planning and advocating for this work by garden staff, especially curators.

The garden as a whole presents a myriad of research opportunities for a host of disciplines from ecology to zoology, sociology to museology, entomology to plant pathology, archaeology to fine arts. The notion of sociologists finding any significance in the collections and facilities of a botanical garden is not so far-fetched. Researchers from nontraditional disciplines can often shed new light on collections, gardens, and botanical garden programs with refreshing results. Consider a fine arts example: "Caravaggio's beautiful paintings, extensively studied and appreciated by artists and historians for their detailed realism, when studied by [a] horticulturist provide in unanticipated fashion a unique glimpse of the crop diversity, pests, and diseases present in the late sixteenth and early seventeenth centuries."[23]

Before going any further in a review of research programs connected to botanical garden collections, I must say in no uncertain terms that the role of the curator in this important segment of botanical garden work is just as you would expect, to curate any and all elements of the collections related to botanical garden research. In doing so, they will bring to bear on that research their breadth and depth of knowledge about the collections to facilitate and assist in its success.

Textbox 7.1. Recommendations for Collections Research

Basic
- The botanical garden has a written research policy.
- The botanical garden undertakes basic documentation of each plant in its collections.
- The botanical garden undertakes research related to its collections and professional operation.
- The results of research are published or made publicly accessible in some form, for example, exhibition, catalog, book, and so forth.
- The botanical garden maintains a general reference library related to its collections for in-house use by staff.
- Visiting researchers are permitted supervised access to the collections.

Intermediate
- The botanical garden undertakes a specific program of summative and applied research based on its collections and programs.
- The botanical garden has the resources and space to accommodate visiting researchers.
- The botanical garden maintains records of visiting researchers and the plants in the collections they have studied.

Advanced
- The botanical garden maintains a lending reference library related to its collections.
- The botanical garden undertakes a specific program in basic research based on its collections and programs.

Research Policy

The relationship between garden research programs and collections should be established within a research policy. This policy outlines the nature, scope, and process of doing research and the garden's research commitments. As an important reminder, policies and plans are usually generated as part of a strategic planning process for the entire institution or as a department or programmatic supplement to an institutional strategic plan. Refer to *The*

Manual of Strategic Planning for Museums by Gail Dexter Lord and Kate Markert.[24]

> A museum with a well-defined statement of purpose and established goals can evolve a research policy from the themes identified and the areas of knowledge represented by its collections. Museums without a clear statement of purpose or goals must do this work before a useful research policy can be developed. It will require assessing the existing collection for its value as a source of knowledge, and identifying themes and subject areas with research potential.[25]

The Alberta Museums Association makes the following recommendations for research policies:

- define the nature and scope of research, e.g., the type of research (summative, applied, basic) and priorities by type and research methods;
- ensure the institution's commitment to research, e.g., budget, space, equipment, materials, staff, staff time;
- address ethical issues, e.g., copyright of research results, acknowledgment, use of museum and archival materials in published research, restrictions to public access;
- provide guidelines for handling public inquiries, e.g., amount of time to devote to requests for information, types of information that can be researched;
- outline the research process, e.g., for what purposes will research be conducted, by whom, what should be the results, how will the results be published;
- ensure commitment to hiring and ongoing training of qualified staff;
- define access and control of research material, e.g., by in-house as well as visiting researchers, who and what type of research is given priority;
- outline the uses of research, e.g., to generate exhibitions and/or publications, to produce new knowledge;
- state policy regarding loan of specimens for research purposes;
- ensure evaluation of research and its effects on the museum's aims and other activities;
- outline the responsibilities and obligations of research participants, e.g., written consent;
- define publication: format, where research will be circulated, will the museum publish a researcher not on staff, time for staff to write, publication of research given priority among other publication activities;

- outline supervision and security requirements for visiting researchers; and
- outline the relationship to other museum policies, e.g., collections management policy, conservation policy, exhibition policy, etc.[26]

Included in the research policy will be ethical guidelines for research. Some ethical concerns related to collections are the following:

- Balancing the needs of research programs with those of collections preservation
- Conducting research within the guidelines of the collections policy
- Conducting field research and adhering to ethical and legal restrictions
- Accuracy in research and interpretation
- Access to research collections[27]

Under the heading of setting research policy, the subject of institutional bias should be addressed. Living museums of all sorts tend to collect and conduct research with a bias for what may be called charismatic organisms. For zoos, that would be "charismatic megafauna," the larger, more iconic members of the animal world. Regarding botanical gardens the tendency is to focus on vascular plants, and within that category, flowering plants.[28] For conservation purposes in particular, the neglect of nonvascular plants is a serious oversight in that "[n]on-vascular species are the living representatives of the first plants to colonise the land." "Within these plants are captured key moments in the early evolutionary history of life on Earth, and they are essential for understanding the evolution of plants."[29] Consider this bias and research need when planning and setting policy for garden research programs.

Notes

1. R. Y. Edwards, "Research: A Museum Cornerstone," in *Museum Collections: Their Roles in Biological Research*, ed. E. H. Miller (Victoria: British Columbia Provincial Museum, 1985), 4.

2. Alden H. Miller, "The Curator as a Research Worker," *Curator* 6, no. 4 (1963): 282.

3. Robert Shalkop, "Research and the Museum," *Museum News* 50, no. 8 (1972): 11.

4. Winker (2004) in Michael S. Dosmann, "Research in the Garden: Averting the Collections Crisis," *Botanical Review* 72, no. 3 (2006): 214.

5. Mori and Mori (1972) in Dosmann, "Research," 219.

6. Pettitt (1994) in Dosmann, "Research," 219.

7. Alberta Museums Association, *Standard Practices Handbook for Museums* (Edmonton, AB: AMA, 1990), 186.

8. *Concise Oxford Dictionary*, 7th ed. (Oxford: Clarendon Press, 1982).

9. Lorin I. Nevling, "On Public Understanding of Museum Research," *Curator* 27, no. 3 (1984): 190; Ontario Ministry of Citizenship and Culture, "Developing a Research Policy for Museums," in *Museum Notes for Community Museums in Ontario* (Toronto: OMCC, 1983), 7:2.

10. American Association of Museums, "Stewards of a Common Wealth," in *Museums for a New Century* (Washington, DC: AAM, 1984), 49.

11. Communications Canada, *Challenges and Choices: Federal Policy and Program Proposals for Canadian Museums* (Ottawa: Minister of Supply and Services Canada, 1988), 48.

12. Dosmann, "Research," 222.

13. Dosmann, "Research," 218.

14. Miller (1963) in Dosmann, "Research," 221. Miller, A. H. 1963. "The Curator as a Research Worker." *Curator* 6(4): 282–286.

15. Dosmann, "Research," 214.

16. Ricklefs (1980) in Dosmann, "Research," 214.

17. Dosmann, "Research," 214.

18. Dosmann, "Research," 213.

19. Rae (1995) in Dosmann, "Research," 210.

20. P. Raven, "Research in Botanical Gardens," *Public Garden* 21, no. 1 (2006): 17.

21. Dosmann, "Research," 218.

22. Dosmann, "Research," 218.

23. Dosmann, "Research," 219.

24. See Gail Dexter Lord and Kate Markert, *The Manual of Strategic Planning for Museums* (Lanham, MD: AltaMira Press, 2007).

25. Alberta Museums Association, *Standard Practices*, 188.

26. Alberta Museums Association, *Standard Practices*, 189.

27. Alberta Museums Association, *Standard Practices*, 191.

28. Michael Gross, "Can Botanic Gardens Save All Plants?" *Current Biology* 28 (2018): R1076.

29. Ross Mounce, Paul Smith, and Samuel Brockington, "Ex-situ Conservation of Plant Diversity in the World's Botanic Gardens," *Nature Plants* 3 (2017): 796.

CHAPTER EIGHT

~

Research and Curatorial Practice

For the botanical garden profession to grow in strength and excellence it must, in addition to conducting collections research, undertake research in botanical garden practice—a branch of museology.

Summative and applied research is necessary to the development and successful functioning of many basic curatorial programs. The planning, successful implementation, and management of these programs embody a certain amount of summative and applied research. Perhaps the most sophisticated part of this research begins when one attempts to establish a context for the plants in our collections, for example, taxonomic, geographic, historic, ecological, cultural, and so forth. To synthesize and interpret the evidence of this research requires knowledge of the discipline on which the interpretation draws, such as taxonomy and systematics, plant ecology, phytogeography, plant ecology, reproductive biology and genetics, conservation biology, and horticulture. This is one of the primary responsibilities of the curator. To maintain their connoisseurship of collections, curators must conduct research on their collections. DeMarie warns of the consequences of insufficient curatorial interest in the collections: "Lack of sufficient interest or awareness of a collection's contents and purpose often means that the collection is not used effectively, which results in diminished efforts to advocate for the collection. Thus, a downward spiral begins, resulting in the loss of valuable collections of plant genotypes."[1]

Collections Acquisition

As was first stated in chapter 3, good collections are the result of thoughtful collecting based upon logical, intelligent planning. The bulk of this research work concerns the development of complete, accurate collections. A basic pursuit of plants for the collection requires a high level of competency in plant recognition, identification, nomenclature, systematics, phytogeography, and ecology. Such competency requires research and investigation. Plants (and plant parts) that need to meet special acquisitions criteria may require additional research and familiarity in the areas of conservation, reproductive biology, and genetics.

First discussed in chapter 3, acquisitions may be based upon a gap analysis oriented to acquiring plants for a specific programmatic theme or specialized program (e.g., conservation collection). A gap analysis is just what it sounds like: the identification of unexamined, overlooked, unclear, or missed elements of a process, plan, program, or group; it is a type of analytical survey. For our purposes, it may apply to the development of a complete collection according to a particular theme or program criterion (e.g., theme: systematics or ecogeography; program criterion: International Union for Conservation of Nature [ICUN] Red List).

Best Practice 8.1: Gap Analysis

A useful and impressive example of a gap analysis is the *Conservation Gap Analysis of Native U.S. Oaks* by Beckman et al. published by the Morton Arboretum as part of a consortium of sponsors.[2] This particular gap analysis is a comprehensive, analytical survey of both the current status and the most urgent needs for *in situ* and *ex situ* conservation of at-risk oak species in the United States. To give you a clearer idea of what a comprehensive gap analysis might entail, the native oak gap analysis had the following major components:

- Examination of oak species richness across the United States using readily accessible databases: USDA Plants and Biota of North American Program (BONAP).
- Survey of *ex situ* collections focused on level III and IV ArbNet-accredited arboreta; BGCI PlantSearch database for *Quercus*; BGCI 2009 global survey of *ex situ* oak collections; APGA oak curatorial group governing Nationally Accredited Quercus Multisite Collection™.
- Identify species of conservation concern by integrated metrics from IUCN Red List of U.S. Oaks; NatureServe; USDA Forest Service Project CAPTURE risk assessment; gap analysis *ex situ* survey data.

- Vulnerability of wild populations: using the data from the preceding bullet point, assess the vulnerability status of wild populations of the U.S. oak species of concern through a scoring matrix that calculates an average vulnerability score for each species.
- Identify threats to wild populations: using literature reviews and expert feedback, threats driving the decline of wild populations were categorized and rated for severity as high, medium, low, or no impact.
- Identify conservation activities for each species of concern: 10 conservation activity categories were formulated for all past, present, and future conservation actions. A conservation action questionnaire was also distributed to institutions holding species of concern.
- Identify priority conservation actions for species of concern.
- A spatial analysis of the populations of species of concern: compile and standardize 10 large and differing spatial point datasets.

Of the 91 native U.S. oaks, we identified 28 species of conservation concern based on extinction risk, vulnerability to climate change, and representation in *ex situ* collections. For each of these 28 species we completed an in-depth analysis of native distribution and ecology, status of wild populations, threats, geographic and ecological coverage of *ex situ* collections, and current conservation actions.[3]

The preceding account is a very abbreviated synopsis of an interesting and important gap analysis. To reiterate a point I made in chapter 3, a similar, although less comprehensive, gap analysis may be used to make decisions about collection priorities, core groups, thematic specialties, and such based on gaps among regional, national, and international collections. The point here, of course, is to identify areas of need in collections development and programming, which is an issue of particular concern for preservation and conservation. Reference is made in the preceding gap analysis to the BGCI PlantSearch database, an item I first discussed in chapter 3. The vital collections databases—GardenSearch, PlantSearch, GlobalTreeSearch, and ThreatSearch—developed and maintained by Botanic Gardens Conservation International will prove useful and important for all sorts of collections referencing and analysis. These databases are reliant on the contributions and participation of all botanical gardens to be complete and relevant.

Other areas of inquiry and research related to acquisitions, as first mentioned in chapter 2, are quarantine protocols and testing of exotic acquisitions for their invasive capabilities. Much of the information produced by this research is useful not only to the institution but also to the public it serves. Let me provide a reminder at this point of the obligations of botanical gardens when it comes to the international policy context for plant collecting and acquisitions. Unless you are fortunate to have funds and institutional mandates for broad-ranging plant collecting programs that keep you abreast of international regulations, you will likely need to research the global multilateral environmental agreements (or MEAs) and resource sharing obligations that shape many national laws and conservation initiatives that govern your collecting and acquisition activities. Of particular importance are the Convention on International Trade in Endangered Species of Wild Fauna and Flora (CITES) and the Convention on Biological Diversity (CBD). There will likely also be country-specific export rules, regulations, and laws that you will need to conform to in your collecting activities.

Collections Documentation

"Metaphorically speaking, collections are icebergs: the living plants represent the proverbial tip, while the vast array of records, previously collected data, reports, and evaluations represent the submerged portion of the iceberg. And as new information is added, the iceberg increases in size and value, becoming more appealing to future researchers, provided they can easily access and manipulate these data and connect them back to specific accessions. The cycle of information mining and generation can potentially continue ad infinitum."[4]

The iceberg metaphor is particularly pertinent to illustrate the value of documentation, especially for research. I hope it is clear to you after reading chapter 4 that a collection's value is proportional to the information it contains and can provide others. Also, "collections" are not just collections of objects but an institutional symbiosis of objects and information. Widrlechner and Burke reported that "users request germplasm based upon knowledge gained about specific accessions through personal experience and by examining the collective results of past evaluation and characterization work."[5] Plant collections with a comprehensive history of research documentation serve as benchmarks against which future results can be compared. Improving the quality of documentation and providing access to it will make collections more useful for research purposes.

On a day-to-day basis, the documentation portion of the collections management program requires the greatest amount of research. Simply put, botanical garden collections must be identified and documented through research. This research starts with an analysis of the plant itself and with information obtained from the source. Then, the documentation used as catalog data requires research that extends beyond examination of the plant to other sources of information.

We can categorize the types of research sources that gardens commonly use in documenting their collections:

Written
- Field notes
- Unpublished correspondence
- Trade catalogs
- Machine-readable records
- Government reports
- Published secondary sources: reference books, periodicals, and so forth

Visual or Pictoria
- Herbarium specimens
- DNA samples
- Photographs
- Maps
- Drawings and plans
- Illustrated books
- Videos
- Compact discs
- Hard drives
- Memory cards

Oral
- Tape recordings
- Compact discs
- Memory cards
- Hard drives
- Interviews

Three-Dimensional
- Garden and museum collections
- Landscape
- Habitat

Much of the written and visual source material will be acquired and contained within the documentation program as supportive documentation. In addition, botanical gardens will often have a reference library related to the plants in their collections along with a herbarium.

Botanical gardens may find it useful to describe various documentation research methods appropriate to their collections in their documentation manual. Here is an example of a historical research method described in the Alberta Museums Association's *Standard Practices Handbook for Museums*:

> The historical method is a process of analysis whereby genuine or authentic objects [plants], written and oral evidence are studied to obtain facts that allow a reconstruction of a part of the past. The basics of the historical research method are:
> - the collection of relevant, authentic written and/or oral materials . . . ;
> - the extraction of credible evidence from the testimony of authentic primary sources; and
> - the synthesis of reliable testimony into a meaningful narrative or exposition.[6]

There are several other important, post-research, documentation steps that gardens often overlook or fail to follow up on:
- Require researchers to complete a material-use form that records the type of research being conducted, information about the user, and a listing of accessions used. This establishes a record of research for those accessions and documents their use for this purpose. Also, researchers may need to reclaim some of that information in the event of loss.
- Be sure that your institution is acknowledged as the source of the research material.
- Be sure that all accession information is referenced in publications; this will provide much-needed background information for research reviewers and future researchers.
- Acquire reprints of any published works resulting from the research. These will become part of the important catalog data for the accessions and serve to further document their use in research.
- If the loaned or donated germplasm is undergoing long-term research, be sure to ask for research updates.[7]

Herbaria were first introduced with a cursory description in chapter 4. Gardens committed to specialized collections research programs will likely also need to make a similar commitment to a supporting herbarium. To be clear, research collections should be accompanied by voucher specimens, be they whole plants, seeds, or tissues. Therefore, it seems pertinent to take a closer look at herbarium operations in connection with research. As also mentioned in chapter 4, *The Herbarium Handbook* by Bridson and Forman is the definitive reference. A herbarium is a collection of carefully dried plant specimens mounted in a prescribed way on archival quality paper with a consistently formatted, identifying label. These herbarium specimens are most likely a type of voucher that provides a verifiable means to identify and distinguish a particular taxon. A herbarium is of particular importance connected to research collections as a means to voucher the identification and other attributes of plants used in a given research program. As you might imagine, herbaria are integral to systematics, genomics, breeding, and conservation biology research.

Herbarium collections may be developed to support a geographic, systematic, ecologic, programmatic, or combination of themes. No matter their overall focus, specimens are individually arranged (filed) according to an accepted taxonomic protocol. The *Index Herbariorum* is a global directory of public herbaria and their staff hosted by the New York Botanical Garden.[8] The herbarium collections are usually made up of wild-collected plants, plants from an institutional collection, the products of plant selection and breeding programs, and specimens acquired through an identification service. The value of herbarium specimens is based on the quality of the sample—it must be complete and well mounted for examination and preservation—and the documentation, which must be complete and accurate.

A herbarium sample is considered complete if it contains all of the diagnostic features that are key to identifying that taxon. A well-mounted specimen is firmly fixed to its herbarium sheet with the diagnostic features prominently displayed. It may be necessary to collect more than one herbarium sample for a given taxon if seasonal features are important for identification (e.g. flowers *and* fruit). Some herbarium sheets may be accompanied by packets of supporting material, such as fruits and seeds, that may be attached to the sheet or stored in a supplementary file. Herbarium sheets should be of archival quality, acid-free paper, and the specimens should be affixed to them with a pH neutral adhesive. Bulky, fleshy, and high-moisture-content plants, such as succulents, palms, bananas, and aquatic plants, will require special handling, preservative, and mounting/storage procedures and equipment. Photographs may be an important supplement for specimens that are difficult to mount and/or may lose important features quickly in storage (e.g., color, coatings, and other anatomical features).

The operation of a herbarium involves ongoing research in systematics, taxonomy, and both botanical and horticultural nomenclature to invest the purpose of the herbarium as a botanical reference center and plant identification vouchering authority. As first described in chapter 4, the World Flora Online and the International Plant Names Index (also online) are valuable resources for valid plant names. The herbarium should have copies of *The International Code of Botanical Nomenclature* and *The International Code of Nomenclature for Cultivated Plants* on hand to guide them in naming plants. The herbarium reference shelf should be stocked with books and monographs on groups of plants, genera, and in the case of horticultural collections, cultivars pertinent to the herbarium's focus. It's not uncommon for herbarium staff to seek counsel with botanical authorities on particular taxa for verification of their collections. In time, through the necessary research of curatorship, the curators of pointed herbarium and botanical garden collections will become authorities in their own right over those collections.

Finally, it is important that collections and collections documentation of institutional herbaria be well coordinated and connected to the living collections of that institution. The several documentation software programs described in chapter 4 will facilitate this coordination as long as they are used with that purpose in mind.

An emerging technology that will likely prove itself of even greater fundamental importance to botanical garden documentation, plant taxonomy, and systematics is DNA barcoding. It goes without saying among botanical garden professionals that "the ability to identify plant species is fundamental to our understanding of the world around us. To conserve plants, their habitats and ecosystems we need to be able to identify and monitor species."[9] DNA barcoding relies on short sections of DNA to identify species; these are often referred to as microsatellites. "The aim of DNA barcoding is to have global agreement on the regions of DNA and protocols used for different groups of living things in order to create an international resource for species identification."[10] Based on the findings of researchers associated with the International Barcode of Life, a research alliance of nations working together to build a DNA bar code set of standards and a reference library, it was determined that vascular plant DNA bar codes should be derived from chloroplast DNA; of course, that may change over time.

Using this standardized approach, several botanical gardens are involved in DNA barcoding projects: the New York Botanical Garden is working on trees of the world, the Atlanta Botanical Garden is DNA barcoding their extensive orchid collection, the Missouri Botanical Garden is barcoding the flora of the Shaw Nature Reserve, the Royal Botanic Garden, Kew is working on barcoding the floras of biodiversity hotspots, and the National Botanical Garden of Wales is working on barcoding both angiosperms and gymnosperms of Wales. The Wales project involves DNA samples taken from herbarium specimens. This approach takes great advantage of existing collections and their documentation without the additional fieldwork and documentation that would otherwise be necessary. However, a special DNA extraction protocol was required to make full use of the herbarium specimens. In order to obtain three samples of each species, some needed to be collected from the wild, and those were accompanied by vouchers. In the case of endangered species, photographs were taken in lieu of herbarium vouchers.[11] DNA banking and sampling is outlined in chapter 10. DNA barcoding adds a new level of precision and reliability to studies of plant taxonomy and systematics, as well as other areas of plant science and conservation.

Collections Preservation

The botanical garden collections preservation program offers many opportunities for applied and basic research in horticulture. Successfully caring for diverse collections of plants will certainly require some trial and error. The results of this type of applied research, if well organized, comprehensively documented, and carefully interpreted, often yields useful horticultural information for the institution and the public. Caution must be exercised during such operations to ensure that valuable collections are not irreparably damaged; after all, the intent is to preserve plants! Basic, potentially damaging research should be carried out on separate research collections that may be propagated from the core collections for this purpose.

Among other practices, there is a great need for information on how to successfully propagate species of wild plants. Little is known about the germination requirements of the seeds of many wild species or how they may best be stored. As part of a conservation program, botanical gardens could conduct research on the viability and germination requirements of the seed of wild plants. The focus of this research would be on viability testing and genetic deterioration in stored samples. In addition, special attention could be focused on the problems associated with recalcitrant seeds. Chapter 10 on *ex situ* conservation collections offers more on this subject. Refer to the section on monitoring and evaluation below and Best Practice 8.2 on the Chicago Botanic Garden plant evaluation program.

Public Programs

Before gardens develop interpretive and exhibit plans or educational programs, research on the collections and the themes to be presented must be conducted. Comprehensive research is critical to the presentation of accurate information. Not surprisingly, this research is facilitated by thorough collections documentation.

Collections Monitoring and Evaluation

"Only through evaluation can the suitability of a plant collection be assessed to address the current needs of an organisation. It represents one of the most important activities undertaken in the curation department, yet it in many botanic gardens this is seldom taken into consideration."[12]

A core and ongoing part of any curatorial program is the monitoring and evaluation of the collection and, in a slightly different vein, audits of curatorial program effectiveness. All of these endeavors should be anchored to the institution's collections management policy for relevance and authority. Monitoring and evaluation are two similar and sometimes confusing collections research tasks. For the purposes of this text, I use the following definitions:

Monitor: to scrutinize or check systematically with a view to collecting certain specified categories of data.[13]
Evaluate: the segment of any research or other program or project when the value or worth of the outcomes or products of that program or project are fixed or ascertained.[14]

Considering these definitions, one may conclude that monitoring and evaluation are often closely linked activities. In fact, it would be difficult to conduct an evaluation without having first monitored the subject. Monitoring is a critical component of any evaluation program. However, we may monitor plants in the collections for, among other things, the purposes of characterization (bloom time, mature height, etc.), without evaluating them—a common occurrence in documentation programs.

Basically, there are two types of evaluation: (1) formative evaluation, which takes place during the planning and implementation phases of a program or project; and (2) summative evaluation, which takes place at the end of a program or project. We may decide to conduct formative evaluations of the mechanics and protocol developed for a plant evaluation program. We may then follow this up with summative evaluations of the overall program and the data collected when the subject plants were monitored. What is common to both of these research activities is documentation. Monitoring and evaluation are most meaningful to botanical gardens if the data and results of this research are well documented. In this regard, evaluation data is critical to the development of catalog records. This work contributes to the complete characterization of taxa in the collections and the collections themselves.

For the most part, collections "evaluation" refers to catalog research to monitor and record the performance, development, phenology, hardiness, or other attributes of plants. In some cases, this work may involve evaluating the data collected during this research. This documentation may then be used for plant introduction and other specialized collections research programs, as well as public education.

Collections-monitoring activities, as mentioned above, most often serve the purposes of characterizing taxa for documentation and collections preservation programs. Collections staff will "monitor" the phenology of plants in the collections and log this data in the catalog records for later evaluation and use by research and public programs. Pest management programs contain a monitoring element that generates data that is evaluated for use in making decisions about pest control.

> In the simplest case, an institution may create a formal monitoring protocol designed to provide environmental and horticultural data with which to improve the care of collections whose primary functions are aesthetic and educational. Gathered systematically over longer periods of time, such data may also yield valuable insights into local trends related to larger environmental variables such as climate. Depending upon the scope of the measured variables, and the quality and duration of the monitoring records, this can yield publishable information that constitutes valuable research.[15]

Data generated through monitoring and other activities is used to evaluate collections against standards established in policies and to ensure that they contribute to the purpose(s) for which they were assembled (fit for purpose). Most important, the curatorial staff will evaluate collections to reconcile them with their collections management policy.

> Most botanic gardens have a collections policy which guides the content and development of their living collections. Rather few, however, check to see if the current content of the collection matches the guidelines specified in the policy. Likewise, it seems that few curators analyse the collection, or even key genera, to check if accession numbers are going up, down or remaining static.[16]

There are some basic steps that should be taken at the beginning of any comprehensive collections monitoring and evaluation programs. There should be a clearly identified goal(s) for the process and an agreed upon set of criteria by which accessions will be evaluated. Make sure the curatorial staff are informed and, if appropriate, engaged in the development and implementation of the program. There may be other institutional stakeholders that should also be informed, for example, the director and collections committee. The target accessions group should be listed and checked for correct identification. Verification itself may be the subject of evaluation, and this may be a preliminary to the proposed evaluation program. Be sure to take note of any special characteristics of the accessions in the evaluation group that may convey a special status relative to the evaluation, for example, endangered plants and/or those with donor qualifications. Finally, ensure that all the important references for this work have been identified and made accessible; this will certainly involve printed and online resources such as the BGCI databases mentioned several times in this text.[17]

Some examples of collections evaluations that occur at botanical gardens include the following:

- Monitoring and evaluation of accessions for hardiness and pest resistance.
  - Accessions of agricultural and horticultural value.
  - Conservation collections.
- Monitoring and evaluation of agricultural and horticultural taxa for potential food, fiber, and other harvest products of commercial value.
- Monitoring and evaluation of taxa for attributes of conservation and research value.
- Monitoring and evaluation of the fitness of the collections within the specifications of the collections policy.
  - Conformance to collections themes and programmatic foci (e.g., taxonomic groups, ecological groups, etc.).

Best Practice 8.2: Collection Evaluation

Botanical Garden Meise, Belgium
In 2006, the largest collection at the Botanical Garden Meise was *Cactaceae*. With 2,507 accessions of 1,642 taxa, this was touted as a significant conservation collection. At that time, two experts in the *Cactaceae* were brought in to evaluate the collection to verify its presumed conservation status. Examination of the documentation revealed that 90 percent of the collection were undocumented, and only 251 accessions were of wild origin.

Cacti experts Dr David Hunt and Dr Nigel Taylor spent two days evaluating and verifying the collection and came to the conclusion: "Two-thirds of the collection could be discarded without any loss in conservation or research value." Despite this sobering outcome, the audit also discovered a jewel in the collection, *Opuntia stenarthra*, a wild-collected Paraguayan species that had not been observed as a living plant for over a century.[18]

Given the results of that evaluation, extraneous accessions were distributed to other gardens with a statement regarding their evaluation and limited value as a conservation collection.

Chicago Botanic Garden

The Chicago Botanic Garden has a long-standing plant evaluation program curated by a permanent member of the scientific staff. "The program evaluates herbaceous and woody plants in comparative trials, ultimately recommending the top performers to gardeners and the horticulture industry. Approximately 900 taxa are currently evaluated in the Bernice E. Lavin Plant Evaluation Garden, the Green Roof Gardens, and various ancillary sites."[19] The goal of the program is to evaluate and recommend horticultural plants of superior performance in the upper Midwest of the United States.

Horticultural qualities and attributes that are evaluated are environmental hardiness, pest resistance, adaptability, and aesthetic attributes. The evaluations are conducted at two different sites in organized, well-marked locations providing a breadth of exposure and environmental conditions. These evaluations occur over varying lengths of time depending upon the type of plant: herbaceous perennials (four years), shrubs (six years), and trees (seven years). The results of these evaluations are published both in *Plant Evaluation Notes* directly available online and in the popular horticultural periodical *Fine Gardening*.[20]

A recent evaluation of available cultivars of *Perovskia atriplicifolia* was completed and reported in *Plant Evaluation Notes*. There is a complete description of this Russian sage and its horticultural applications illustrated with detailed color images. The context and organization of the evaluation is described, and all of the cultivars in the test are listed indicating that three plants of each were included in the evaluation. The top-rated cultivars are fully described highlighting their most noteworthy characteristics followed by a summary of the study. The evaluation program and an available list of all the published *Plant Evaluation Notes* and links to the corresponding magazine articles are available on the Chicago Botanic Garden website.

The Chicago Botanic Garden evaluation program described here is a formalized part of the garden's overall research program and, in a sense, is multipurpose in terms of the information on the collection it generates. It is also useful by conducting less formal monitoring and evaluations for similar phenological, developmental, and performance traits as part of collections catalog documentation.

Textbox 8.1. Recommendations for Research and Curatorial Practice

Basic
- Botanical gardens develop and maintain a comprehensive desideratum for building and maintaining their collections.
- Botanical gardens gather and maintain basic documentation on their collections.
- Botanical gardens selectively implement, fully document, and evaluate the results of new or innovative collections preservation practices.
- Botanical gardens systematically monitor, evaluate, and document the conformance of their collections to policy standards.

Intermediate
- Botanical gardens have a written collections development plan.
- Botanical gardens systematically monitor and evaluate their collections for the development of catalog documentation and the fulfillment of specific summative and applied research programs.
- Botanical gardens have a written evaluation policy.

Advanced
- Botanical gardens have a research division that engages in various types of research projects and makes public the results.

Notes

1. DeMarie (1996) in Michael S. Dosmann, "Research in the Garden: Averting the Collections Crisis," *Botanical Review* 72, no. 3 (2006): 216.
2. E. Beckman, A. Meyer, A. Denvir, D. Gill, G. Man, D. Pivorunas, K. Shaw, and M. Westwood, *Conservation Gap Analysis of Native U.S. Oaks* (Lisle, IL: Morton Arboretum, 2019).
3. Beckman et al., *Conservation Gap Analysis*, 4.
4. Dosmann, "Research," 226.
5. Widrlechner and Burke (2003) in Dosmann, "Research," 225.
6. Alberta Museums Association, *Standard Practices Handbook for Museums* (Edmonton, AB: AMA, 1990), 196.
7. Dosmann, "Research," 224.

8. New York Botanical Garden, *Index Herbariorum*, http://sweetgum.nybg.org/science/ih/.

9. Natasha de Vere, "Barcode Wales," *BG Journal* 9, no. 1 (2012): 7.

10. de Vere, "Barcode Wales," 7.

11. de Vere, "Barcode Wales," 8.

12. J. Gratzfeld, ed., *From Idea to Realisation: BGCI's Manual on Planning, Developing and Managing Botanic Gardens* (Richmond, Surrey, UK: Botanic Gardens Conservation International, 2016), 61.

13. *American Heritage Dictionary*, 2nd college ed. (Boston: Houghton Mifflin, 1982), 810.

14. *American Heritage Dictionary*, 469.

15. R. E. Cook, "Botanical Collections as a Resource for Research," *Public Garden* 21, no. 1 (2006): 19.

16. David Rae, "Fit for Purpose? The Value of Checking Collections Statistics," *Sibbaldia*, no. 2 (2004): 61.

17. Gratzfeld, *From Idea to Realisation*, 61.

18. Gratzfeld, *From Idea to Realisation*, 62.

19. Chicago Botanic Garden, "Plant Evaluation," https://www.chicagobotanic.org/collections/ornamental_plant_research/plant_evaluation.

20. Chicago Botanical Garden, *Plant Evaluation Notes*, https://www.chicagobotanic.org/collections/ornamental_plant_research/plant_evaluation

CHAPTER NINE

Conservation

"No biodiversity is dispensable or redundant—every population of every species, in fact all of nature, is worth conserving."[1]

"Through their great experience in growing plants, botanic gardens are the most suitable organizations in the world to rescue and conserve individual plant species. Even though each botanic garden may focus on only a small number of plant species, it should see its activities as part of a wider global effort to preserve biological diversity."[2]

There is a role for every botanical garden in plant conservation. Indeed, it is often stated as a programmatic goal in the mission statements of these institutions. Based on a 1986 survey by the American Public Gardens Association (APGA) Plant Conservation Committee, "botanical gardens and arboreta practice and promote plant conservation in a variety of innovative ways."[3] These efforts have intensified as our awareness and understanding of the growing threats to species and their extinction rates has increased. Plant conservation programs may be categorized as *in situ*, with preservation and restoration taking place in the wild, or *ex situ*, with preservation taking place outside or away from the wild, usually on the property of the garden.

There is also emerging a series of innovative, intermediate techniques that bridge the gap and occupy the programmatic space between *in situ* and *ex situ* conservation. The *"inter situs"* approach involves "the reintroduction of native species outside their current range but within the recent past range of that species."[4] The other intermediate strategy is called the "quasi

in situ" approach. Diverse germplasm collections are established in reserves
similar in ecology to the target species where adequate populations may be
stewarded for preservation and propagation; the propagules are then used in
reinforcement and/or reintroduction programs.[5] Institutions may choose to
work and conduct research in any one or a combination of these conservation approaches.

Obviously, cultivating a few specimens of a threatened plant is only the
beginning of a comprehensive conservation program that will open up possibilities for research, education, and the reintroduction of plants into natural
ecosystems. "If the natural world itself remained healthy and robust around
us, if the air and water remained clean, and if the impact of the growing
human population did not threaten our own quality of life and that of our
children, perhaps we could focus narrowly within the garden walls, formulating in that isolation the themes, policies and practices we need to operate."[6]
However, we know differently.

What have botanical gardens been doing in the way of conservation?
Acquiring and preserving rare and endangered plants, learning to propagate
them, promoting habitat preservation, and training staff, guides, and volunteers are the major ways that gardens have been involved in conservation.
A commitment to plant conservation in botanical gardens involves many
facets of the entire garden operation from administration and collections
management to public programs. "Conservation is difficult both scientifically
and technically, it is expensive and requires properly trained staff, space and
facilities, and it requires long-term commitment."[7]

From the standpoint of collections management and conservation, there
are two principal types of collections: (1) conservation collections of rare,
gene-pool-representative germplasm that serve to prevent extinction and as
source material in ecological restorations, and (2) collections of conservation
value that contain rare plants that are the subjects of research and education programs.[8] Conservation collections may serve both of the purposes just
described. Before going any further, a comprehensive definition and description of conservation collections is required, and I prefer this one from the
Center for Plant Conservation (CPC):

> A conservation collection is an *ex situ* (offsite) collection of seeds, plant tis-
> sues, or whole plants that supports species' survival and reduces the extinction
> risk of globally and/or regionally rare species. A conservation collection has
> accurate records of provenance, maternal lines differentiated, and diverse
> genetic representation of a species' wild populations. To be most useful for spe-
> cies survival in the wild, a conservation collection should have depth, meaning

that it contains seeds, tissues, or whole plants of at least 50 unrelated mother plants, and breadth, meaning it consists of accessions from multiple populations across the range of the species. Conservation collections should have tests of initial germination and viability, cultivation protocols developed, and periodic testing of long-term viability. A conservation collection differs from a horticultural collection, which may have few genetically unique individuals, or is solely comprised of unusual appearing forms.[9]

However, gardens may be most effective if their conservation work is integrated as a combination of *in situ* collaboration with nature conservation agencies and consortia of other gardens as well as *ex situ* programs on site. Conservation agencies, such as The Nature Conservancy (TNC), are often responsible for a protected area network and may have a general responsibility to conserve habitats. Over the last several years, more gardens have been working together in consortia and networks devoted to plant conservation. Several "metacollections"—the combined holdings of a group of collections—have been formed to assist in this goal.

[M]etacollections are envisioned as common resources held by separate institutions but stewarded collaboratively for research and conservation purposes. The American Public Gardens Association's Nationally Accredited Multisite Collections™, BGCI's Global Conservation Consortia and the CPC National Collection are established examples of metacollections. Like any collection, a metacollection can be of any scope or taxonomic level.[10]

In order to preserve plant genetic diversity, gardens may wish to acquire and preserve whole plants. Many authorities now agree that this may not be the most efficacious strategy for reaching this goal.

History indicates that living botanical collections by themselves serve as unreliable long-term alternatives to *in situ* conservation. In the majority of taxa, which are genetically variable, even when reduced to small numbers, the expense of maintaining adequate genetic representation, combined with the impossibility of simulating natural selection, must inevitably limit the value of gardens in species conservation for periods exceeding one generation.[11]

Be that as it may and as pointed out in chapter 3, the impacts of climate change on the world's ecosystems are occurring at such a rapid pace that *in situ* approaches to conservation are threatened by the loss of supporting habitat. The emerging intermediate approaches to conservation I described above, *inter situs* and quasi *in situ*, are a response to the changing circumstances. Still, in the face of these increasingly dire circumstances for *in situ* conservation, many authorities do agree that field bank collections of plant ecotypes, lines, and clones are useful and productive.

Although maintaining and restoring healthy populations of plants in the wild is the condition we strive for, preservation in cultivation is greatly preferable to extinction. However, *ex situ* collections are more than just an insurance policy. They are the subjects of research and education, and they can become important sources for the reestablishment of endangered species. The most useful *ex situ* conservation measures available to gardens for preserving genetic diversity are referred to as gene banks.

As defined in chapter 3, "gene bank" refers to collections of propagules such as seed, pollen, modified roots and shoots such as tubers and rhizomes, and tissues grown as thallus cultures or plantlets. Collections of whole plants for the purposes of genetic diversity are referred to as "field gene banks." The storage, or "banking," of propagules involves the use of low temperatures and may require low humidity, both of which slow down metabolic processes within the living material. Regardless of the type, gene banks perform three major functions:

- Preserve rare or threatened plant material against possible extinction
- Preserve the genetic diversity of wild species with large stocks of stored living seeds and other tissues
- Provide access to plant material that would otherwise be difficult for researchers to obtain without impacting wild populations

Preserving this material in gene banks implies that it will be used in the restoration of species and populations to agriculture or the wild if necessary. It is the *restorative role* of conservation programs that makes them unique.

Before examining some specific conservation research programs, I want to highlight a broad set of possibilities in conservation research suggested by the targets of the *International Agenda for Botanic Gardens in Conservation*.[12] "The International Agenda is a policy document for botanical gardens that links to the broad range of international biodiversity initiatives and particularly closely to the GSPC [Global Strategy for Plant Conservation]."[13] Any discussion of conservation research at botanical gardens must begin with plant diversity. Plant diversity must now become part of the raison d'être of plant collecting at botanical gardens and, therefore, a mainstay of any botanical garden research. Plant diversity research is facilitated by laboratories, herbaria, greenhouses and growth chambers with controlled conditions, field experimental areas, climatic and weather stations, data management systems, and advanced equipment for molecular and genetic studies. Botanical gardens have much of this infrastructure already in service to other parts of their program.

Any talk of research at botanical gardens nearly always turns first to plant taxonomy and systematics, and with good reason. What could be a more poignant area of research in conservation to help us recognize, know, and understand what we are conserving, as well as the extent of diversity among those unknowns? An expansive example is the well-known work of the Royal Botanic Garden, Kew and Missouri Botanical Garden partnership in producing the World Flora Online[14] and the Plant List,[15] two important taxonomic compilations and resources toward meeting target 1 of the GSPC. These two old and venerable institutions have been at the center of botanical work in plant systematics and taxonomy for most of their long histories. There are many smaller botanical gardens with vital systematics and taxonomic research programs on specific ecological and/or taxonomic groups of plants, such as the Montgomery Botanical Center's work on cycads, the Marie Selby Botanical Garden's study of epiphytes, and the Betty Ford Alpine Garden's research on alpine plant diversity and conservation. Contributing to the important work of "red listing" threatened species in your area is research that all gardens can do. "Be involved in the preparation of floras, field guides, taxonomic monographs, identification keys and manuals, handbooks, other reference works and publications that assist in the identification, monitoring and recovery of plant diversity."[16]

The urgent necessity for botanical gardens to become involved in plant conservation is due in large part to climate change. There is a lot of work to do and many opportunities for botanical gardens of all kinds to participate in research on the impacts of climate change on their collections and the wild plants within their service areas. By way of their catalog and other record keeping, many gardens may already have valuable phenological and other data important for use in the analysis of climate change impacts. Phenological studies are a natural for public outreach and citizen science projects that could yield valuable information and valuable public contact and education. For more on citizen science, see chapter 6. There are also many research needs in the areas of carbon sequestration and assisted migration, just to name two.

Another broad area of need in conservation research includes the biotechnology requirements of gene bank collections and their applications in plant propagation and biosecurity. "Major areas of activity in biotechnology in botanic gardens include *in vitro* propagation and multiplication, tissue and cell culture, recombinant DNA technology, molecular and genetic research, plant breeding and disease elimination."[17] As of this writing, there are only two botanical gardens in the United States actively conducting research on cryopreservation; there are, however, several U.S. Department of Agriculture (USDA) Germplasm Repositories.

The following is a select list of other areas of potential conservation research for botanical gardens:

- Collaborate on conservation research with other stakeholders in your area involved in protected area and natural ecosystem conservation.
- Conduct research in areas that support *in situ* conservation such as conservation biology, restoration ecology, and horticulture.
- Commit to and develop *ex situ* conservation collections aligned with your collections policy and the conservation demands of your region.
- Expand your plant propagation program to include research on protocols for threatened species allied to your collections and/or native to your area. Share this work as a member of the International Plant Exchange Network (IPEN).
- Focus you collection development and acquisition plans on wild-collected accessions of maximum genetic diversity.
- Conduct research on the successful reinforcement and reintroduction of threatened plants and join the Ecological Research Alliance of Botanic Gardens.[18]
- Research and contribute to the development of sustainable practices for the growth and production of plants that are of cultural, subsistence, or economic value, including species listed by the Convention on International Trade in Endangered Species of Wild Fauna and Flora (CITES).
- Research, monitor, and evaluate plant collections and commercial introductions for invasive qualities and share that information.

Not surprisingly, many conservation research groups and organizations recommend that botanical gardens concentrate on their local floras. Within this general recommendation, gardens should focus on the following:

Wild Species
1. Rare and endangered species (at the local, national, regional, and global level)
2. Economically important species, particularly
   - minor food crops and crop wild relatives: "a wild plant taxon that has an indirect use derived from its relatively close genetic relationship to a crop"[19]
   - products other than food
   - medicinal plants

3. Species required for the restoration and rehabilitation of ecosystems
4. Keystone species—that is, those that are known to be of particular significance in the maintenance and stability of ecosystems
5. Taxonomically isolated species whose loss would be serious from a scientific point of view

Cultivated Species
1. Primitive cultivars (land races)
2. Semidomesticates[20]

Gardens with greater capacity might expand their conservation collections beyond their local floras to other areas of the world using the above criteria.

Above all, keep in mind that botanical gardens must ensure that their approach to plant conservation is justified and governed by their collections management policy. If not, seek to change the governing policy to conform to this new direction. From there, build on your strengths so that conservation work is consistent with the core vision, policies, and programs of the institution. Conservation collections may naturally be an offshoot of the institution's core collections; for example, those institutions participating in the APGA Nationally Accredited *Quercus* Multisite Collection™ (or metacollection) have a long-standing, policy-driven institutional commitment to that genus. Further, institutions with greater capacity may expand to conservation collections of species ranging from similar ecoregions. The following is a description of a multifaceted conservation research program conducted at the North Carolina Botanical Garden, a leading institution in conservation programming in the United States:

The Garden's Conservation Programs conduct plant conservation at multiple levels. We conduct both pure and applied research on the ecology and restoration of rare plants as well as curate a seed bank for the protection of genetic material for some of the rarest plants in the Southeastern US. We also have begun a Native Plant Materials Development Program to increase the diversity of common species available for ecological restoration. These materials will not only benefit our own natural areas, but those of our partners throughout the region. Our program also actively works in land conservation and in applying the best management practices to our natural areas. In all of our work, we seek to engage the public and to work with partners to advance the protection and conservation of our native plants and natural communities.[21]

Conservation Assessment

Specific conservation programming choices should be based on a general conservation assessment. A conservation assessment involves gathering information on the natural distribution of species, the threats they face in the wild, and their extinction risk to help reveal the types of programming and the taxa involved that will be most suitable to the interests and capabilities of your institution. Some botanical gardens, based on the results of conservation assessments, focus on select families and genera of plants as a major research initiative. They become the important providers of information and justification for "red listing" species with the International Union for Conservation of Nature and Natural Resources (IUCN).

For the purpose I have in mind here, much of the information you will need for a simple conservation assessment is available if you know where to look. Beginning with a target list of taxa derived from your core collections and local floras following the criteria listed above, the IUCN Red List of threatened species is an invaluable resource for refining this list. The IUCN Red List provides information on both a national and global scale. Another very useful resource (and partner) for North American gardens, and those interested in the threatened flora of North America, is NatureServe, a nonprofit organization that maintains an online database of the status of biodiversity in the United States and Canada. The Botanic Gardens Conservation International (BGCI) databases, in particular ThreatSearch and GlobalTreeSearch, are important, as well as the Center for Plant Conservation's Rare Plant Finder tool.[22] The information available from these sources may then motivate a more refined examination of conservation needs through queries and visits to other gardens and work in the field. The result of these investigations should help you decide where to target your conservation efforts and by what means.

Diversity Analysis

The ability to conduct reasonably accurate and rapid genetic diversity analyses is of growing importance for biodiversity conservation and may be part of a comprehensive conservation assessment.

> Adaptive genetic diversity is a primary focus of conservation genetics because of its importance over the long-term, during which it allows for the maintenance of adaptive evolutionary potential, and over the short-term, during which it is associated with the maintenance of reproductive fitness. Quantifying adaptive genetic diversity within and across the ranges of tree species

is therefore critical for achieving the objectives of both *in situ* and *ex situ* conservation.[23]

At an earlier time when an understanding of population biology and diversity was investigated without the acute pressures of global extinction, lengthy field gene bank trials were conducted to quantify the adaptive traits significant to the fitness of a species to its habitat. Now, however, we need to make such determinations within a shorter time frame. Through DNA microbiology, conservation biologists can access DNA microsatellites as a means for estimating genetic diversity. First mentioned in the previous chapter, microsatellites are repeating bits of DNA in an organism's genome made up of two or three combinations of nucleotide bases (A, C, G, T). Microsatellite sequences within a species genome vary between different individuals. This characteristic makes it possible to fingerprint DNA within a particular individual and, in so doing with many individuals, create a portrait of the genetic diversity within a population.

> Range-wide studies using molecular markers can identify populations harboring relatively high or low levels of genetic variation and offer insights into the existence of intra-species evolutionary lineages that may have undergone differential natural selection and thus may possess unique adaptive traits. This information should be highly applicable to conservation planning and management decision-making.[24]

I don't expect that garden curators will take it upon themselves to learn the intricacies of genetic analysis using the molecular approach (ecological genomics) I've described here. However, they will certainly benefit from an understanding of its usefulness and perhaps employ the skills of a trained technician to help in building and curating genetically diverse conservation collections.

Before considering the details of specific types of conservation programs and research, it should be made clear that reaching the goal of global biodiversity conservation will only be successful through integrated approaches that take advantage of our collective knowledge and capability among the larger community of conservation organizations, institutions, and other entities. Those gardens with the necessary resources and programmatic capabilities should first consider integrated programs of more than one conservation strategy involving both *in situ* and *ex situ* conservation strategies. Smaller gardens or those with fewer options should seek to work within a larger, integrated network or consortium. Figure 9.1 is a helpful illustration of the components and relationships of an integrated conservation program.

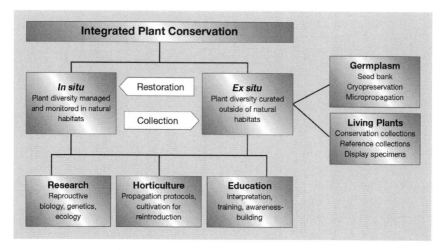

Figure 9.1. Integrated plant conservation combines *in situ* (on-site) and *ex situ* (off-site) conservation approaches to support species survival.
Source: Botanic Gardens Conservation International, *International Agenda for Botanic Gardens in Conservation*, 2nd ed. (Richmond, Surrey, UK: BGCI, 2012).

In Situ Conservation Programs

"Plant conservation is largely dependent in most countries on the creation of a system of protected areas. This is complemented by both *in situ* and *ex situ* actions at the species and population level, notably species recovery actions, reintroductions and conservation translocations and the creation of genebanks for storing germplasm such as seed, pollen, cell and tissue cultures. Also, much effort is now being placed on ecological restoration."[25]

In situ conservation refers to efforts to identify, protect, manage, monitor, and restore plant species in their native habitats. This type of conservation is preferable, provided supportive native habitat still exists, because it offers the opportunity to maintain the integrity of natural populations and to allow them to continue to evolve. Because human impacts have become so pervasive, the scope of *in situ* conservation is being expanded to include the intermediate approaches I described earlier in this chapter. These intermediary applications have found utility in urban and suburban areas, locations that heretofore were often ignored within the natural range of a threatened species. This requires a perspective that includes the full spectrum of landscapes from natural to domestic. "The continuum of land uses from the urban centre to the heart of the forest provides a number of locations where both spontaneous regrowth and planted 'wild' species could be used to promote *in situ* conservation."[26]

Protected Areas and Reserves

"[B]ecause the primary threat for most terrestrial and freshwater species is the destruction of their habitats, the establishment of protected areas for these species has emerged as one of our most important and effective tools to safeguard biodiversity."[27]

Many countries have set aside protected areas under the aegis of various departments and agencies in their government. Some, although not nearly enough, have been created for the express purpose of protecting plants; most have not. At this point in time, most endangered species of plants exist outside of protected areas, especially in the tropics.[28] *In situ* conservation of endangered species is more commonly practiced in reserves held by public, nonprofit organizations and private institutions or individuals. In fact, there is an emerging trend in tropical countries for local communities to cocreate and comanage protected areas. Reserves and protected areas for endangered plants must be identified and mapped out based on ecogeographical surveys and an assessment of the threats to species of concern within those surveyed areas. This is a critical area of research for botanical gardens in helping to describe and establish protected areas for *in situ* conservation. The most important facet of this work is climate modeling to ensure that protected areas will continue to provide protection amid shifting climate impacts.[29] Chapter 11 offers more information on a particular type of climate modeling.

Some large botanical gardens have formed effective partnerships with governments and conservation organizations to create large conservation reserves for wild populations of keystone and endangered plant species. These gardens may also conduct basic research programs in conservation biology within the confines of such reserves and at other locations where such work is important. And, at the risk of sounding like a broken record, smaller botanical gardens may play a role in the support of such activities as part of a conservation network, consortium, or other joint avenue for participation.

Other gardens have established and manage small reserves with an emphasis on the conservation of native species and populations of plants. These small reserves, usually about 5 to 50 hectares, can be an essential part of a protected areas network if they can be adequately buffered and protected from disturbance. Some gardens, such as the Holden Arboretum, the North

Carolina Botanical Garden, the Native Plant Trust Garden in the Woods, and the Xishuangbanna Tropical Botanic Garden maintain large reserves.

Before taking on the ownership and preservation of a reserve property, be sure that it is of adequate size to effectively conserve the target species and populations of interest to your conservation efforts. A common oversight among the holders and proprietors of protected areas and reserves is a lack of attention for the ongoing monitoring and stewardship of endangered species within those protected areas; simply holding those protected areas is not enough. If those species are imperiled within the protected areas by invasive competitors, declining environmental conditions, or other deleterious factors, actions must be taken to preserve those species.[30] Under such conditions, endangered species may be recovered for propagation and multiplication at a facility in the reserve and, when stocks of plants have been sufficiently increased, reintroduced to a habitat in the reserve where it has been lost after the limiting factor(s) in that habitat have been mitigated. In any case, each garden should conduct and maintain a plant inventory of its reserves that identifies endangered species.

Botanical gardens contemplating the acquisition of a large property for the purpose of a reserve need to assess that property for its potential as a key biodiversity area using the IUCN criteria and gap analysis protocol found in the IUCN protected areas best practices series handbook *Identification and Gap Analysis of Key Biodiversity Areas*.[31] Key biodiversity areas often encompass large tracts, an approach that may only be possible for large institutions or consortia of botanical gardens pooling their resources. Small areas containing the principal or only remaining population of an endangered species should be within the capability of smaller gardens for preservation and stewardship. The following are some very basic protected area management guidelines used by the North Carolina Botanical Garden:

- Allow and encourage the function of natural processes to the greatest extent possible.
- Rehabilitate sites of impaired ecological function
- Actively manage areas that can benefit from human intervention
- Maintain habitat diversity

Within the above framework, successful *in situ* species conservation will be achieved by the following:

- Maintaining and protecting the habitats in which they occur
- Identifying any threats to the species

- Taking steps to remove or contain these threats, and
- Monitoring the results.[32]

Curators and protected area managers must be very familiar with the flora within a protected area if they are to be protected and conserved. This base of knowledge begins with a survey and status check of species when a candidate area is being considered for protection. What comes next is an ecogeographical study of the area. "Ecogeographic studies provide critical information on plant genetic resources (PGR) to assess their current conservation status and prioritize areas for conservation."[33] "An eco-geographical study is an ecological, geographic, and taxonomic information gathering and synthesis project. The results are predictive and can be used to assist in formulation of collection and conservation priorities."[34] Such a study may be undertaken with varying degrees of breadth, although these studies are usually targeted to species identified as endangered. An ecogeographical study includes information from the following areas derived from both office work and fieldwork:

- Taxonomy and nomenclature: all the plants of concern must be correctly identified and named.
- Ecology of each species: habit, growth rate, reproduction mechanisms, pollination, seed dispersal, population structure, seed storage behavior, predators, diseases, and such.
- Genetic information: population size and variation, breeding behavior, gene flow and breeding anomalies, evidence of clonal propagation.
- Habitat preferences: habitat type, distribution, grove mycorrhizal networks, autecology.
- Horticulture requirements: environmental controls, control of biological competition, growth regulation, propagation strategies.
- Human interactions: ethnobotany, specific human threats and competitions.[35]

The fieldwork portion of an ecogeographic study requires many of the measurements and other data collection along with herbarium samples typical of field collecting for plant acquisitions. Great technological advances in the available hardware, software, and data in the areas of geography, the environment, and biodiversity have made the use of geographic information systems (GIS) for ecogeographic studies in the field an important tool. A typical ecogeographic field record will include the following:

- Documentation information: project number, target species, name of recorder.
- Site characterization, including vegetation type and species composition.
- Types of threats: population growth and resource consumption, climate change, habitat destruction, biological competition, overexploitation, environmental degradation.
- Population information: plant demographics.
- Samples: herbarium samples, seeds, tissues (DNA samples).
- Photo checklist.[36]

See the sample data collection form in the BGCI *and IABG's Species Recovery Manual.*[37]

South Africa, through its South African Biodiversity Institute, Threatened Species Program, which includes the expertise of the Kirstenbosch Botanical Garden, surveyed and collected spatial data for 2,345 of its 2,576 threatened plant species as of 2017. Sixty-six percent of these have at least one record within a protected area. This exemplifies the type of survey work and data collection botanical gardens are well positioned to undertake for *in situ* conservation.

Ecogeographic studies are an integral part of an endangered species management plan, several of which may be prepared for any one reserve or protected area. For more details on ecogeographic studies, see chapter 15/16 of *Collecting Plant Genetic Diversity: Technical Guidelines—2011 Update* available online, as well as the BGCI *and IABG's Species Recovery Manual* available from Botanic Garden Conservation International.[38]

In situ conservation may also be practiced by botanical gardens in collaboration with public reserve and protected area holders, such as governmental agencies and nongovernmental organizations (NGOs). In such cases, botanical garden staffs will bring their expertise to bear to assist in conservation programs, such as the ecogeographic studies (or one of their elements) described above, germplasm preservation (seeds and tissues), propagation studies, and other critical elements.

Biodiversity Monitoring

In addition to ecogeographic studies and floristic surveys related to their reserves and those of other institutions, botanical gardens have an *in situ* research role to play in biodiversity monitoring. As I mentioned earlier in this section, many of the existing reserves and protected areas are insufficiently monitored to track the status of the threatened species they are meant to protect. The Smithsonian Tropical Research Institute has developed a global network (Tropical Forest Earth Observatory) and protocols for monitoring populations of threatened trees within forest plots measuring from 2 to 60 hectares in size. It monitors the growth and survival of six million trees of 10,000 species within this extensive network. However, botanical gardens can be highly effective working with much smaller networks and plots. The North Carolina Botanical Garden has been monitoring, gene mapping, seed banking, and conducting propagation studies on the only known population of Venus flytrap, *Dionaea muscipula*, scattered in a 90-mile radius of Wilmington, North Carolina. Whether on a global scale like the Smithsonian Tropical Forest Earth Observatory project, or the highly localized conservation research on the Venus flytrap by the North Carolina Botanical Garden, statistical techniques and spatial distribution modeling applications are used to monitor these populations to obtain representative samples.

Species Recovery Programs and Ecological Restoration

"Ecological restoration, the process of assisting the recovery of an ecosystem that has been degraded, damaged or destroyed, is becoming ever more necessary to ensure that resilient ecological communities, including the biodiversity that they support and the ecosystem services they provide, are maintained."[39]

Botanical gardens are well suited for species recovery programs and ecological restoration with their full complement of plant identification, taxonomy, ecology, horticultural, and social skills. This critically important element of *in situ* conservation allows botanical gardens to bridge their research programming between *in situ* and *ex situ* contexts, making full use of their capabilities and collections. Species recovery is often confused with restoration. Notice in the definition above, restoration is the recovery of an ecosystem, while species recovery is the process of returning an endangered species, or species population, to a stable and self-supporting condition. Both of these types of conservation research and programming are essentially *in situ* in nature, but because they also have *ex situ* components and are of such significance in conservation work, they will both be covered in greater detail in chapter 11.

Assisted Migration

Another important and controversial area of *in situ* conservation research is assisted migration. Plants have the ability to alter their natural ranges and migrate into new areas based on the dispersal capabilities of their seeds and other propagules along with the prevailing conditions into which their propagules travel. This process may prove to be an important survival mechanism for many plants with extensive dispersal capabilities in the face of deteriorating environmental conditions due to climate change. Other plants with more restricted dispersal mechanisms may not have the migratory potential to establish themselves in new, more accommodating habitats. In this case, human-assisted migration is one solution. Carefully considered assisted migration requires a great deal of the ecological and horticultural knowledge and skills held by botanical gardens. A modified and smaller scale version of assisted migration is assisted colonization—moving plants to a new and suitable habitat—a practice that usually takes place in a reserve or protected area. Assisted migration is an emerging and potentially important solution to climate change, species extinction, and habitat restoration and will be taken up in more detail in chapter 11.

Local and Indigenous Populations

Local and indigenous populations must be approached as partners in any species recovery and ecological restoration projects. Among other good reasons for these partnerships, important and well-meaning *in situ* conservation research programs and protected areas may create conflicts for indigenous and local populations of people with a history of dependence on the target plants and use of the protected areas. Avoiding and/or mitigating such conflicts is the principal goal of target 13 of the Global Strategy for Plant Conservation. We touched on this subject in chapter 3 on field collecting, documenting, and crediting indigenous knowledge as well as benefit sharing. At the outset of any planning for *in situ* conservation, the possible ethnobotanical importance of any populations of plants under consideration must be assessed. Botanical gardens with a history of ethnobotanical research and programming will be well suited for assessments and surveys of the ethnobotanical significance of *in situ* conservation programs. It will be important to partner with indigenous and local peoples in that their knowledge about plants and the environment will be crucial toward successfully conserving any threatened plants. Documenting indigenous knowledge meets target 9 of the Global Strategy for Plant Conservation, and *Documenting Traditional Knowledge: A Toolkit* is a useful resource.[40]

Indigenous groups can have a critical role to play as part of the *in situ* conservation team and as land stewards; in fact, it may be the only way to facilitate any effective conservation. The following is a checklist of ways botanical gardens can assist local communities to benefit from *in situ* conservation taken from "The Role of Botanic Gardens in In Situ Conservation" in *Plant Conservation Science and Practice*:

- Provide technical support and training to local people for harvesting [plants] sustainably from protected areas.
- Assist with marketing of sustainably harvested wild products.
- Support local people and their rights in negotiations with protected area authorities.
- Help with promotion of ecotourism and ensuring that local people's interests are protected.
- Conduct outreach [to interpret] . . . ecosystem service, health and social well-being brought by protected areas to local people.[41]

I would add to this list and reiterate something I suggested in the previous paragraph: train local people to be directly involved in protected area preservation and management. Also, credit them with their contributions in all the project documentation and publicity.

The following is a synoptic list (in no particular order) of *in situ* research opportunities for botanical gardens:

- Identification and delineation of viable *in situ* reserves and protected areas.
- *In situ* investment of local and indigenous people.
- *In situ* population monitoring and habitat management practices.
- Inter-reserve and protected area pollen dispersal and gene flow dynamics between populations.
- Preservation of *in situ* ecosystem processes and functions.
- The development of innovative active management solutions (e.g., assisted colonization, thinning, burns, weed and pest management) and the conditions and triggers for using them.
- Potential and best use of gene banks within and in association with *in situ* conservation programs.
- Identification and delineation of viable "near site" (*situ*) and mixed-use areas for limited *in situ* conservation (e.g., corridors, forestry areas, recreational lands).
- Development of adaptive management practices for *circa situ* conservation areas (e.g., preservation of structural and genetic diversity.

Best Practice 9.1: *In Situ* Conservation

North Carolina Botanical Garden

The North Carolina Botanical Garden (NCBG), in addition to managing several protected areas, conducts *in situ* conservation research in partnership with other landowners, usually on a specific endangered species. One of those projects is in collaboration with the U.S. Department of Defense on the army's Fort Bragg installation, a very large military base in the Sand Hills region of the state of North Carolina. This property contains several rare and threatened species, one of those being the endemic legume *Astragalus michauxii*, the Sandhills milkvetch. In order to preserve and restore populations of the endangered milkvetch, research was needed to develop protocols for germinating and successfully growing the plant. The staff at NCBG collected seed of *A. michauxii* and determined that germination was inhibited by a mechanical dormancy that can be broken by scarification, a treatment that then yielded a 94 percent germination rate. Previous work on germination made use of petri dishes, but the NCBG staff also determined that a better protocol after scarification is direct sowing of seed into prepared media, which yielded a 69 percent survival rate for seedlings. This work has vastly improved the conservation, reintroduction, and recovery of *Astragalus michauxii* in North Carolina.[42]

Native Plant Trust, Garden in the Woods

Like the North Carolina Botanical Garden, the Native Plant Trust (NPT) Garden in the Woods carries out a diverse program in plant conservation and conservation research. One of their contributions to *in situ* conservation has been the creation of conservation and recovery plans for a range of threatened New England species. These are collaborative efforts involving garden staff and local experts working together to produce individual taxonomic, ecologic, and conservation profiles and recommendations for 117 endangered species. Each species is fully described, including a dichotomous key for its identification; range and habitat descriptions are supplied, including specific details regarding the population(s); the natural history of the plant is described, including reproductive biology; and, of course, there is a detailed description of the causes of endangerment and recommendations for mitigating and otherwise lessening those conditions and protecting the species.

Chicago Botanic Garden, Miami University (Ohio), Morton Arboretum, University of Illinois, and U.S. Department of Agriculture Forest Service Collaboration

This collaborative selected a genetically diverse subset of white ash trees (*Fraxinus americana*) within 17 distinct populations in the Alleghany National Forest of northeastern Pennsylvania for *in situ* conservation treatments of a systemic insecticide to protect them against emerald ash borer (*Agrilus planipennis*).

> *In situ* preservation will help maintain breeding populations of large trees in the landscape, thereby protecting ecosystems services, helping to ensure future ash reintroduction, and allowing continued natural selection and adaptation. Evidence from early *in situ Fraxinus* conservation efforts suggest a degree of associational protection, in which insecticide treatment of a small number of ash trees in a stand can promote the health of untreated trees.[43]

The authors used eight microsatellites to genotype the selected trees toward obtaining a genetically diverse sample. This research demonstrated that greater genetic diversity was obtained by treating more populations of fewer trees rather than more trees in fewer populations.

Textbox 9.1. Recommendations for Conservation Research and Collections

Basic
- Conservation research and programming is guided by the institutional mission and policies.
- Begin by collaborating with experienced institutions, agencies, and nongovernmental organizations (NGOs).

Intermediate
- Independent conservation research and interventions are guided by conservation assessments.
- Collections of conservation value should conform to collections policies.

Advanced
- Garden establishes and manages a preservation and conservation reserve.
- Garden develops and preserves gene bank conservation collections.
- Garden engages in species recovery and ecological restoration.

Notes

1. P. F. Langhammer et al., *Identification and Gap Analysis of Key Biodiversity Areas: Targets for Comprehensive Protected Area Systems* (Gland, Switzerland: IUCN, 2007), 24.

2. International Union for the Conservation of Nature and Natural Resources, Botanic Gardens Secretariat, *The Botanic Gardens Conservation Strategy* (London: IUCN, 1989), ix.

3. American Public Gardens Association (APGA) Plant Conservation Committee survey (1986).

4. Sergei Volis, "Complementarities of Two Existing Intermediate Conservation Approaches," *Plant Diversity* 39 (2017): 380.

5. Volis, "Complementarities," 380.

6. P. S. White, "In Search of the Conservation Garden," *Public Garden* 11, no. 2 (1996): 11.

7. V. H. Heywood, "The Future of Plant Conservation and the Role of Botanic Gardens," *Plant Diversity* 39 (2017): 312.

8. B. A. Meilleur, introduction to "Profiles: Conservation Collections versus Collections with Conservation Values," *Public Garden* 12, no. 2 (April 1997): 38.

9. Center for Plant Conservation, *CPC Best Plant Conservation Practices to Support Species Survival in the Wild* (Escondido, CA: CPC, 2019), xiv.

10. M. Patrick Griffith et al., *Toward the Metacollection: Safeguarding Plant Diversity and Coordinating Conservation Collections* (San Marino, CA: Botanic Gardens Conservation International, 2019), 2.

11. Peter Ashton, "Biological Considerations in *In Situ* vs. *Ex Situ* Plant Conservation," in *Botanic Gardens and the World Conservation Strategy*, ed. D. Bramwell et al. (London: Academic Press, 1987), 117.

12. Botanic Gardens Conservation International, *International Agenda for Botanic Gardens in Conservation*, 2nd ed. (Richmond, Surrey, UK: BGCI, 2012).

13. Botanic Gardens Conservation International, *International Agenda*, 3.

14. World Flora Online, "Home," http://www.worldfloraonline.org.

15. The Plant List, "Home," http://www.theplantlist.org.

16. Botanic Gardens Conservation International, *International Agenda*, 21.

17. Botanic Gardens Conservation International, *International Agenda*, 25.

18. Botanic Gardens Conservation International, "Ecological Restoration Alliance of Botanic Gardens," https://www.bgci.org/our-work/projects-and-case-studies/ecological-restoration-alliance-of-botanic-gardens/.

19. Nigel Maxted and Shelagh Kell, "A Role for Botanic Gardens in CWR Conservation for Food Security," *BG Journal* 10, no. 2 (2013): 32.

20. Botanic Gardens Conservation International, *International Agenda*, 28.

21. North Carolina Botanical Garden, "Plant Conservation Programs," https://ncbg.unc.edu/research/plant-conservation/.

22. See Botanic Gardens Conservation International, ThreatSearch (database), https://www.bgci.org/threat_search.php; Botanic Gardens Conservation International, GlobalTreeSearch (database), https://tools.bgci.org/global_tree_search.php; and the Center for Plant Conservation's Rare Plant Finder tool, "Rare Plants," https://saveplants.org/search/?category=plants/.

23. Kevin M. Potter et al., "Banking on the Future: Progress, Challenges and Opportunities for the Genetic Conservation of Forest Trees," *New Forests* 48 (2017): 157.

24. Potter et al., "Banking on the Future," 157.

25. Heywood, "The Future of Plant Conservation," 310.

26. J. Chen, R. T. Corlett, and C. Cannon, "The Role of Botanic Gardens in In Situ Conservation," in *Plant Conservation Science and Practice*, ed. S. Blackmore and S. Oldfield (New York: Cambridge University Press, 2017), 77.

27. Langhammer et al., *Identification and Gap Analysis*, xiii.

28. Heywood, "The Future of Plant Conservation," 310.

29. Chen, Corlett, and Cannon, "The Role of Botanic Gardens," 78.

30. Chen, Corlett, and Cannon, "The Role of Botanic Gardens," 78.

31. See Langhammer et al., *Identification and Gap Analysis*.

32. V. H. Heywood, K. Shaw, Y. Harvey-Brown, and P. Smith, eds., *BGCI and IABG's Species Recovery Manual* (Richmond, Surrey, UK: Botanic Gardens Conservation International, 2018), 21, https://www.bgci.org/wp/wp-content/uploads/2019/04/Species_Recovery_Manual.pdf.

33. M. van Zonneveld, E. Thomas, G. Galluzzi, and X. Scheldeman, "Mapping the Ecogeographic Distribution of Biodiversity and GIS Tools for Plant Germplasm Collectors," in *Collecting Plant Genetic Diversity: Technical Guidelines—2011 Update*, ed. L. Guarino et al. (Rome, Italy: Bioversity International, 2011), ch. 15/16, 1.

34. N. Maxted and L. Guarino, "Ecogeographic Surveys," in *Plant Genetic Conservation*, ed. N. Maxted, B. V. Ford-Lloyd, and J. G. Hawkes (Dordrecht: Springer, 1997), https://doi.org/10.1007/978-94-009-1437-7_4.

35. Heywood et al., *Species Recovery Manual*, 47.

36. Heywood et al., *Species Recovery Manual*, 51.

37. Heywood et al., *Species Recovery Manual*, 51.

38. See Heywood et al., *Species Recovery Manual*; L. Guarino et al., eds., *Collecting Plant Genetic Diversity: Technical Guidelines—2011 Update* (Rome, Italy: Bioversity International, 2011).

39. K. Havens, "The Role of Botanical Gardens and Arboreta in Restoring Plants," in *Plant Conservation Science and Practice*, ed. S. Blackmore and S. Oldfield (New York: Cambridge University Press, 2017), 134.

40. World Intellectual Property Organization, *Documenting Traditional Knowledge: A Toolkit* (Geneva: WIPO, 2017).

41. Chen, Corlett, and Cannon, "The Role of Botanic Gardens," 91.

42. Michael Kunz et al., "Germination and Propagation of *Astragalus michauxii*, a Rare Southeastern US Endemic Legume," *Native Plants Journal* 17, no. 1 (2016): 47.

43. Charles E. Flower et al., "Optimizing Conservation Strategies for a Threatened Tree Species: In Situ Conservation of White Ash (*Fraxinus americana* L.) Genetic Diversity through Insecticide Treatment," *Forests* 9 (2018): 2.

CHAPTER TEN

Ex Situ Conservation

"As a professional community, botanic gardens and arboreta conserve and manage a far greater proportion of known plant species diversity than forestry, agriculture or any other sector."[1]

The purpose of *ex situ* conservation is to preserve plants as part of an overall strategy to ensure that species ultimately survive in the wild. In other words, the *ex situ* preservation of species is not an end in itself and only has real conservation value if viewed as a means to an end.[2] This type of program is most effective, as I've stated previously, if it is part of an overall, integrated program complemented and reinforced by *in situ* conservation. *Ex situ* conservation collections with direct connections to *in situ* species recovery programs will be of great value in completing a cycle of species conservation. This point will be underscored later in the next chapter on species recovery programs and ecological restoration. This may require that many botanical gardens, certainly the smaller ones, form partnerships with other gardens and agencies to pool their conservation resources.

Ex situ collections may be categorized as "conservation collections" or "collections with conservation value" as described in the previous chapter on conservation programs. "*Ex situ* conservation of plant germplasm provides a vital backup in the event of species extinction in the wild."[3] Of all the conservation programming and research presented in this section, *ex situ* conservation is a natural for botanical gardens as these are institutions that collect and preserve plants. There is a programmatic niche for all botanical gardens in *ex situ* conservation research and programming toward achieving target 8 of the Global Strategy for Plant Conservation.

Before diving into specific types of *ex situ* conservation research and collections, I want to begin this process of analysis with a general list of considerations from the Montgomery Botanical Center's guide for building conservation collections:

- Define the purpose and scope of a conservation collection.
  - Consistent with the institution's mission and collections policy.
  - Build on existing collection strengths.
  - Focus on threatened species.
  - Join and/or build partnerships of integrated programs.

- Build the collection thoughtfully.
  - Curate first generation, wild-collected germplasm from well-documented sources.
  - Curate genetic diversity; maternal lines are preferable.
  - Precise and accurate documentation is critical beginning with passport data.[4]

In the first set of bulleted items above is a recommendation for conservation program partnerships. A great opportunity for partnerships exists through the American Public Gardens Association and United States Forest Service Tree Gene Partnership Program.[5] From Pamela Allenstein, Plant Collections Network Manager at APGA, "we offer matching grants to projects targeting at-risk exceptional tree species. Funds support scouting and collecting trips, encourage collaborations, and propagules to be distributed among several public gardens for safeguarding in living collections."[6]

No matter the specific kind of collection, "in order to maximize its value, a *conservation collection* has accurate records of provenance, differentiated *maternal lines*, and diverse genetic representation of the species."[7] This goal alone presents a large array of conservation research questions and opportunities for botanical gardens, for example, how to capture the greatest genetic diversity in a collection; how to maintain that diversity *ex situ* over an extended period of time. Keep in mind that the more information we have about the ecology, breeding biology, and genetics of a target species, the better our sampling and management of these accessions will be. This understanding forms the basis for a great deal of conservation research related to *ex situ* conservation collections. I'll offer one last generic item related to conservation documentation: make sure that all conservation collections are

Table 10.1. The Relative Conservation Value of *Ex Situ* Approaches

Type of ex situ collection	Genetic diversity	Longevity	Relative costs per individual	Relative conservation value	Notes
Seed bank	High (if proper protocols followed)	High (with proper storage)	Low (if facilities exist)	Reintroduction: high; research: high; education: low	Seed storage is not possible for many tree species, particularly those of the humid tropics
Cryopreservation	High (if proper protocols followed)	High (with proper storage)	Intermediate (if facilities exist)	Reintroduction: high; research: high; education: low	Techniques for most tree species not yet available
Tissue culture	High (if proper protocols followed)	Intermediate (with proper storage)	Intermediate (if facilities exist)	Reintroduction: high; research: high; education: low	Techniques for many tree species not yet available
Conservation collection / field gene bank	Intermediate	Short (species' generation length)	High	Reintroduction: intermediate; research: high; education: high	Cultivation is the only option for many tree species; adaptation to cultivation and hybridization may be a concern
Reference living collection	Low*	Short (species' generation length)	High	Reintroduction: low*; research: intermediate*; education: high	Source may be unknown, often one or few individuals, likely adaptation to cultivation
Display living collection	Low*	Short (species' generation length)	High	Reintroduction: low*; research: low*; education: high	Source often unknown; often one or few individuals, likely adaptation to cultivation

* May have higher genetic diversity or conservation and research value if material is wild collected and maintained as multiple genetically diverse accessions, although adaptation to cultivation and hybridization is a concern.

Source: S. Oldfield and A. C. Newton, *Integrated Conservation of Tree Species by Botanic Gardens: A Reference Manual* (Richmond, Surrey, UK: Botanic Gardens Conservation International, 2012).

geo-referenced; providing geographic information system (GIS) coordinates for collections is sufficient.

First of all, before embarking on a program of *ex situ* conservation collecting and research, some developmental research is in order involving, among other things, the conservation assessment described in the previous chapter. Table 10.1 lists a nice synopsis of attributes associated with the principal types of *ex situ* conservation collections that should help you to evaluate your programmatic options.

Combine the comparative information in table 10.1 with your answers to the following questions (adapted from the Center for Plant Conservation [CPC] manual, *CPC Best Plant Conservation Practices to Support Species Survival in the Wild*):

- What is the purpose of the collection (consistent with Montgomery above)?
- Will it benefit the species survival, reduce the risk of extinction, and not harm the wild population?
- What is the population status of the species in the wild?
- What is the species' natural history: breeding biology and dispersal mechanism, ecology?
- What are the horticultural and propagation requirements of the species?
- What is the best long-term preservation mechanism (whole plant, seed, tissue)?
- Will the accession be used for *in situ* conservation?[8]

The research necessary to answer the above questions, research that will likely require a conservation assessment, will help you decide the basic nature of any conservation collection you might pursue. Be prepared to address the two most important curatorial challenges of these collections: obtaining optimum genetic diversity and maintaining the *ex situ* integrity of that diversity. To help determine the most efficient way to preserve germplasm long term, the CPC has fashioned a helpful decision flowchart (see figure 10.1). The flowchart may be relatively simple, but some of the information necessary to use it is not—it requires the specific curatorial research and knowledge indicated in the above list of questions.

Before taking up research associated with particular types of *ex situ* conservation collections, I want to advocate for a category of plants worthy of much greater conservation attention: crop wild relatives. "The global value of the introduction of new genes from CWR to crops is estimated to be $115 billion annually. However, CWR taxa cannot be used by plant breeders to sustain

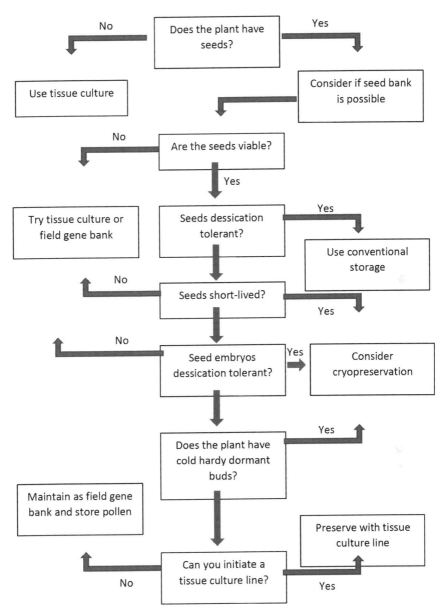

Figure 10.1. Gene Bank Decision Flowchart from the Center for Plant Conservation
Source: Center for Plant Conservation, *CPC Best Plant Conservation Practices to Support Species Survival in the Wild* (Escondido, CA: CPC, 2019), 1–12.

food security if they are not conserved and available for utilization."[9] Conservation efforts, as I've mentioned previously, are most intensely focused on charismatic species and landscapes, an unfortunate priority in the face of potentially disastrous food insecurity. Agricultural gene banks naturally focus on crops that make up the core of agricultural production, but in doing so they largely ignore their wild relatives. Gardens in the United States could work to support the U.S. Department of Agriculture (USDA) national plant germplasm repositories with sister conservation collections of crop wild relatives. The repositories can provide helpful guidance in what genera and species of plants are most in need of *ex situ* preservation. It's very likely that taxa of current conservation interest to botanical gardens contain crop wild relatives that may be included in those collections.

One last general item that applies to all types of *ex situ* conservation collections is local and indigenous group partnership. It is likely that the target species has some importance and/or historical significance to local or indigenous groups in the area that must be brought into the project. No matter how large or small their contribution as partners, be sure that they are fully acknowledged and credited and that they share in benefits accrued from the project.

Finally, recall that collection preservation (care and maintenance), including gene banks and such, is outlined in chapter 5.

Conservation Collections: Seed Gene Bank

The preferred type of gene bank for the *ex situ* conservation of genetic diversity, given our current level of technology, is the seed gene bank. Seed bank technology has been in use to preserve agricultural crops for decades, perhaps for centuries, with an increasing level of competence for long-term seed preservation. Nowadays, the purpose of a seed bank is (1) the long-term storage of germplasm for conservation purposes in case taxa are lost in the wild and/ or in the landscape and (2) for research and use in breeding, restoration, and other programs.[10]

Botanical gardens have an important role to play in seed banking to fulfill both of these purposes. In fact, some of the largest and most comprehensive seed banks are found at botanical gardens; the Millennium Seed Bank of the Royal Botanic Garden, Kew is a primary example. Within their two main purposes, comprehensive seed bank collections must hold the broadest possible genetic base of their subject taxa. This strategy drives the collecting and storage practices used at seed banks.[11] Your institution's mission, capability, and expertise will help determine which types of *ex situ* conservation are suitable to your institution, be that a seed bank, some other *ex situ* strategy, or a combination of these.

If a seed bank is a viable option, I encourage you to look into the Millennium Seed Bank Partnership (MSBP) for institutional mentoring and networking opportunities within the seed bank community. The MSBP has become a huge network of botanical gardens, forestry institutes, and agricultural organizations in more than 80 countries training thousands of people in seed bank technologies.[12]

The next stage in your research will be to determine what taxa to collect, assuming that they produce viable seeds. Participating institutions can search for priority species for their region in the CPC's portal on the "Rare Plant Finder" tab.[13] Another valuable resource for helping you get started developing a seed bank is the Botanic Gardens Conservation International (BGCI) Global Seed Conservation Challenge program. You will find a set of helpful learning modules on their program webpage to walk you through the basic process.[14] Take note that as of this writing, there is a serious shortage of taxa within seed banks that are useful in ecological restoration.

Once you've composed a conservation desiderata of threatened, seed-producing species, you will need to consider the seed storage requirements of the target taxa to help plan the seed bank infrastructure required to accommodate those seeds. Refer again to the flowchart in figure 10.1 showing some basic issues and questions pertaining to seed collections. For seeds to remain viable in storage, they must tolerate desiccation. If so, they are labeled "orthodox" seed; if not, they are considered "recalcitrant." There is also an intermediate seed storage category known as "exceptional" that is closely aligned with recalcitrant seed and is not viable for seed banking. We will return to recalcitrant and exceptional seed preservation in a later section on tissue gene banks. To determine the seed storage classification of any taxon, check online using the scientific name and "seed storage" as key words. Also, the USDA-ARS National Laboratory for Genetic Resources Preservation (NLGRP) in Fort Collins, Colorado, and the Millennium Seed Bank, Royal Botanic Garden, Kew are good resources. The CPC offers some useful generalities to help you ascertain if the plants in question have an orthodox seed storage tolerance:

- Plants are not wetland adapted, particularly xerophytes.
- Seeds are subject in nature to a drying maturation and hard freezes.
- Season of seed production is other than spring.
- Plant life form is other than a tree.
- Viable seeds are persistent in nature.
- Seeds are dormant when dispersed.
- Seeds are neither very large nor very small.[15]

You may find that the taxa you have chosen to work with have a relatively unknown seed biology and little helpful information is available. In that case, you will either move on to a new target species with documented storage requirements or decide to conduct your own applied research to discover the best conditions for the storage of that species.

Seed collecting must follow a sampling strategy designed to capture the greatest diversity with the smallest sample size and number. Referring back to the sampling strategy first discussed in chapter 3, diversity must be sampled in three ways: (1) within-population diversity, (2) between-population diversity, and (3) ecogeographical diversity. In addition to the samples collected using the CPC strategy outlined in chapter 3, collectors may also want to sample any interesting variants. These samples are kept separate from the rest and given individual collection numbers. Refer to chapter 3 regarding the handling and documentation of seed collections.

In some cases, the size of the population and/or the fruit set may be quite limited necessitating a special effort to obtain an adequately diverse sample. In these cases, a maternal line collection should be established.

> For research purposes, and for the conservation of rare species that occur in populations of fewer than 50 individuals, as well as for less fecund common species, where collections will result in fewer than 3,000 seeds, we recommend collecting seeds along maternal lines. In a maternal line collection, seeds from each individual plant (maternal line) are collected and bagged separately (as opposed to bulking the seed collection of multiple plants into one or two bags as we do for regular seed bank collections). This ensures the greatest genetic diversity is available in a small collection.[16]

When a maternal line collection is used for translocations or reintroductions of species, the seeds must be selected in equal proportion from each maternal line for the propagation of new plants.

The taxa you may have chosen to focus on for a conservation collection of seed may be so rare that you can't meet any of the recommended standard CPC collecting protocol numbers for obtaining optimum genetic diversity, including maternal line collections. First of all, be sure to do as much research on the natural history of the targeted rare species as possible. This should include field investigations of the ecology, breeding biology, population dynamics, and fecundity of the target species. Fortunately, the CPC has taken the likelihood of institutions seeking to conserve very small, rare populations of plants into consideration with some of the following recommendations:

- Plan to make multiyear, maternal line seed collections and tag or mark individuals to track representation.

- Do germination tests and growing-on trials of small samples.
- Expand the collection through regeneration by adding wild-collected seed and seed from second-generation plants grown in cultivation (field gene bank).
  - "Because *genetic drift* and artificial selection are possible in a cultivated setting, strive to have at least 30 *randomly chosen* individuals of a fully *outbreeding* sexual species (*outcrossing*) or 59 randomly chosen individuals of a self-fertilizing species as the seed-bearing plants as source material."[17]
  - "Because genetic drift and artificial selection are possible with every generation, use the most original *sample* to regenerate accessions (prioritize wild first, then the first generation (F_1), then second generation (F_2) to grow for increasing the total number of seeds."[18]
  - Controlled pollination will facilitate out-crossing and must be distinguished by labeling and in the documentation.

- Clearly distinguish all the maternal lines as well as wild and regenerated germplasm among the samples and in the documentation. Indicate the generation number for the regenerated seeds (F_1, F_2, etc.) and note their growing conditions in the field gene bank.
- Follow the standard viability testing protocols.[19]

I'll offer a word of caution here about biosecurity. In consideration of your seed-collecting desiderata, and while making field observations during collecting, apply the information available through the Plant Sentinel Network and weed risk assessment literature, including the information in chapter 3 on invasive plants, to your target species. Then take advantage of your time field collecting to observe the target plants for signs and symptoms of pernicious pest problems and evidence of invasiveness.

In addition to containing as much genetic diversity as possible, storage of seeds demands that the genetic integrity of the samples, as well as the seeds themselves, be preserved over long periods of time—several decades or longer. The techniques for accomplishing this are outlined in chapter 5. If, however, you have decided to acquire seed of an unknown storage capacity and do the necessary research to try to identify the tolerance of that seed, begin with the assumption that the seed is recalcitrant. With this protocol, you will preserve the viability of the seed regardless of its storage type while conducting some limited germination tests. If the seed gives every indication of being dormant and/or otherwise orthodox, revert to the standard CPC guidelines in chapter 5 for seed storage. Further research may indicate that the long-term storage of this particular species will benefit from slight deviations from the CPC specifications.

All the necessary steps for preparing seeds for long-term storage—cleaning, drying, packaging, and such—are comprehensively described and illustrated in the CPC *Best Plant Conservation Practices* manual. Every one of these important operations are open for applied botanical garden research, and some may be in need of basic research to improve and refine our ability to successfully store seeds for extended periods of time.

To review from chapter 5, there are two basic types of seed bank collections. A *base collection* is stored under optimal, long-term conditions and is not altered unless indicated by viability testing. An *active collection* is stored under short-term conditions for the purposes of access and distribution, for example, *Index Seminum*.[20]

The operation of a seed bank may be easily illustrated with a flowchart. Figure 10.2 is adapted from "A Strategy for Seed Banking in Botanic Gardens" by J. G. Hawkes.[21] Each step in that flowchart represents a node of potential botanical garden research in seed bank operation and seed preservation.

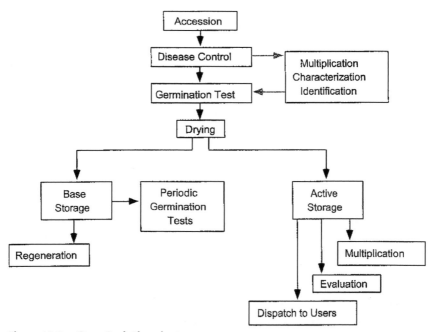

Figure 10.2.   Gene Bank Flowchart
Source: J. G. Hawkes, "A Strategy for Seed Banking in Botanical Gardens," in *Botanic Gardens and the World Conservation Strategy*, ed. D. Bramwell et al. (London: Academic Press, 1987), 144.

Apart from the base and active collections illustrated here and described in greater detail in chapter 5, a duplicate of the base collection should be established at another location. It is worth reiterating that, even though the CPC has put together an extremely valuable best practices manual on conservation collections with specific protocols for seed and other types of banking operations, a great deal of research still needs to be done on the viability, germination, and genetic deterioration of seeds of wild plants in order to refine our approach to this important work.

Before moving on to other types of gene banks, I'll add a few comments on a technology allied to seed banks: pollen and spore banks. Pollen banks are used as a small-scale means of *ex situ* conservation of species with either recalcitrant or poor-quality seeds. Spore banks may be used for the *ex situ* conservation of pteridophytes and their relatives. Like seeds, pollen and spores may be segregated according to their storage tolerance as orthodox or recalcitrant. Some spores are so tolerant of desiccation that they may be extracted from old herbarium specimens and germinated.[22] It is possible to grow some whole haploid plants from pollen grains. Of course, all ferns from spores are naturally haploid.

Whole anthers and individual pollen grains may be stored following cryo-preservation standards in liquid nitrogen.[23] Some pollen grains may require drying before freezing while others do not require any pretreatments. Both pollen and spores may be stored temporarily in vials. For long-term storage, pollen needs to be embedded in gelatin capsules, paper pouches, or packed on a paper strip.[24] For regeneration into viable plants, pollen may require a crossing parent to obtain the desired diploid whole plant. There is a great deal of research that needs to be done on the handling and long-term storage of pollen for *ex situ* conservation.

Best Practice 10.1: Seed Bank
The two examples that follow are good models of small-scale, comprehensive seed banks within the technological and financial reach of many botanical gardens.

Fairchild Tropical Botanic Garden
The Fairchild Tropical Botanic Garden Connect to Protect program is a wonderful example of an integrated conservation program that includes both *in situ* and *ex situ* conservation, as well as translocation of species. The Connect to Protect program is structured around the *in situ* conservation of pine rocklands habitat in South Florida and the Caribbean area; the *ex*

situ seed banking of threatened plants found in the pine rocklands; and the translocation of threatened plants to pine rocklands produced from collected and stored seeds.

Seeds of threatened plants are collected and tested for their tolerance to seed bank storage. Germination, desiccation, and freezing trials indicate that most of the pine rocklands species are orthodox and tolerant of long-term storage. With the exception of temporary holdings for translocation plantings, the program's seed collections are stored at the National Center for Genetic Resources Preservation in Fort Collins, Colorado.

University of Washington Botanic Garden (UWBG) Miller Seed Vault
The Miller Seed Vault is part of the UWBG Rare Care program, an integrated conservation program involving both *in situ* and *ex situ* conservation focused on the flora of Washington State. The vault is a state-of-the-art, environmentally controlled, 150-square-foot multifunctional facility with a workroom, short-term storage room, and a long-term storage freezer. The ambient conditions in the work and short-term storage rooms are maintained at 57°F and 22 percent relative humidity to maintain the integrity of seeds held for propagation and project work. Seed accessions for long-term *ex situ* preservation are dried to 15–25 percent moisture content and stored at –18°C in heat-sealed, foil laminate bags.

Conservation Collections: Tissue Gene Banks

Tissue banking, as first defined in chapter 5, is an alternative preservation and conservation strategy for plants that produce recalcitrant seeds; species classified as "exceptional" due to conditions that exclude seed banking as a conservation strategy; species that are not responsive to vegetative propagation; and species that may be short-lived or difficult to grow as *ex situ* whole plants in field gene banks. DNA banking may be considered another type of tissue gene bank in that the generic requirements for the preservation of DNA samples are the same as those for other tissues. Tissue-banking technology utilizes *in vitro* and/or cryopreservation methods for long-term storage depending upon the tolerances of the taxa to be preserved and the specific physiological status of recalcitrant seeds (or their embryos) and tissues.[25] Cryopreservation is often dependent upon *in vitro* technologies in some intermediate steps and in the recovery of frozen germplasm for propagation purposes.

Before going any further, it should be clear from the preceding descriptions that tissue banking is fraught with a host of variable and poorly known factors related to exceptional species and intraspecies variations. This presents a matrix of research opportunities for botanical gardens pursuing tissue-banking programs. Writing on the role of botanic gardens in *ex situ* conservation, Paul Smith and Valerie Pence pose three statements of need for conserving threatened exceptional species:

- Information Challenges: there is incomplete information on which threatened species are exceptional and who is working with this group.
- Identify Research Priorities: the biology of exceptional species is poorly understood impeding the development of efficacious conservation technologies: *in vitro* and cryopreservation techniques.
- Funding, Communication and Coordination Challenges: the methods required for conservation of recalcitrant and exceptional species are significantly more expensive than standard seed banking. Shared technologies and coordinated banking operations will improve practice and mitigate costs.[26]

Something else that I would add to the preceding second bullet item is that the *in vitro* and cryopreservation techniques for long-term storage are only useful conservation solutions if the tissues can be successfully restored to whole plants for transplantation to the wild.[27] This is an example of another research opportunity.

Before getting into the details of these methods, be reminded that tissues should be acquired using the same genetic diversity and ethical protocols used for seed collections (some may in fact begin as seed collections). That means keeping accurate records of provenance and establishing and maintaining differentiated maternal lines as well as diverse genetic representation. It is important to remember that most *in vitro* conservation collections and some cryopreserved conservation collections are established using vegetative, clonal tissues that must be acquired and curated in genetically distinct lines of sufficient quantity to obtain the necessary diversity within the collection. There are some other general practices that must be followed: all material must be in good condition—seeds viable and tissues vigorous, unblemished and uninfected with pests—and of consistent maturity (seeds mature and tissues either juvenile or mature). Transport time from the field to accessioning must be minimized, and all material must be disinfected before being preserved. See chapter 3 for information on handling and transporting recalcitrant seeds and tissues as well as information on passport data.

Both cryopreservation and *in vitro* conservation collections require some common facilities and equipment for their preparation. I know of some individuals and small institutions that have achieved success with *in vitro* micropropagation with ordinary facilities: any small room that may be thoroughly cleaned; simple countertop workspace; a small box open on one side to serve as a simple, aseptic workspace with minimal air exchange; simple agar media prepared with hot plates, firmed at room temperature and stored with explants on LED light shelves; and all facilities sterilized with alcohol sprays. The following concise recommendation for a more comprehensive *in vitro*/cryopreservation lab comes from the *Technical Guidelines for the Management of Field and In Vitro Germplasm Collections*:

> Equipment needed includes laminar flow benches, pH meters, balances, sterilization equipment (either autoclave or pressure cookers), hot plates or stove, magnetic stirrer, appropriate chemicals, analytic and culture glassware, refrigerator, and freezer. Protective clothing, gloves and safety devices such as showers, eyewash, and fire extinguishers are also recommended. Arrangement of the equipment and supplies in a logical fashion will allow for the most efficient use of time and should minimize mistakes. A flow chart of lab activities might be used to organize lab equipment and supplies. For cryopreservation, a reliable source of liquid nitrogen (LN), tanks, dewar vials, and specific chemicals will be needed in addition to tissue culture facilities.[28]

In addition to the preceding list, you will need to have on hand the chemical constituents of *in vitro* growth media as well as sterilants and disinfestation materials. You will also need a good source of single- or double-distilled, deionized, and filtered water for preparing the medium as well as rinsing seeds and explants. The air in both the culture lab and the grow rooms/chambers should be HEPA filtered. Grow room and growth chamber lighting is usually provided by LED fixtures or florescent tubes to provide illumination in a range "from 10 to 1000 $\mu mol \cdot s^{-1} \cdot m^{-2}$ but most plant cultures require 50-200 $\mu mol \cdot s^{-1} \cdot m^{-2}$."[29] Ambient temperatures in these spaces may range from 22°C to 28°C, depending upon the requirements of the taxa being grown. Finally, grow rooms and growth chambers should be outfitted with security mechanisms to warn of system failures. I will offer more on materials and equipment related to *in vitro* conservation later in this section.

Before looking more closely at cryopreservation, I'll add a few words about recalcitrant seed storage. Recalcitrant seeds must be kept at the same hydration levels during handling and in any temporary storage as they had at dispersal. Generally speaking, recalcitrant species from temperate climates are chilling tolerant, while those from subtropical and tropical climates are not. Chilling recalcitrant seeds from temperate areas while maintaining hydration levels will extend their viability.[30] If you are uncertain as to the storage classification of the collected seeds and there are enough of them, test a small sample for their tolerance to drying using a desiccator and silica gel. Consult the CPC *Best Plant Conservation Practices* manual for specific recommendations on temperatures and relative humidity for drying seeds.[31] A measure of water content is a helpful indicator. High water content at the point of dispersal is a good indicator of exceptional seed.

Next we'll take a closer look at cryopreservation technologies and practices, followed by those for *in vitro* preservation and, finally, DNA banking. Banking of species with recalcitrant and exceptional seeds is most often accomplished through the cryopreservation, or deep freezing, of *in vitro* cultured embryo axes, shoots, or plantlets. However, some exceptional seeds may be cryopreserved intact if they are desiccation tolerant. It is important to recognize that some species with exceptional seeds require a symbiont partner for germination and growth (e.g., mycorrhizal fungi). The tissue bank may need to preserve these organisms as well to ensure that the exceptional taxa may be successfully restored in the future to whole plants.

"Four steps are essential in any cryopreservation protocol: (i) selection, (ii) preculture, (iii) cryopreservation techniques, (iv) retrieval from storage, and, (v) seedling or plantlet establishment."[32] Recalcitrant seeds are commonly cryopreserved by removing their embryos and trimming them down to their axis under a microscope. If there are adequate supplies, set aside a control group of embryo axes for viability testing after each step in the cryopreservation process. If the embryos prove to be untenable due to size or intolerances to any of the cryopreservation steps, shoot tip meristems[33] and buds[34] are suitable substitutes. Next, sterilize (disinfest) the tissue as per *in vitro* practice and place in an osmoprotectant, such as sucrose, for two days. Osmoprotectants assist tissue cells to reach an osmotic balance that allows them to tolerate extreme drying. This is another element in the process that may need testing and verification. Next, flash dry the embryos with air and treat with a cryoprotectant—a substance that prevents tissues from freezing or that prevents damage to cells during freezing. Flash drying embryo axes facilitates rapid cooling with less concern for potential freeze damage. At this juncture, you may want to test some embryos from the control group to see if they survived the treatments. Finally, place the embryos in cryovials, label, wrap in aluminum foil, and immerse in liquid nitrogen.[35]

In the above process, the following variables may not be well documented necessitating trials and experimentation to find the best application and most reliable outcomes:

- Disinfestation: determine the correct solution ratio to avoid necrosis.

- Use of osmoprotectants: determine osmotic dehydration time and its rate of treatment.
- Determine the air desiccation time.
- Determine the correct cryoprotectant to avoid toxicity.
- Place in liquid nitrogen: determine the correct cooling rate to avoid crystallization (flash drying mitigates this concern for embryo axes).

Tissues should be periodically thawed in a warm water bath (40°C) to test for viability.[36] To successfully recover viable tissues from cryopreservation, they should be thawed as above and placed in vitro for the formation of plantlets (see the in vitro protocols that follow). The addition of an antioxidant, such as cathodic water, may reduce tissue browning and necrosis during the recovery process. Cathodic water is a commonly used antioxidant (an electrolyzed dilute solution of calcium chloride and magnesium chloride).[37] Like many other steps in the cryopreservation protocol, the successful recovery and "germination" of embryo axes in vitro are areas where research is needed to refine the processes.

As I mentioned earlier, some exceptional seeds may be cryopreserved intact if they are desiccation tolerant. If they are not desiccation tolerant, you will have to utilize other tissues, such as shoot tips or buds, that can be frozen and/or preserved in vitro. Generally speaking, dormant buds from cold climate, woody plants are most suitable for use as cryopreserved substitutes for seeds.[38]

Cryopreservation, like the early days of in vitro propagation and preservation, involves a great deal of empirical research. The principal concern is to achieve an acceptable level of success to capture an adequate amount of genetic diversity among a viable inventory of preserved tissues. This is difficult to predict when working your way through an experimental process. Fortunately, "probabilistic tools are available which facilitate calculation of the number of propagules to store and retrieve, depending on the objectives, survival after cryostorage, and other parameters."[39] The U.S. Department of Agriculture–Agricultural Research Service (USDA-ARS), Agricultural Genetic Resources Preservation Research lab in Fort Collins, Colorado, described later in Best Practice 10.2, aims to meet a 40/60 standard for cryopreservation. That means that to achieve 60 viable propagules after cryopreservation at a 40 percent viability level, at least 150 propagules should be processed.[40] This is considered a good rate of return. Finally, keep in mind that these tissue collections must be well curated from the time of acquisition through the banking process to maintain the integrity of provenance, maternal lines, and their overall genetic diversity.

Best Practice 10.2: Cryopreserved Conservation Collections

Cincinnati Zoo & Botanical Garden
Perhaps the first and longest-running cryopreservation program among botanical gardens, the Center for Conservation and Research of Endangered Wildlife (CREW) at the Cincinnati Zoo & Botanical Garden seeks to propagate and preserve endangered plants, particularly those for which traditional methods are not adequate. Their innovative program uses a combination of *in vitro* and cryopreservation for recalcitrant and exceptional species of threatened plants.

This work centers on the CryoBioBank (CBB), CREW's liquid nitrogen storage facility. The plant collections, or "Frozen Garden" within the Cryo-BioBank, includes seeds, spores, and tissues of about 150 plant species in four distinct collections, all stored at –196°C (–320°F) in liquid nitrogen. "The Regional Seed Bank includes seeds of state and federally endangered species from Kentucky, Indiana, and Ohio. . . . Because not all endangered plants are seed plants, there is a Pteridophyte (Fern) Bank and a Bryophyte (Moss) Bank, which contain spores, as well as tissues of gametophytes and sporophytes. The Endangered Plant Tissue Bank contains samples of tissues from *in vitro* cultures of endangered plants that are grown at CREW."[41]

U.S. Department of Agriculture–Agricultural Research Service (USDA-ARS), Agricultural Genetic Resources Preservation Research (AGRPR): Fort Collins, Colorado
The AGRPR lab at Colorado State University in Fort Collins, Colorado, is part of the extensive Agricultural Research Service of the U.S. Department of Agriculture. Its mission states:

> To support U.S. agriculture in producing high-quality food, feed, and fiber, the Agricultural Genetic Resources Preservation Research Unit provides genetic security by focusing on: acquiring, evaluating, preserving, and distributing critical genetic resources including plant, animal, insect, and microbial material for industry and the research community.[42]

A key operative term in this mission is "genetic security," meaning that a great deal of the work at this lab is dedicated to gene banking. Among the wide array of genetic resources preserved by this lab are seed and tissue samples from the U.S. National Plant Germplasm System (NPGS) of 15 repositories. The collection contains genetic diversity from temperate, subtropical, and tropical fruits; industrial crops; nuts; ornamentals; specialty crops; vegetables; and wild relatives constituting nearly a million accessions—primarily seeds but also *in vitro* and cryopreserved germplasm. The Genetic Resources lab has been researching and utilizing cryopreservation technology since the early 1990s. Currently, their focus is on the cryopreservation of dormant buds and meristem shoot tips.

Dormant budwood is received from the NPGS repositories and then pre-pared and cryopreserved at the Genetic Resources lab. Meristem shoot tips are either submitted to the lab *in vitro* from the NPGS repositories for trans-fer and cryopreservation or plant material is sent to the Genetic Resources lab from the repositories for sterile preparation of shoot tips and transfer to cryopreservation. "Collections of dormant buds of temperate trees, shoot tips of *in vitro* cultures of many crops, and embryonic axes of some large seeded or recalcitrant seeded plants are all part of the clonal backup storage system."[43]

The skilled and secure services of the Agricultural Genetic Resources Pres-ervation Research laboratory in Fort Collins are available to botanical gar-dens in the United States that wish to have plant genetic diversity preserved as original collections or duplicates. For more specific information on their cryopreservation practices in general and on specific crops, go to the USDA-ARS, Agricultural Genetic Resources Preservation Research webpage.[44]

The Huntington Botanical Garden Plant Tissue Culture and
Cryopreservation Lab
A newcomer to the field of cryopreservation at the time of this writing, the Huntington Botanical Garden has created an ambitious *in vitro* and cryopreservation conservation research and collections program. For more information, consult the Botanical Conservation Program online at https://www.huntington.org/botanical-conservation-program.

Turning now to *in vitro* conservation collections, as many of you know from a study of basic botany, "the basis for this technology and practice relies on the fact that many plant cells have the ability to regenerate a whole plant (totipotency). Single cells, plant cells without cell walls (protoplasts), pieces of leaves, stems or roots can often be used to generate a new plant on culture media given the required nutrients and plant hormones."[45] A huge industry has developed over the last 40 years in the production of all kinds of plants for agriculture, horticulture, and industry through tissue culture micropropa-gation techniques based on the principle of totipotency. We can now take advantage of that understanding and the tissue culture technologies that have sprung up from it to conserve threatened plants through *in vitro* culture and preservation.

Tissues intolerant of desiccation and/or freezing—in other words, those that cannot be cryopreserved—may be preserved as *in vitro* cultures for an intermediate time period (months to years). Transfer of *in vitro* tissue is much easier and more secure than cryopreserved tissue. These cultures are used in two ways for *ex situ* conservation: (1) they may be subcultured indefinitely as a backup for critically endangered wild species and/or as material for restoration and/or research work and (2) they may be used as source material for cryopreservation. These are the costliest of conservation collections, but, at the same time, they may conserve more species diversity per square foot of preservation space than traditional whole plant collections. The expense comes from the near constant requirement for the input of labor, energy, and other resources to operate and maintain the facilities and preserve and develop the tissues. In addition, staff must be on guard for possible contaminants, genetic alterations, and unconscious selections.

In vitro cultures used as conservation duplicates for critically endangered species may be held in a slow growth condition that will spread out the subculturing intervals and reduce labor and resource inputs. Be reminded that regardless of the purpose of *in vitro* culture or the particular process chosen, the tissues involved are clonal, meaning that genetically distinct lines must be acquired and curated in sufficient quantity to develop a diverse collection.[46]

Plant *in vitro*, or tissue, culture is a three- or four-stage process, depending upon how one groups the procedures. The first stage is initiation or establishment of the tissue successfully into aseptic culture. The second stage is multiplication of those initial tissues by dividing existing cultures and using those divisions to establish additional cultures (subculturing). The third stage involves the transfer of shoot cultures and some plantlet cultures to an aseptic rooting culture. Many authorities consider these three stages as the complete tissue culture process, although some consider the establishment of fully developed cultures of *ex vitro* potted, soil-based, nursery-grown plants to be a fourth stage in the process.

There are a number of factors to consider when material (explant) is being selected for *in vitro* culture: type of tissue, physiological condition, and sanitation. "In vitro culture and storage facilities are quite variable. There are many ways to achieve the desired level of asepsis, and each laboratory will need to do what is necessary for the security of their cultures."[47] Establishing aseptic cultures is a pivotal step to *in vitro* culture, and the key to that is training and practice. Staff members that develop a rapid and aseptic technique are invaluable for *in vitro* culture. I recommend staging the transfer to aseptic culture and carefully choreographing the steps before actually beginning the process.

A basic reference library on *in vitro* techniques for working with plants should include the following:

- P. C. Debergh and R. H. Zimmerman, eds., *Micropropagation Technology and Application* (Dordrecht: Kluwer Academic, 1991).
- E. F. George, *Plant Propagation by Tissue Culture: Part 1* (Westbury, UK: Exergetics, 1993).
- E. F. George, *Plant Propagation by Tissue Culture: Part 2* (Westbury, UK: Exergetics, 1996).
- L. A. Kyte and J. Klein, *Plants from Test Tubes. An Introduction to Micropropagation*, 3rd ed. (Portland, OR: Timber Press, 2001).

Among these references, a good introduction to tissue culture and *in vitro* practice can be found in *Plants from Test Tubes*.

The common types of explants acquired and accessioned for *in vitro* conservation collections are meristematic stem tips that give rise to shoots and/or tissues that then give rise to whole plantlets. Many plants that multiply vegetatively may be tissue cultured to produce plantlets, such as strawberries (*Fragaria* spp.), a product that often streamlines the tissue culture process. You may find an explant recommendation for the taxa you wish to preserve, or you may need to research and experiment with more than one type before identifying the best choice.

> The wide diversity of genetic resources available in genebanks require a similar diversity of tissue culture media and growing conditions. Standard methods may not be applicable to all accessions of a genus, or even all cultivars of a species. A thorough literature review and library research before beginning culture is extremely important for determining a starting point. Modifications to published techniques will be required in most cases and new techniques must be developed for other genera.[48]

This quote is very pertinent regarding the formulation of *in vitro* and tissue culture media for the tissues in your conservation collection. A formula will need to be researched and specified before any *in vitro* culturing can begin. A basic tissue culture formula for the macro and micro nutrients needed was developed in 1962 by Murashige and Skoog (MS) using a culture base of agar medium. Agar is a gelatinous substance obtained from various kinds of red seaweed that is commonly used as the medium for *in vitro* tissue cultures. Customized MS formulations are mixed with the agar medium and placed in various types of small vessels, such as petri dishes, vials, and test tubes, to house the *in vitro* explants during culture and preservation.

The basic MS solution is often made up on a monthly schedule and refrigerated until customized for use with a particular taxon. These custom formulations are fiercely guarded by commercial horticulture tissue culture labs. However, they might be shared with botanical gardens by legal agreement that includes a nondisclosure clause. Otherwise, botanical gardens will need to dedicate research time to developing a successful media formula. The skills of a laboratory trained microbiologist or chemist are usually required. A curator's insights into the ecological and cultural requirements of the taxa concerned often prove helpful to crafting the right formulas. Available formulas for taxonomic and ecologic affiliates to the target species are often good starting points.

Once an *in vitro* medium has been formulated, make sure it is clearly recorded in a standard format on a medium, or recipe, document that is held in duplicate. Then, record all the stages of *in vitro* culture and all the necessary procedures at each stage in a separate database. At first it is difficult to capture all the details, with some seeming nearly intuitive. This is critically important for maintaining, among other things, the required level of sanitation throughout the *in vitro* process. Speaking of documentation, you will need to document the *in vitro* conservation collection as thoroughly as any other. This will include accession labels with script and machine readable symbols for each individual plate, vial, or test tube of an accession. The documentation file will also include passport data, field data, and *in vitro* data that includes dating of all the *in vitro* stages, as well as developmental and other attribute data.[49]

Beginning the establishment stage of *in vitro* preservation involves a careful process of infestation and sterilization of plant material, trimmed explants, equipment, working surfaces, and such. This is well described in the references I've listed earlier as well as in the *Technical Guidelines for the Management of Field and In Vitro Germplasm Collections* in section II.2.5 "Introduce Plant Material into Tissue Culture."[50]

This first step—the initiation stage–is often the most troubling for tissue culture neophytes. As I suggested earlier, set up the workspace for this stage with all the necessary equipment and supplies for visualizing and choreographing the steps before actually beginning to work; this will assist you in doing the work quickly and carefully. A difficulty I had to overcome at the beginning is a propensity to be too conservative in the use of disinfestants and sterilizing agents out of fear of killing the explants. It's important to keep in mind during this step that chemically damaged tissue will often still respond and grow *in vitro*, while contaminated tissue will ultimately fail. Trials designed to determine the best aseptic practices for unknown taxa might best be started with disinfectants at higher levels of concentration then diluted to find the right ratio as quickly as possible. Mites and thrips are insidious and pernicious insect pests in vitro. The disinfestation and disinfection work should be targeted to their possible presence on explants as well as a larger concern for excluding them from the workspace. Also, proper sealing of in vitro containers is critical to keeping them free of mites and thrips.

As mentioned earlier, the multiplication stage is exactly what it sounds like: it is the increase of developing explant inventory by subculturing established tissues (division and initiation) to reach the desired inventory of a given accession. The transition from the initiation stage to the multiplication stage involves diligent maintenance of aseptic conditions in vitro and ex vitro. A part of this practice involves monitoring the in vitro accessions for signs and symptoms of contamination. This concern was first raised in chapter 5, and it will involve staff training to recognize contamination in the containers (petri dishes, vials, test tubes, etc.) of in vitro cultures. Contaminated cultures should be autoclaved before being discarded. Staff will also need training to recognize the less common but also important phenotypic changes in explants that may be the result of undesirable genetic mutations. Finally, characterizing documentation should be keeping pace with the in vitro process.

Once the necessary inventory of explants for the in vitro conservation collection has been achieved, two trajectories are followed based on the specific purpose of that accession. If the accession is for restoration or reintroduction, then normal in vitro production will continue through the remaining stages toward the development of transplants for use in the field. I will offer more on that scenario later. If the in vitro collection is a backup for seriously threatened wild taxa, then the accessions are subjected to slow growth conditions to achieve the longest possible preservation before starting new cultures. As you might expect, this process involves a cutting back on conditions that would normally be prescribed for optimal growth to those that will achieve a lower, preservation level of growth, in some cases one close to a state of dormancy. Some species are naturally slow growing in vitro, in which case standard tissue culture protocols may be followed. Although experimentation is currently being done on slow growth protocols for in vitro conservation collections, it is likely that you will need to conduct your own research and experiments to achieve successful results with your taxa. Published information on similar taxa may be a useful guide for your experimentation.

Slow growth tissue culture protocols, like standard approaches, are based on the Murashige and Skoog standard solution on agar medium. It's quite possible that all, or a large portion, of the chemical constituents of the medium may need modification to achieve the best slow growth condition. This involves the nutrient content, the sugar or sucrose content, the type and concentration of growth regulators, and other materials. Not surprisingly, temperature regimes will be lower than normal. "Optimal storage temperatures for cold-tolerant species may be from 0 to 5°C or somewhat higher; for material of tropical provenance the lowest temperatures tolerated may be in the range from 15 to 20°C, depending on the species."[51] Cultures should be acclimated to these temperatures gradually over the course of one to two weeks. Light requirements—intensity, quality, and duration—will vary by taxa. In some cases, the type, volume, and means of sealing the growth vessel may be an important variable that also requires experimentation.[52] As I've mentioned before, curatorial knowledge of the ecological and horticultural requirements for growth of a taxon will be useful in helping to surmise the optimal conditions for slow growth *in vitro* preservation.

Some issues and problems to watch out for within a slow growth regimen are pests and pathogens, anatomical anomalies, and genetic deformations. Woody plants may produce polyphenolics that can lead to deterioration of the explants and also hyper-hydricity in their tissues causing necrosis. Bacterial infection is always a concern and may occur in a covert and chronic way that appears later in the tissue culture process requiring removal of the vessel, autoclaving, and disposal. Like all *in vitro* cultures, slow growth collections will require close monitoring for anatomical and other growth changes that may be a sign of somaclonal and other genetic changes—the appearance of callus growth may be such an indicator.[53]

Whether you are working with a standard inventory of *in vitro* cultures for use in restoration and reintroduction programs, or slow growth cultures for conservation backup, there are some steps pertinent to both of these collection types that should be taken at the point of subculturing them. Subculturing may be done to expand or rejuvenate an *in vitro* conservation collection or to create a safety duplicate collection. Subdivide the total number of cultures and subculture them one at a time to ensure that each subgroup is reestablished *in vitro* successfully before moving to the next group. Be alert to the possibility of artificial selection in returning the new subcultures to storage.[54]

To use *in vitro* tissues for growing whole plants, they must be stimulated for further development from their cultured condition as shoots or plantlets. Shoot cultures will need to be transferred to a new medium that will stimulate rooting. A notable change in the medium formulation will be the presence of root-promoting growth regulators. Some plantlet cultures may require rooting as well. This transfer will require diligence in maintaining asepsis and, like establishment and subculturing, requires the same laboratory conditions and much of the same equipment. Once the plants are rooted, they are transferred to an *ex vitro* environment in small nursery containers of soilless media and kept under transparent cover to retain high humidity. Tissue-cultured plants have very thin cuticles and are exceptionally vulnerable to desiccation. Plants are irrigated with tempered water and slowly acclimated to lower humidity by periodic exposure until they can be left uncovered. Once fully hardened off, the plants are cultured as is consistent with container nursery production standards. Counterintuitively, xerophytes may require a longer acclimation/hardening-off period.[55]

Finally, and regarding tissue banking as a whole, there is a growing literature on cryopreservation and an even more thorough one on *in vitro* techniques for the preservation of conservation collections of specific taxa. However, you are still likely to find a dearth of information on species and clones of interest to your institution. Here is where a great deal of botanical garden conservation research is required. Slight changes to generic protocols may lead to successful outcomes and, when well documented, become important references for the entire field. It is critically important that all successful protocols be well documented and that technical staff be well trained in those protocols.

Best Practice 10.3: *In Vitro* Conservation Collection

Kings Park and Botanic Garden, West Perth, Australia

The discovery of two specimens of *Symonanthus bancroftii*, a once thought to be extinct member of the dwindling natural flora of Western Australia's wheat belt, by staff members of the Kings Park and Botanic Garden motivated an innovative approach to conservation. The Conservation Biotechnology research group developed an *in vitro* micropropagation protocol as the only viable means of producing a conservation collection of Bancroft's symonanthus. The team was challenged with *in vitro* setbacks working on meristematic shoot cultures that declined in process. The protocol was modified and improved for a second wave of *in vitro* accessions from the other remaining specimen of symonanthus. As luck would have it, they salvaged male and female stock plants. Three years later, female micropropagated clones were artificially pollinated using pollen from micropropagated male clones resulting in the first crop of seed from symonanthus in 50 years.

A year later, field trials were set up utilizing the 2,400 tissue-cultured plants. Of these, 90 plants were selected as a stable population from which seed was collected two years later. Trials with symonanthus then moved to their germination requirements to ensure adequate populations of seed grown plants to replace the need for the more resource intensive *in vitro* approach. Fortunately, during this extended period of seed plant reintroduction and preservation, the *in vitro* protocols developed by Kings Park and Botanic Garden team provide a safety net for the preservation and conservation of *Symonanthus bancroftii*. This conservation research program involving the *in vitro* preservation and propagation of Bancroft's symonanthus, and its reintroduction, is a fine example of an integrated program that utilizes several strategies for species conservation.[56]

Cincinnati Zoo and Botanical Garden

The Center for Conservation and Research of Endangered Wildlife (CREW) at the Cincinnati Zoo & Botanical Garden cryopreservation program was profiled in Best Practice 10.2. In addition, they use their *in vitro* tissue culture not only to supply the cryopreservation program but also as a means to preserve threatened species that are intolerant of desiccation and freezing.

> Of all the types of material collected for conservation of diversity, DNA is undoubtedly the latest and likely the last. After traditional collections of genetic resources, DNA banks provide a new resource for the ready availability of DNA with great potential for the characterization and utilization of biodiversity.[57]

DNA collections may serve as the foundation for a wide array of botanical, conservation biology, and ecological research and study as well as a "resource for biotechnical applications."[58] What distinguishes DNA tissue samples from those described earlier for cryopreservation and *in vitro* preservation is a higher level of quality to ensure good DNA extraction. That same concern for quality applies to the associated documentation.

DNA collections are most commonly derived from leaf samples. The collection of these samples is usually combined with the collection of herbarium vouchers for that taxon. These joint samples must be well labeled to indicate their relationship, and the documentation must also make that linkage clear. Leaf samples (whole leaves or leaf fragments) are commonly stored in one or both of two different ways: (1) dried in silica gel and frozen or(2) placed in containers for storage in liquid nitrogen (LN). All samples should be young but fully formed leaves or pieces from these leaves.[59]

For the silica gel method, place samples in white paper, ULINE S-11485, labeled envelopes—known as "coin envelopes." These are then placed in sealed plastic bags (Ziploc®, etc.) with silica gel beads of a volume calibrated to dry the sample in no more than 48 hours to 30 percent relative humidity. The envelopes of dried samples are then packaged with a relative humidity indicator card and stored in freezers at –80° C. For samples to be stored in LN, they should be placed into labeled 8-milliliter, externally threaded cryovials with o-ring seals, labeled and wrapped in aluminum foil and then placed into cryogenic storage containers in liquid nitrogen. The processes described above must be carried out with some haste to ensure that the samples do not degrade. Hence, field collections are often processed for silica gel drying since LN is not usually available. Where LN preservation is preferred, germplasm may be transported to the LN storage facility where whole plants are propagated and grown to provide samples for DNA storage.[60]

I hope this brief synopsis of DNA banking and collections provides you with a sufficient introduction and orientation to the practice and technology to help you decide how this work might relate to the conservation research and collections programming at your institution. I recommend two very good sources of information here as well as in chapter 3: the Global Genome Biodiversity Network[61] and Gemeinholzer et al. (2010).[62] The Royal Botanic Garden, Kew curates a mind-boggling DNA and tissue bank of 60,000 samples (at the time of this writing) representing nearly all the families and more than half the genera of flowering plants. They make accessible a DNA tissue databank for, among other things, requesting samples.

Conservation Collections: Field Gene Bank

Species that produce few if any seeds or those that produce recalcitrant seeds may be grown in field gene banks. This type of conservation collection may be complementary to other gene bank collections. For example, both seed bank and tissue bank collections may require duplication, reestablishment, and other developmental requirements that call for them to be regrown as whole plants as part of a field gene bank. These conservation collections are grown in plantations or in large "nursery blocks" for efficient care, documentation, tracking, and genetic control. This technique is common in forestry, and those field gene banks are often referred to as seed orchards. There is an emerging use for a type of modified field gene bank called an *inter situ* field gene bank. These are established outside of botanical gardens in semiprotected locations in settings that provide a place where plants experience some natural climatic variation while still receiving supplemental care when necessary. These are often not far from a site where translocation or reintroduction of field bank species will occur.[63]

Before proceeding on the subject of field gene bank conservation collections, recall that there is both introductory and preservation information about them in chapter 5. The subject of best preservation protocols and practices for field gene bank conservation collections is one for botanical garden research and experimentation. For many species, trees in particular, preserving a broad base of genetic diversity would require an immense amount of space. Therefore, most field gene banks will concentrate on the growth of ecotypes, lines, or clones of species. In any case, choosing an appropriate agro-ecological site for field gene bank collections is of high priority. Varied collections may require multiple growing sites, a situation that I discourage unless conservation priorities and program funding allow for it. Some collections are best grown in open field space, some within protected spaces such as screenhouses and/or greenhouses, a situation dictated by ecotypic variation, gene pool preservation, limited hardiness, or other conditions.

Keep in mind that additional space may be needed for expanding the collection as it grows, for rotations related to soil health and pest avoidance, and for safety duplication. "When accessions from different eco-geographical origins are planted in one location, careful attention by the curatorial staff is required to monitor the reproductive phenology and seed production, and identify and transfer poorly adapted accessions to possible alternative sites, greenhouses, or *in vitro* culture to avoid genetic loss."[64] More specifically, inbreeding and out-crossing may require tracking and control; I will offer more on that later in the chapter. Also keep in mind when considering field sites that these collections are vulnerable to natural disasters such as fires and pest outbreaks.

Knowing how many accessions to establish in a field gene bank conservation collection depends on several variables: genetic diversity, need for characterization, cultural space, and safety duplication. We can start with the generalization that vegetatively reproduced species may be well represented by a smaller field gene bank collection. Seed reproduced species will require a larger representative population of accessions.[65] For specific recommendations on collecting for genetic diversity, refer to chapter 3. Regarding field gene banks specifically, I want to reiterate three basic Center for Plant Conservation generalizations on collecting genetic diversity: (1) collect across the population's spatial expanse including the center and edge, (2) collect from morphological maternal plants of different sizes, and (3) collect a range of seed appearance.[66] Such collections may need to be undertaken in multiple years for species that flower sporadically. Use recommendations 1 and 2 when collecting vegetative material for propagation and be sure to collect multiple samples from maternal plants (maternal lines).

Some field gene bank clonal lines may only be successfully propagated and preserved by grafting. In this case, rootstocks must be selected that will support the growth of the clonal scion without generating chimeric growth, suckering shoots, or triggering types of interference with the growth and/or genetic diversity of the preserved clone.

[Accessions] that require cross-pollination should be planted in groups by bloom date. In dioecious species, a suitable number of male/female plants should be planted. For self-incompatible species asexually propagated, the curator has to know which self-incompatibility (SI) system is presented by the species and the allelic combination in order to have a good field collection and to guarantee fruit or seeds formation.[67]

A principal curatorial goal for a field gene bank collection is to minimize genetic drift and artificial selection. A basic task toward meeting this goal is to rogue out all volunteer seedlings that appear in the planting blocks. Grow different accessions (from different populations) of the same species and different species of the same genera in widely separated blocks. If seed will be collected from these conspecifics and congenerics, establish a controlled pollination regimen, for example, bag flowers and hand pollinate or surround the breeding population with a netted cage. Further, consider pollen exchanges with other institutions preserving duplicate collections. Maintain maternal lines at stable populations, grow them in clearly identified subsections of accession planting blocks, and make sure they are clearly identified on labels and in documentation.[68]

Artificial selection forms the basis for plant breeding programs where conscious choices are made in keeping and discarding plants according to established criteria. Those same criteria may drive breeding programs among plants with desirable traits. However, artificial selection threatens the genetic diversity of conservation collections when it occurs unconsciously through the careless or biased activities of curators and their staffs. For instance, there is a natural proclivity for curators to select the most vigorous individuals within a conservation collection accession for replication. This may inadvertently eliminate some maternal lines from the collection, a form of artificial selection. It is important to maintain all the founding maternal lines within a field gene bank collection. As genetic analysis improves, genetic studies of field gene bank and other types of conservation collections will help determine genetic diversity with greater precision.

Field gene bank conservation collections will require propagation and regeneration. Standards for regeneration and propagation of field gene bank accessions must take the above protocols for genetic integrity into consideration. The best method to accomplish this is through vegetative propagation to maintain each maternal line (clone). There are many sources of propagation information; the FAO crop calendars are particularly useful for economic crops. However, this is also another area requiring research and experimentation. If it is necessary to propagate and regenerate out-crossing accessions by seed, then they must be isolated by netting or hand pollination. Whatever propagation method you choose, make sure the process and the plants involved are carefully documented.

Field gene bank conservation collections should be characterized and evaluated as part of a routine scrutiny and monitoring of these valuable resources. The curator, or their associates, will need to establish standards that can be incorporated into a checklist or evaluation form. From then on, I advise using the same individual(s) for this work in order to establish greater efficacy and continuity.

> Characterization is the description of plant germplasm, and a tool for the description and fingerprinting of the accessions, confirmation of their trueness to type, and identification of duplicates in a collection.[69]

Characterization involves a representative sample of an accession and documents noteworthy and heritable morphological, anatomical, phenological, hardiness, and other essential information that curators use to distinguish accessions and that researchers and others will find useful. Such work, perhaps in a less detailed and exacting way, is often a routine part of collections management and documentation conducted by curators as a means of adding research and education value to the collections. The characterization process may be prescribed through the use of international descriptor lists for phenotypic, physiological, morphological, and breeding system traits for the species of concern. DNA barcoding samples may be taken as part of the characterization process. How representative a characterization sample should be is based on its diversity. The Food and Agriculture Organization (FAO) recommends that there should be a minimum of three plants for diverse accessions and one to two for clonal plants. They also recommend that species prone to mutations, such as citrus, have characterizations done on an annual basis.[70]

Closely connected to characterization is evaluation. "Evaluation data on field genebank accessions should be obtained for traits of interest and in accordance with internationally used descriptor lists where available."[71] In many ways, an evaluation of field gene bank conservation collections is much like the collections monitoring and evaluation described in chapter 8. There is a common goal for all collections evaluations: an assessment of traits of interest, as described above. As you can see from the above quote, there are descriptor lists available to assist in and help standardize evaluations. For economic crops, these lists focus on commercial, food quality attributes. For reintroductions and restorations, the evaluations may focus on genetic, environmental, and reproductive fitness. Evaluations are more exacting and time consuming than characterizations and are often reserved for high-priority collections, in other words, those with specific and/or critical applications. Evaluations must be done with greater specificity and continuity than characterizations, and the data must be even more reliable. For example, "variations in the incidences of pests and diseases, the severity of abiotic stresses and the fluctuations in environmental and climatic factors in the field impact on the accuracy of data and should be mitigated through reasonably replicated, multi-locational, multi-season and multi-year evaluations."[72]

Best Practice 10.4: Field Gene Bank Conservation Collection

U.S. Department of Agriculture–Agricultural Research Service (USDA-ARS), U.S. National Plant Germplasm System
I couldn't resist profiling the USDA-ARS germplasm system again because they have such an extensive and comprehensive conservation program involving a broad range of conservation technologies. Although they conduct seed and tissue banking (cryopreservation and *in vitro*) programs, they may be best known for their field gene banks. From their webpage: "The U.S. National Plant Germplasm System (NPGS) is collaborative effort to safeguard the genetic diversity of agriculturally important plants." Their mission involves

- acquiring crop germplasm,
- conserving crop germplasm,
- evaluating and characterizing crop germplasm,
- documenting crop germplasm, and
- distributing crop germplasm.[73]

The NPGS is a network of 15 germplasm repositories, 9 with extensive field gene banks. I am most familiar with the NPGS facility in Corvallis, Oregon, which contains several field bank collections of commercially important trees, bushes, and other fruits. All of the field gene bank collections at the Corvallis facility will serve as useful models for the establishment, organization, preservation, documentation, and distribution of field gene bank accessions. I'm sure the others are of equal instruction. The United States National Arboretum and the North Central Regional Plant Introduction Station in Ames, Iowa, are part of this system, and both have noteworthy track records of partnership with the botanical garden community.

The Dawes Arboretum
The USDA NPGS germplasm collections are huge but are not the only model of successful field gene banking for botanical gardens. The Dawes Arboretum of Mentor, Ohio, curates a field gene bank of dawn redwood (*Metasequoia glyptostroboides*) that contains 500 trees representing 132 accessions—96 of them with documented wild origin. This is one of, if not the largest, representation of genetic diversity in dawn redwood outside of China, its country of origin. Of these accessions, there are also 29 named cultivars of horticultural interest. This field gene bank of dawn redwood is preserved on 8 acres within the Arboretum and is available for viewing by visitors.[74]

The Brenton Arboretum
The Breton Arboretum of Dallas Center, Iowa, was founded in 1997 and might be considered still in its infancy, but that hasn't stood in the way of its commitment to tree preservation and conservation. In 2004, Arboretum manager and director of horticulture Andy Schmitz started looking for a signature collection for the Arboretum that would also qualify for accreditation through the American Public Gardens Association, Plant Collections Network. Having always appreciated the Kentucky coffee tree (*Gymnocladus dioicus*) as a beautiful but underappreciated and underutilized native tree, Andy began extensive fieldwork that resulted in 111 geo-referenced accessions collected from 16 states containing a large portion of the gene pool of *Gymnocladus dioicus*.

Partnering with the USDA North Central Regional Plant Introduction Station in Ames, Iowa, this became a dual-purpose collection for both the Breton Arboretum and the USDA. Now there is an extensive germplasm collection of Kentucky coffee tree seed in the USDA system and a genetically diverse *ex situ* living collection at the Breton Arboretum.

Collections with Conservation Value

The Dawes Arboretum and the Brenton Arboretum collections profiled above are not traditional field gene bank collections like those curated at the USDA Germplasm Repositories. However, they are certainly comprehensive collections of conservation value given their genetic diversity, curatorial attention, and public display. Nowadays, most botanical gardens collect and curate collections of conservation value, in part because there has been a stronger emphasis put on the acquisition of collections of wild origin for just this purpose. However, limited collections of rare or endangered plants may serve many other research goals besides germplasm preservation and ecological restoration. These collections could be very useful for research in, among other things, the breeding systems of plants. Dr. Peter Raven concludes:

> We don't know about the breeding systems in most kinds of plants in the world, especially trees. Who's breeding trees, who's breeding tropical trees to know what kind of reproduction barriers there are? We're all befuddled by a dumb biological species concept that was invented by somebody's fantasy about birds in the 1930s and assume that plants all behave the same way, that they are all biological species and they are all interfertile within but not fertile without. But what do we really know? Anything we want to find out about those organisms must be discovered sooner rather than later because we are driving them to extinction.[75]

Many of the research and experimental opportunities described so far in this chapter also apply to collections with conservation value, albeit on a more limited basis. Their value may be expanded and amplified if networked as part of a larger metacollection. I can imagine rare and unusual accessions in limited numbers providing the missing link in helping to complete national and global *ex situ* metacollections. Not only that, but collections with conservation value also present a number of educational opportunities to create and amplify public awareness and appreciation for the conservation of biological diversity.

Cycad collections serve as a good example of botanical garden collections of increasing conservation value. The current circumstances of over-collection of wild plants for the horticultural trade, as well as habitat destruction and invasive insect species, make it imperative that cycads be preserved in botanical gardens. This need is underscored by "low seed germination, low seedling survival rates, and long generation times that exacerbate cycad decline as natural regeneration cannot keep up with losses in nature."[76] The need to preserve living cycads as opposed to relying on other types of *ex situ* banking technologies is based on the fact that *in vitro*, cryopreservation, and traditional seed banking technology is currently not effective for most cycads. The needs of security duplication as well as provenance surveys and integrated preservation research dictates that *ex situ* collections of cycads be shared among a number of botanical gardens around the globe as part of a cycad metacollection.

Metacollection: The combined holdings of a group of collections. For gardens, metacollections are envisioned as common resources held by separate institutions but stewarded collaboratively for research and conservation purposes. Networking multiple collections into a single metacollection increases potential coverage within a group, allows broader access to greater diversity, dilutes risk of loss, and can reduce maintenance costs. The American Public Gardens Association's Multisite Collections, BGCI's Global Conservation Consortia and the CPC National Collection are established examples of metacollections. Like any collection, a metacollection can be of any scope or taxonomic level.[77]

The above example of a shared conservation metacollection centered on an entire division of the plant kingdom, *Cycadophyta*, may be paired down to individual populations of a single species, such as *Quercus oglethorpensis*. This evergreen oak is endangered across its range in the southeast United States by land use changes, competition, and the chestnut blight fungus *Cryphonectria parasitica*. It survives in the face of these threats in a series of disjunct populations that may harbor unique genetic diversity or adaptive variation.[78] Since it produces recalcitrant seeds, it may not be easily gene banked and is currently preserved in a population-based metacollection of living plants as part of the American Public Gardens Association Nationally Accredited *Quercus* Multisite Collection™.

Seed and scions of *Quercus oglethorpensis* were collected from five counties and several populations with accompanying herbarium vouchers and propagated at the Morton Arboretum. Seedlings and grafted trees were then distributed to five other gardens: Chicago Botanic Garden, Starhill Forest and Arboretum, Holden Arboretum, Donald E. Davis Arboretum of Auburn University, and Moore Farms Botanical Garden. In addition, herbarium voucher specimens were also collected. Fortunately, this tree has proven quite adaptable with a large horticultural range that allows it to be successfully cultivated and preserved in locations outside of its natural range. These collections may now be studied and used in reintroduction programs.

For more information on collections of conservation value, networked collections, and metacollections, see the online information on this subject at Botanic Garden Conservation International, the American Public Gardens Association, and the Center for Plant Conservation.

Before moving on to the subject of species recovery programs in the next chapter, I want to leave the subject of *ex situ* conservation collections with some suggestions for areas of investigation and research that may have been overlooked in the above review or deserve reiteration. This is not an all-inclusive list, and I'm sure you may have already identified some areas of research interest based on your garden's mission and purpose.

- Preservation of genetic diversity during propagation, restocking, and duplication of conservation collections.
- Allied to the above: coordinated germplasm exchange and breeding to improve genetic diversity.
- Species-specific research on breeding biology of threatened species.
- Research on the syn- and autecologies of threatened species, including morphological differences between groups and phenotypic differences among populations.
- Research and investigations on sampling from single-source populations versus mixed-source populations.
- Research on genetic rescue effects for declining populations of collections.
- Studies on the length of seed viability and how it may be extended.
- Research on improving species-specific post-harvest handling and storage of seed, especially new protocols for recalcitrant and exceptional seeds.
- Research species-specific seed stratification and germination requirements, especially for trees.
- Identify gene bank specific pests and their natural histories; develop pest management protocols.
- Research on species-specific tissue-banking protocols.
- Research on managing genetic decline in tissue gene banks and field gene banks, such as genetic drift, mutation accumulation, and inbreeding depression, as well as some unique ones, such as artificial selection and outbreeding depression.
- Research and develop species-specific field sampling guidelines for capturing maximum genetic diversity.
- Conduct research on improved molecular and genomic studies on endangered species.
- Research ways to better assess the genetic diversity of wild-collected germplasm.
- Commit to the preservation and conservation of imperiled nonvascular plants and the research necessary to effectively preserve them.
- Investigate and trial new ways to coordinate both *ex situ* and *in situ* preservation and conservation work.
- Research and trialing for improved and specific horticulture practice for all conservation collections.

Textbox 10.1. Recommendations for *Ex Situ* Collections

Basic
- Define the purpose and scope of a conservation collection consistent with the garden mission and collections policy.
- Build on existing collections strengths; focus on threatened species.
- Join or build partnerships of integrated programs (ex *situ* and *in situ*).
- Begin with collections of conservation value.
- Precise and accurate documentation is critical, beginning with passport data.[1]

Intermediate
- Conduct a conservation assessment.
- Determine the best long-term preservation mechanism (whole plant, seed, tissue?).
- If appropriate, begin with a seed gene bank base collection.
- Curate first-generation, wild-collected germplasm from well-documented sources for genetic diversity; maternal lines are preferable.

Advanced
- Add an active collection to the seed gene bank for recovery and restoration.
- Preserve recalcitrant and difficult species in a tissue gene bank.

Notes

1. Botanic Gardens Conservation International, *Building Living Plant Collections to Support Conservation: A Guide for Public Gardens* (Richmond, Surrey, UK: BGCI, 2014), 2–3.

Notes

1. Paul Smith and Valerie Pence, "The Role of Botanic Gardens in Ex Situ Conservation," in *Plant Conservation Science and Practice*, ed. S. Blackmore and S. Oldfield (New York: Cambridge University Press, 2017), 102.

2. International Union for the Conservation of Nature and Natural Resources, Botanic Gardens Secretariat, *The Botanic Gardens Conservation Strategy* (London: IUCN, 1989), 21.

3. J. Gratzfeld, ed., *From Idea to Realisation: BGCI's Manual on Planning, Developing and Managing Botanic Gardens* (Richmond, Surrey, UK: Botanic Gardens Conservation International, 2016), 145.

4. Botanic Gardens Conservation International, *Building Living Plant Collections to Support Conservation: A Guide for Public Gardens* (Richmond, Surrey, UK: BGCI, 2014), 2–3.

5. American Public Gardens Association, Tree Gene Partnership Program, https://www.publicgardens.org/programs/plant-collections-network/tree-gene-conservation-partnership.

6. Pamela Allenstein, personal communication, October 2021.

7. Center for Plant Conservation, *CPC Best Plant Conservation Practices to Support Species Survival in the Wild* (Escondido, CA: Center for Plant Conservation, 2019), 3-3.

8. Center for Plant Conservation, *CPC Best Plant Conservation Practices*, 3-4.

9. Nigel Maxted and Shelagh Kell, "A Role for Botanic Gardens in CWR Conservation for Food Security," *BG Journal* 10, no. 2 (2013): 32.

10. Smith and Pence, "The Role of Botanic Gardens," 102.

11. J. G. Hawkes, "A Strategy for Seed Banking in Botanic Gardens," in *Botanic Gardens and the World Conservation Strategy*, ed. D. Bramwell et al. (London: Academic Press, 1987), 132.

12. Smith and Pence, "The Role of Botanic Gardens," 109–10.

13. See Center for Plant Conservation, "Rare Plants," https://saveplants.org/search/?category=plants/.

14. Botanic Gardens Conservation International, "Global Seed Conservation Challenge," https://www.bgci.org/our-work/plant-conservation/seed-conservation/global-seed-conservation-challenge/.

15. Center for Plant Conservation, *CPC Best Plant Conservation Practices*, 1-21.

16. Center for Plant Conservation, *CPC Best Plant Conservation Practices*, 1-22.

17. Center for Plant Conservation, *CPC Best Plant Conservation Practices*, 1-44.

18. Center for Plant Conservation, *CPC Best Plant Conservation Practices*, 1-44.

19. Center for Plant Conservation, *CPC Best Plant Conservation Practices*, 1-44.

20. Hawkes, "Seed Banking," 137.

21. Hawkes, "Seed Banking," 145.

22. Gunnar Øvstebø, Alex Twyford, and Tina Westerlund, "Propagation of Dry Habitat Fern Species Using Spore Collections from Historic Herbarium Specimens," *Sibbaldia* 9 (2011): 44.

23. Food and Agriculture Organization, *Genebank Standards for Plant Genetic Resources for Food and Agriculture*, rev. ed. (Rome, Italy: FAO, 2014), 140.

24. Food and Agriculture Organization, *Genebank Standards*, 140.

25. Food and Agriculture Organization, *Genebank Standards*, 118.

26. Smith and Pence, "The Role of Botanic Gardens," 113.

27. Center for Plant Conservation, *CPC Best Plant Conservation Practices*, 2-4.

28. B. M. Reed, F. Engelmann, M. E. Dulloo, and J. M. M. Engels, *Technical Guidelines for the Management of Field and In Vitro Germplasm Collections*, IPGRI Handbooks for Gene banks 7 (Rome, Italy: International Plant Genetic Resources Institute, 2004), 21.

29. Reed et al., *Technical Guidelines*, 23.

30. Food and Agriculture Organization, *Genebank Standards*, 118.

31. Center for Plant Conservation, *CPC Best Plant Conservation Practices*, 1-37.

32. Food and Agriculture Organization, *Genebank Standards*, 139.

33. Food and Agriculture Organization, *Genebank Standards*, 140.

34. Smith and Pence, "The Role of Botanic Gardens," 114.

35. Center for Plant Conservation, *CPC Best Plant Conservation Practices*, 2-14.

36. Center for Plant Conservation, *CPC Best Plant Conservation Practices*, 2-15.

37. Food and Agriculture Organization, *Genebank Standards*, 143.

38. Smith and Pence, "The Role of Botanic Gardens," 116.

39. Food and Agriculture Organization, *Genebank Standards*, 154.

40. Maria M. Jenderek and Barbara M. Reed, "Cryopreserved Storage of Clonal Germplasm in the USDA National Plant Germplasm System," *In Vitro Cellular & Developmental Biology—Plant* 4 (2017): 303.

41. Valerie Pence, "From Freezing to the Field: In Vitro Methods Assisting Plant Conservation," *BG Journal* 9, no. 1 (2012): 16.

42. U.S. Department of Agriculture–Agricultural Research Service, "Agricultural Genetic Resources Preservation Research: Fort Collins, CO," https://www.ars.usda .gov/plains-area/fort-collins-co/center-for-agricultural-resources-research/paagrpru/.

43. Jenderek and Reed, "Cryopreserved Storage," 299.

44. U.S. Department of Agriculture–Agricultural Research Service, "Agricultural Genetic Resources Preservation Research: Fort Collins, CO: Plants," https://www .ars.usda.gov/plains-area/fort-collins-co/center-for-agricultural-resources-research/ paagrpru/docs/plants/plant-science-at-the-national-laboratory-for-genetic-resources -preservation/.

45. *Wikipedia*, "Plant Tissue Culture," last updated May 17, 2021, https://en.wiki-pedia.org/wiki/Plant_tissue_culture.

46. Smith and Pence, "The Role of Botanic Gardens," 116–17.

47. Reed et al., *Technical Guidelines*, 21.

48. Reed et al., *Technical Guidelines*, 27.

49. Reed et al., *Technical Guidelines*, 29.

50. Reed et al., *Technical Guidelines*, 45.

51. Food and Agriculture Organization, *Genebank Standards*, 135.

52. Food and Agriculture Organization, *Genebank Standards*, 136.

53. Food and Agriculture Organization, *Genebank Standards*, 136–37.

54. Reed, *Technical Guidelines*, 136–37.

55. Center for Plant Conservation, *CPC Best Plant Conservation Practices*, 2-12.

56. Eric Bunn and Kingsley Dixon, "Botanic Gardens: Meeting the Restoration Challenge with Critically Endangered Plants: A Case History, *Symonanthus bancroftii* (Solanaceae)," *BG Journal* 6, no. 1 (2009).

57. M. C. de Vicente, "Collecting DNA for Conservation," in *Collecting Plant Genetic Diversity: Technical Guidelines—2011 Update*, ed. L. Guarino et al. (Rome, Italy: Bioversity International, 2011), 1.

58. de Vicente, "Collecting DNA," 1.

59. Vicki A. Funk et al., "Guidelines for Collecting Vouchers and Tissues Intended for Genomic Work (Smithsonian Institution): Botany Best Practices," *Biodiversity Data Journal* 5 (2017): 11–12.

60. Funk et al., "Guidelines for Collecting," 12–13.

61. Global Genome Biodiversity Network, "The GGBN Data Portal," http://www.ggbn.org/ggbn_portal/.

62. B. Gemeinholzer et al., "Organizing Specimen and Tissue Preservation in the Field for Subsequent Molecular Analyses," in *Manual on Field Recording Techniques and Protocols for All Taxa Biodiversity Inventories*, ed. J. Eymann et al., ABCTaxa 8 (n.p.: Belgian Development Cooperation, 2010), 129.

63. Center for Plant Conservation, *CPC Best Plant Conservation Practices*, 2-22.

64. Food and Agriculture Organization, *Genebank Standards*, 69.

65. Food and Agriculture Organization, *Genebank Standards*, 77.

66. Center for Plant Conservation, *CPC Best Plant Conservation Practices*, 2-24.

67. Food and Agriculture Organization, *Genebank Standards*, 79–80.

68. Center for Plant Conservation, *CPC Best Plant Conservation Practices*, 2-29.

69. Food and Agriculture Organization, *Genebank Standards*, 91.

70. Food and Agriculture Organization, *Genebank Standards*, 93.

71. Food and Agriculture Organization, *Genebank Standards*, 96.

72. Food and Agriculture Organization, *Genebank Standards*, 99.

73. U.S. Department of Agriculture–Agricultural Research Service, "U.S. National Plant Germplasm System," https://www.ars-grin.gov/npgs/.

74. American Public Gardens Association, "*Metasequoia glyptostroboides*," https://www.publicgardens.org/programs/plant-collections-network/collections-showcase/metasequoia-glyptostroboides.

75. P. Raven, "A Look at the Big Picture," *Public Garden* 12, no. 2 (1997): 10.

76. Botanic Gardens Conservation International, *Cycads: A Model Group for Ex Situ Plant Conservation* (Richmond, Surrey, UK: BGCI, 2015), 2.

77. M. Patrick Griffith et al., *Toward the Metacollection: Safeguarding Plant Diversity and Coordinating Conservation Collections* (San Marino, CA: Botanic Gardens Conservation International, 2019), 2.

78. Matthew S. Lobdell and Patrick G. Thompson, "*Ex-situ* Conservation of *Quercus oglethorpensis* in Living Collections of Arboreta and Botanical Gardens," in *Proceedings of Workshop on Gene Conservation of Tree Species—Banking on the Future* (Chicago: U.S. Department of Agriculture, 2016), 144.

CHAPTER ELEVEN

Species Recovery Programs and Ecological Restoration

Throughout the world, wild plants and their habitats are under increasing threat. Clearly, the conservation of habitats and species *in situ* must be seen to be preferable to *ex situ* measures, but the scale of habitat destruction in many regions has too often denied or reduced the option of extensive *in situ* conservation. Thus, the reintroduction of individual species into protected sites in the wild (i.e., native or seminatural habitats)—and, in some cases, the restoration or reconstruction of whole communities—will become essential measures to conserve threatened plants.[1]

> Species recovery refers to the procedures whereby species as a whole, or targeted populations of species that have become threatened, for example through loss of habitat, decrease in population size, or loss of genetic variability, are recovered to a state where they are able to maintain themselves without further human intervention.[2]

The principal steps to be taken for species under serious threat must be *ex situ* collection, threat control, and habitat management.[3] We may interpret these steps as the three pillars of species recovery. My primary focus in this chapter is primarily on the connection between *ex situ* conservation collections and species recovery interventions that depend on those collections because of its relevance to curatorial practice. Botanical gardens that engage in *in situ* and *ex situ* conservation programs will recognize the synergy in these complementary activities and may choose to integrate them as part of species recovery programs that involve plant reintroduction, reinforcement, translocation, and habitat restoration interventions. Species recovery programs will draw heavily upon the conservation/curatorial practices described in the previous chapters. What is not covered here are species management and conservation plans that gardens holding conservation reserves may need to formulate and implement. These specify active preservation monitoring and interventions to remove or mitigate factors that threaten species within these reserves. This subject was briefly addressed in chapter 9, and I urge you to seek out further details from other texts on the subject.

Plants from *ex situ* conservation collections may be used to reinforce—add to—declining and degraded wild populations. They may also be used in reintroduction schemes where the species has completely disappeared from a part of its range. Under more dire circumstances, accessions from *ex situ* conservation collections may be used to restore completely degraded or destroyed habitats or ecosystems. And finally, "in times of rapid climate change and transforming ecosystems, botanic gardens may also hold plant material of species that likely require translocation and introduction to climatically more suitable habitats, which will lead to new species combinations that have not occurred before."[4]

Clearly, *in situ* and *ex situ* programming should be integrated for a complementary and synergistic impact on species preservation and conservation. This presumes that when it comes to involving a botanical garden in any number of species recovery programs, serious thought will have already been given to this prospect at the point when the institution dedicated itself to conservation research. Using a recent example in the previous chapter on collections with conservation value, the gardens that chose to participate in the *ex situ* conservation of *Quercus oglethorpensis* expect that this germplasm will likely be used in an *in situ* program of reinforcement, reintroduction, translocation, or habitat restoration. We can surmise that the culminating purpose of most, if not all, *ex situ* conservation collections is their use *in situ* for species recovery (or crop plant reintroduction and/or breeding).

This is a good point to reiterate that the definition of *in situ* has expanded in recent years to include intermediary conditions that include *quasi in situ* (*ex situ* collections preserved in a seminatural state); *circa situm* (*ex situ* collections preserved within the species natural range); forest seed plots (forest tree seed plots established *in situ*); dynamic conservation units (active genetic management of forest plots); and plant micro-reserves (managing vegetation fragments of threatened species as small-scale reserves). These intermediary conservation sites may be recipients of *ex situ* collection via their progeny: F_1, F_2, F_3 . . . generations.

A species recovery program includes the rescue, documentation, and restoration of a particular species. Botanical gardens should partner with appropriate land management agencies and other conservation organizations that will ultimately assume long-term management responsibility. Choose a species of local conservation interest as a candidate for a species recovery program. These may be identified in national species recovery plans. At this time, the majority of species recovery plans are developed in Australia, Canada, China, Europe, New Zealand, South Africa, and the United States.

Because species recovery programs are costly and require a long-term commitment, I want to reiterate that botanical gardens work with government agencies, nongovernmental organizations (NGOs), and other botanical gardens currently developing and implementing species recovery plans. I would expect that these entities will also be working within the Global Strategy for Plant Conservation (GSPC) and the Aichi Biodiversity Targets of the Strategic Plan for Biodiversity. Joining a partnership that has already established and works within these useful frameworks will save a great deal of time and expense and establish a focus for your contribution to the overall recovery plan. Those gardens that intend to undertake their own species recovery project will need to develop a species recovery plan if one does not currently exist or is insufficiently detailed. There are three essential components to a recovery plan:

- An evaluation of the current status of the species, including a thorough analysis of the threats.
- The aims and objectives of the plan.
- The actions proposed.

For more on writing a species recovery plan, see BGCI and IABG's Species Recovery Manual.[5]

Whether a garden is creating a species recovery plan or following an existing one, there are some important curatorial questions associated with them:

- How many plants are required to establish a viable population?
- What plant life stage is most viable for reintroduction (e.g., seeds, plugs of young plants, etc.)?
- What threats to the reintroduced population are there?
- What special horticultural requirements are there to establish the new population?
- How will reintroductions be monitored?

- How will failures be assessed and replacements initiated?
- What are the responsibilities of each curatorial staff member?
- What are the documentation requirements and procedures?[6]

These are a sampling of the kinds of concerns and issues that curators will need to address as part of any species recovery program. These issues and concerns are compounded for multispecies recovery plans and programs; the programs may be shared and coordinated among gardens each having a single *ex situ* species among several needed for use in the recovery effort. In multispecies recovery programs, extra diligence is needed to ensure the viability of individual species recovery.[7] When considering your institutional conservation research and collection priorities, give serious consideration to an exceptional habitat (e.g., serpentine outcrops, barrens, balds, etc.) or plant community for multispecies *ex situ* conservation collections and *in situ* species recovery programs. Multispecies recovery programs may have greater efficacy and success managed by a single institution, although the expense will certainly be greater.

No matter the type of species recovery program you are involved in, a recognition and an understanding of critical habitat for the species in question is imperative. Curators should come to the recovery effort armed with this knowledge in as much as it is also important to the *ex situ* establishment and preservation of the target species. This information also becomes a major part of any species recovery plan. In addition to the BGCI *and IABG's Species Recovery Manual*, another valuable resource is the International Union for Conservation of Nature and Natural Resources (IUCN) *Strategic Planning for Species Conservation* handbook.[8] This reference pertains to both animals and plants but is still a useful overall strategic planning guide for recovery and conservation.

A concern for acquiring and preserving genetic diversity within collections has been expressed throughout this text. When it comes to species recovery programs, the same concern applies in reverse: the reintroduction of genetic diversity back into the wild. Interestingly, we are still faced with a similar set of questions: how many individuals and how many populations are sufficient for successful species recovery? Unfortunately, a recent survey of species recovery projects shows a distinct lack of concern for genetic diversity in the recovery of species in the wild.[9] Clearly there is no point in undertaking expensive species recovery interventions unless the species can be established in large and diverse enough populations to persist and adapt. This underscores the importance of having an adequate *ex situ* conservation collection available for recovery in terms of both numbers of plants and their genetic diversity. It should be clear that the use of *ex situ* propagules and plants is common for species recovery interventions.

U.S. federal agencies charged with administering the Endangered Species Act have adopted a three Rs approach for species recovery:

- Representation: Representation requires the protection of populations across the full range of ecological settings of a species' range.
- Resiliency: Local populations of a species are large enough, have sufficient genetic variation, and are sufficiently mixed with respect to the age and sex of individuals to persist in the face of periodic threats such as drought, wildfire, and disease.
- Redundancy: Establish multiple populations to reduce the risk of extinction and increase viability.[10]

The most common forms of *ex situ* material used for species recovery interventions are seeds and whole plants. Whole plants are propagated and grown to a prescribed size from conservation collections of seed; *in vitro* or cryopreservation tissue collections; or field gene bank supplied seeds, cuttings, or divisions. The *BGCI and IABG's Species Recovery Manual* points out the needs and benefits of *ex situ* seed collections available for species recovery:

- Knowing the amount and quality of seed available
- Time and ability to confirm identification from a voucher specimen
- Having seed available in seasons or years when low or no wild seed is produced
- The ability to store extra after seasons of high wild seed production
- Seed collected for recovery should be of high quality and purity (percentage of target species' seed)
- Be of known provenance (location of origin) and contain sufficient genetic variation.[11]

Finally, to help you identify and justify your target species for recovery, the Center for Plant Conservation (CPC) has a convenient checklist for use in justifying species reintroductions, reinforcements, and translocations:

Species is extinct in the wild, OR;
There are few, small, and declining populations; AND
Alternative management options have been deemed ineffective; AND
Threats have been identified; AND

Threats from habitat destruction, invasive species, land conversion and/or climate change are imminent and uncontrollable.

There is also a companion checklist of criteria to indicate when species introductions are inappropriate until conditions change:

Reintroduction will undermine the imperative to protect existing sites.

Previous tests indicate that it has not been possible to propagate plants or germinate seeds.

High-quality, diverse source material is not available.

Existing threats have not been minimized or managed.

The reintroduced species may negatively impact species in the site.

The reintroduced taxon may have other threatened or endangered species or conflict with their management.

Suitable habitat is not available or understood.[12]

Both of these lists indicate that the species recovery sites must have some form of protection, which, among other things, could simply be isolation. Also, no matter which species recovery intervention is used, the recipient site is certain to require some preparation before the introductions are planted. Controlling biological competition is of major concern, particularly weeds and invasive species. Other issues that must be addressed include the following:

- Are pollinators known and available?
- Will plants need protection from herbivory?
- Have other threats been reduced or eliminated?
- What is the experimental design?
- What aftercare will be needed?
- How will monitoring be accomplished?
- Are local people involved?[13]

All forms of species recovery interventions with *ex situ* collections will likely require some experimentation. Unless you are following a proven protocol (which is unlikely for a given species), consider designing the entire project as an experiment. In this case, project team members will need to represent all the individual facets of the experiment, such as someone familiar with experimental design. Here are some research and experimental ideas to consider from the CPC for a species recovery plan built around reintroductions or translocations; the first is this: how can this project be designed to address unknowns? Some others include the following:

- Is this project an appropriate venue for testing ecological theory related to species recovery, such as founder effects, competition, and other aspects of population dynamics?
- Can you use the reintroductions as a cohort to examine natural variation, mortality, and recruitment against ecological factors?
- Will the reintroduction test key habitat gradients such as light, moisture, temperature, and elevation?
- Will traits be monitored, and how will they be analyzed?
- How will the reintroduction further our knowledge of key principles related to rare species ability to cope with climate change?[14]

See a larger list of possibilities in the *CPC Best Plant Conservation Practices to Support Species Survival in the Wild* manual on page 4-11. Above all else, be sure to have your species recovery plan reviewed by experienced professionals.

New species recovery projects are sure to require a number of permissions, permits, and other legal documents. These requirements, along with your entire project, may be expedited with local participation and support, particularly that of indigenous groups. Facilitating these contacts will involve announcements, meetings, and updates. Involving local and indigenous people in the real aspects of the project may have significant payoffs in the way of project monitoring and security, just to name two. Make it clear to everyone involved in the recovery that success may not be achieved for several years; expectation management will be important for maintaining commitment. The same goes for funding and time lines for funding; budgets need to be carefully calibrated to meet the long time lines for species recovery projects.

In consideration of the challenges to species introductions presented by the recipient site, give careful consideration to the factors that have negatively impacted the native flora. In addition to mitigating those as part of site preparation, also consider various acclimatizing and toning regimens for the transplants to help condition them for the new habitat. Certainly, the new plants must be weaned off nursery practices and conditions that promote the kind of growth and physiological condition that the *in situ* site will not support. For example, progressive cutbacks on nutrition and irrigation may be necessary to harden off the introductions.

Introductions that involve new recipient sites, such as translocations, should be assessed for environmental consistencies with sites currently supporting existing, healthy populations. Those sites should also be evaluated

for suitability under various climate change scenarios. Consider translocations to multiple sites as a fail-safe strategy. There is a very helpful set of
tables for evaluating potential restoration sites in the CPC *Best Plant Conservation Practices* manual, pages 4–18 and 4–19.

Once a site has been chosen, it is time to turn your attention to the
specific qualifications of the *ex situ* source material for introduction. For
instance, there are several factors to consider when it comes to choosing
what will be the "founder population" of introductions. First, you will want
to choose a genetically diverse founding population to optimize the chances
for success, especially for recipient populations that show signs of inbreeding depression. Within that parameter and if available, choose from *ex situ*
populations derived from the population being augmented. Within that
same guideline, make equal choices from the maternal lines of introduced
seeds or plants. A familiarity with the breeding system of the species under
consideration is important such that if, for example, the species is an obligate out-crosser and is locally adapted to the site, then mixing populations
may cause outbreeding depression. Also, be conscious of a tendency toward
artificial selection during this process, especially if you find yourself choosing
plants that stand out as "high performers" in the *ex situ* environment. Again,
concentrating on maternal lines equally will help avoid artificial selections.[15]
Then, planting logistics and demographics will need to be clearly mapped
out and marked.

There are several considerations that enter into the choice of the right
propagule for introduction, be they seeds, rooted cuttings, or container
plants, for example. For populations of annual and short-lived species, seeds
are commonly used for reinforcement or reintroduction. Keep in mind that
seed germination in the wild is usually quite low (CPC estimates 1–10 percent) requiring a large number of seeds to generate the necessary population
of new plants. Aftercare will involve preserving the seedlings with supplemental irrigation and protection from grazers, among other things. Recipient populations of herbaceous perennials are best augmented with whole,
nursery grown plants as close to maturity as possible. They, too, will likely
need supplemental irrigation and protection from biological competitors,
including weed control. For both herbaceous and woody long-lived plants,
consider introducing plants of various ages and life stages for a more diverse
population structure, especially if you see microsite differences among the
recipient population.[16] For tree introductions, use seedlings grown in deep
rooted band pots, plant at the beginning or during the rainy and/or cool
season, and protect the young trees from biological competitors—they may
require sleeves to protect them from browsing animals.

Like so many of the previous steps and decisions, deciding on what form of *ex situ* reinforcement material will work best may also require experimentation. In an unusual example, a species recovery reinforcement project for *Magnolia longipeduculata* in southern China trialed the use of seed grown versus grafted stock (on *M. kwangtunensis* rootstock). The seed grown and grafted plants were trialed at *inter situ* sites where it was shown that the grafted plants developed with greater vigor and adaptability.[17] Don't forget about crucial symbiont organisms for species that require them. A means of incorporating them in the new area along with the new plant material must be devised and implemented. Finally, count on an extended interval of horticultural care and successive replanting for introduced plants before they reach a point of established stability.

No matter which type of material is used for species recovery interventions, be sure to retain the original *ex situ* founding conservation collection. This requires you to duplicate collections in anticipation of their use *in situ*. These may be duplicate collections of seeds (F_1 generations) or vegetatively propagated clones. These duplicates must be well documented to make sure that lineages are clearly maintained and that the duplicates are genetically diverse.

As I pointed out at the beginning of this chapter, common forms of intervention that draw from *ex situ* conservation collections are reinforcement, reintroduction, and translocation. The IUCN defines species reinforcement as

> the intentional movement and release of an organism into an existing population of conspecifics, and involves reintroduction of plants, propagules or seeds into pre-existing habitat or populations in an area that is currently known to contain the taxon, i.e., within the indigenous range of the species. This approach is likely to be appropriate for species or populations with small or declining populations or ranges and/or high probabilities of extinction.[18]

The goal of reinforcement is to bolster a population(s) using genetically diverse and, if possible, similarly adapted propagules or plants so that it remains viable and self-supporting. A focus on genetic diversity and the introduction of new alleles into a small target population suffering from genetic depletion is a process known as "genetic rescue."[19]

> Botanic gardens are uniquely placed to conduct successful population reinforcement action, especially for plants of urgent conservation concern. Based on their expertise to prioritise species, collect propagules in the wild, grow

in horticultural trials and establish sizable numbers of individuals, they are equipped with the vital fundamentals to reinforce wild populations and monitor their survival over time.[20]

Population reintroduction involves planting and establishing a population of plants in a segment of its natural range where it has disappeared, or to establish an entirely new population in its former range after it has become extinct in the wild. Botanical gardens have played an important role in reintroductions using *ex situ* collections of plants that had become extinct in the wild and are now recovering due to successful reintroductions. Two iconic examples are the cycad *Encephalartos woodii* in South Africa and the leguminous tree *Sophora toromiro* on Easter Island.

Species translocations are defined in the *BGCI and IABG's Species Recovery Manual* as "[t]he deliberate movement and release of a living organism from one location to another with the purpose of improving its conservation status."[21] Unlike reinforcements and reintroductions, species translocations will likely require more extensive trialing in order to achieve success. Informed assumptions must be made as to the environmental factors necessary for the success of a translocated species, but these assumptions must be carefully examined and that usually takes place through trials in carefully selected test sites, such as the intermediary *inter situ* sites described earlier.

Assisted Migration

On the subject of species translocations, an emerging and controversial practice at the time of this writing is assisted or stewarded migration. The impact of climate change will be devastating for species incapable of migrating to more suitable habitats. Many species of plants fall into this category simply because they cannot disperse and migrate quickly enough. This includes plants at the boundaries of habitable ranges (e.g., alpine species) that will disappear.

For species with very specific habitat needs or ranges limited by physical barriers, such as fragmentation or geographic features, this may mean that the entire species could be at risk of extinction or extirpation due to climate change. In the case of widespread species, complete loss of the species may not be a risk; however, climate change may result in large-scale mortality and population extirpation due to maladaptation of populations.[22]

Assisted migration, when plants are moved to new habitats outside their historic range, is one possible solution to this problem. Studies and programs involving reciprocal, successful transplants of trees and other types of plants over long distances among agencies and institutions, such as the U.S. Forest Service, as well as between botanical gardens, have shown that assisted migration is possible to the benefit of the plants involved. In addition to establishing founder populations in areas of predictably suitable habitat, less widespread migrations may be implemented to bring genetic diversity and climate adaptability to at-risk populations. The U.S. Forest Service recognizes more than one type of assisted migration, each with their own risks, ecological implications, and policy considerations:

- Assisted population migration (also assisted genetic migration or assisted gene flow)—moving seed sources or populations to new locations within the historical species range
- Assisted range expansion—moving seed sources or populations from their current range to suitable areas just beyond the historical species range, facilitating or mimicking natural dispersal
- Assisted species migration (also species rescue, managed relocation, or assisted long-distance migration)—moving seed sources or populations to a location far outside the historical species range, beyond locations accessible by natural dispersal.[23]

A challenge in planning and carrying out assisted migration, be it any of the above three types, is that the impact of climate change on species distribution is neither straightforward nor entirely predictable. For example, a study of the impact of several climate change scenarios on the distribution of a selection of palms native to the southeast United States revealed significant variability. For two of the five species in the study, their range of suitable habitat will extend further north as the climate changes, which is a seemingly predictable scenario. But for the remaining three, their range will become more restricted showing no extension of suitable habitat further north or in any direction. Climate change factors other than the expected warmer temperatures come into play to limit the natural ranges of these other palms: annual rainfall amounts and seasonal differences, changes in diurnal temperatures, and such.[24] This condition underscores some of the difficulties of predicting suitable future habitats for assisted migration. Fortunately, there are sophisticated models to assist in making these predictions with increasing accuracy.

My reasons for including the subject of assisted migration in this chapter is that a primary purpose of this type of species translocation is recovery and conservation, although it is not the only purpose. Any intent to get involved in assisted migration programs must be given serious examination that shows this strategy to be a logical one. As I mentioned earlier about the challenge of misguided assumptions, assisted migration has other risks, including the following:

- Introducing a species that becomes invasive, thereby threatening other species.
- A vector for insect and other pests that also threaten other species.
- Hybridization with other species that dilutes the collective gene pool.[25]

These risks have been evidenced among botanical gardens that have shared a great many plants among themselves through the *Index Seminum* and other exchange programs. A number of risk assessment and management frameworks have been brought forward by advocates of assisted migration that will help minimize, but probably not eliminate, these risks. On the flip side, there are some distinct opportunities with assisted migration:

- Assisted species migration (species rescue) can help "lifeboat" a species or population that is at critical risk
- Assisted migration can help maintain crucial ecosystem functions (wildlife habitat, carbon sequestration, etc.), particularly when local species are already declining or are anticipated to decline in the future
- Assisted migration can help ensure that a species occurs in many redundant locations or across a range of conditions, which helps reduce risk from uncertain climate impacts
- Assisted migration can help populations and species move across ecological barriers in fragmented landscapes[26]

Botanical gardens planning an assisted migration species recovery intervention should prioritize rare species and those given highly threatened IUCN designations, those with limited genetic variation, and/or those with low dispersal rates. From a population standpoint, priority might also be given to small and specialized populations, those in fragmented landscapes, and populations on the trailing edge of their range. These priorities may also be tied to established collection priorities, such as the earlier example of gardens involved in the American Public Gardens Association (APGA)

Nationally Accredited *Quercus* Multisite Collection™ having a preexisting collection policy focus on oaks. Botanical gardens have been dabbling with assisted migration for most of their histories by collecting taxa from around the world and testing their suitability for growth and display on their home ground.

A formalized approach to assisted migration at botanical gardens could take the form of "chaperoned" assisted migration. In this scheme, proposed by Adam Smith and Matthew Albrecht of the Missouri Botanic Garden and Abby Hird of Botanic Gardens Conservation International (BGCI), gardens will serve as way points for transferred species; I think of their scheme as a form of *ex situ* field gene bank. Here is what the concept entails:

- Moving species outside of their historic distributions;
- Growing species in regularly managed *ex situ* settings like those provided by botanical gardens;
- Moving species within their potential dispersal envelopes and evolutionary/ecological context;
- Curating species as separate wild-collected specimens;
- Screening species on a regular basis for invasiveness, pests, diseases, and hybridization; and
- Ensuring species' survival as climate changes.[27]

Under this arrangement, these conservation collections may be curated, as I indicated earlier, much like field gene banks with all the preservation and documentation advantages. The risks that I listed above may be more easily and effectively mitigated in the care of the curatorial staff of the host garden. Also like a field gene bank, this *ex situ* collection has a greater purpose than simply taking up residence at a given botanical garden. As the climate changes, accessions and/or their progeny may be transferred to botanical gardens in more suitable locations. I would also suggest that propagules be trialed *in situ* for adaptability and propensity to stabilize and naturalize in the projected new habitat.

Currently, the largest and most comprehensive assisted migration projects in North America are conducted by both the U.S. and Canadian forest services toward the preservation of commercially and ecologically important forest tree species. Involving both private and public agencies, nonprofit organizations, and many individuals, Torreya Guardians is an unusual assisted migration project focused on the rare conifer *Torreya taxifolia*. Sometimes known as Florida nutmeg, *T. taxifolia* became one of the first federally

listed endangered plant species in the United States in 1984, and the IUCN
has listed the species as critically endangered since 1998. In 2010, it was
determined that a new species of *Fusarium* blight had killed most of the
adult trees within an already tiny population on the Florida-Georgia border.[28]
Assisted migration efforts are underway to establish populations of torreya
north of their natural range in North Carolina and other states.[29]

A common component of all the species recovery programs described here
is a commitment to the successful establishment of the introduced taxa and
an increase in the overall health of the wild population. This will be articulated as part of a monitoring plan that must be developed by and shared with
all the responsible parties. At this point, botanical garden personnel may
remain involved in the project to implement and oversee the horticultural
practices necessary to establish the new plants. Monitoring responsibilities
may also be shared with plant ecologists, population biologists, and others
focused on the various aspects of intraspecies recruitment, synecology, plant
breeding, and population demographics (of particular importance to species
recovery projects). Monitoring plans should contain benchmarks for success
based on plant survival, reproduction, dispersal, and recruitment.[30] For more
detailed information on monitoring and project documentation, refer to the
CPC Best Plant Conservation Practices to Support Species Survival in the Wild,
sections 4 and 5.

Best Practice 11.1: Species Recovery

The following are three "integrated" species recovery projects that involved
research and partnerships.

Bok Tower Gardens

Bok Tower Gardens in Lake Wales, Florida, is a holder of several accessions
in the Center for Plant Conservation's National Collection, including *Conradina glabra* (Apalachicola Rosemary). This plant is critically endangered
and the few surviving populations of *C. glabra* grow on land owned by a paper
company. Bok Tower Garden has preserved the plant in a field gene bank
collection and translocated it to a new site only a few kilometers away within
its historic range on Nature Conservancy land.

Fortunately, *Conradina glabra* proved remarkably easy to propagate. Shoot-
tip cuttings root readily, and a collection of 48 clones was established at Bok
Tower Gardens. From these, more plants were readily produced and 1,300
individuals were made available for reintroduction. The Nature Conservancy

agreed to receive the plants from Bok Tower Gardens, and establish and care for them to ensure their survival as part of their overall restoration plan for the site.

The plants more than doubled in size during the first summer. After the first flowering, only nine seedlings were found, but by the second season, many hundreds had appeared. This robust natural reproduction is a reassuring signal that the plants have established successfully.[31]

Native Plant Trust, Garden in the Woods
Robbins' cinquefoil (*Potentilla robbinsiana*) is a critically endangered member of a rare flora: the alpine flora of the few New England mountains in the eastern United States tall enough to support it. A half-hectare site on top of Mount Washington in New Hampshire is home to 95 percent of the world's population of Robbins' cinquefoil. While efforts were being made to stabilize that population by the staffs of the U.S. Fish and Wildlife Service, White Mountain National Forest, and members of the Appalachian Mountain Club, the Native Plant Trust developed protocols for germinating seed and propagating the plants. Transplants were used to reinforce the two existing populations and to establish satellite subpopulations at four additional sites. Today, more than 14,000 plants inhabit the original site, with 300 others reproducing at a site at Franconia Notch. The species was officially delisted as an endangered species in 2002.[32]

Atlanta Botanical Garden
At the time of this writing, the Atlanta Botanical Garden curates the world's largest collection of orchids. Consistent with such a comprehensive collection is a concomitant effort at orchid research and conservation. A good example is the cigar orchid, *Cyrtopodium punctatum*, a native of south Florida where it has been nearly extirpated by habitat loss and collecting.

> In 2007, the Garden was asked by the state of Florida to assist in recovery efforts of the Cigar Orchid within the Fakahatchee Strand Preserve State Park. At the time, fewer than 20 orchids were growing in the park. In 2009, the first orchid capsule was harvested and sent to the tissue culture lab where staff and dedicated volunteers propagated plants for recovery. Since then, the recovery team has established nearly 1,000 new Cigar Orchids into the preserve where it conducts research to monitor survival and map the project's success.[33]

Textbox 11.1. Recommendations for Species Recovery

Basic
- Gardens preserve and curate *ex situ* conservation collections for species recovery.
- Choose a species of local conservation interest identified in a national species recovery plan.
- Partner with land management agencies and other conservation organizations that will ultimately assume long-term management responsibility.
- Projects involve species reinforcement.
- Aim for infraspecific diversity—functional, phylogenetic, and phenological—in species reinforcement and other recovery interventions.

Intermediate
- Plan and conduct an independent species recovery project based on an existing species recovery plan.
- The garden's species recovery plan targets an exceptional habitat (e.g., serpentine outcrop, barrens, balds, etc.) or plant community.
- Projects involve species reinforcement and reintroduction.

Advanced
- Projects involve reintroductions, translocations, and assisted migration.
- Species recovery programs involve multispecies *ex situ* conservation collections and *in situ* species recovery programs.
- Projects incorporate an experimental design as part of species recovery research.

Ecological Restoration

"Ecological restoration is the process of assisting the recovery of an eco-system that has been degraded, damaged, or destroyed."[34]

Botanical gardens are most often involved in ecological restoration as the species recovery partner that supplies important plants as part of species reintroductions or translocations. However, many gardens are engaged in ecological restoration projects that are necessary on their home ground or on newly acquired land. Also, with their growing experience in species recovery, it's only natural that botanical gardens would find scaling up to plant community restorations and from there to habitat and ecosystem restoration, an obvious extension of their experience and expertise. In any case, botanic gardens are well suited to carry out ecological restoration in that they have the necessary expertise with both the practitioners and researchers within the same organization. They also possess the infrastructure and, in some cases, resources that are necessary. Their structure, personnel, and mission facilitates cutting-edge, interdisciplinary scientific research to solve applied problems in restoration.

> Botanic gardens in the Midwestern United States have a long history of research and practice for prairie restoration. Curtis Prairie, the earliest documented prairie restoration, was established in 1935 on post-agricultural land at the University of Wisconsin Arboretum using plant materials gleaned from intact remnant prairies (Cottam and Wilson, 1966). In 1962 the Schulenberg Prairie was initiated at the Morton Arboretum in Lisle, Illinois, by painstakingly planting individual forb and grass plants (Bowles et al., 2012). Many of the restoration and management methods used today to restore diverse prairies, such as prescribed fire and soil inoculation, were developed though research and management at botanic gardens (Cottam and Wilson, 1966; Bowles et al., 2012). By sustaining restorations far beyond the time of any one researcher or practitioner, botanic gardens have helped to write the history of what it means to restore a prairie.[35]

Over the years, botanical gardens have engaged in a range of restoration projects of various sizes and technical challenges on lands in their care. This work has informed best practices around important elements of this work such as the use of native plants and the particular requirements of individual habitats. The prairie projects highlighted here serve as perfect examples. Many gardens have branched out from the restoration requirements of their own properties to the needs of the surrounding community and bioregion. Nearby cloud forest restoration has become a focus of the Francisco Javier Clavijero Botanic Garden in Veracruz, Mexico, and regional subtropical forest restoration has been taken up by the Kadoorie Farm and Botanic Garden in Hong Kong. Botanical gardens with a global programmatic reach are working on habitat restoration in far-flung locations, such as the Missouri Botanic Garden's work in Madagascar, a global biodiversity hotspot under tremendous land use pressure.[36]

Any botanical garden intending to become involved in ecological restoration should join the Ecological Restoration Alliance of Botanical Gardens (ERA),[37] first mentioned in chapter 9. The ERA has the following goals:

1. Work with local partners to set up, maintain, and document a series of long-term sustainable exemplar restoration projects in diverse biophysical, political, and cultural contexts around the globe that provide training and demonstrate the value of a carefully designed, science-driven approach to sustainable ecological restoration.
2. Improve the quality and volume of science-based ecological restoration practice by deploying scientific and horticultural skills to applied work on the ground.
3. Conduct ecological restoration research, to develop an enhanced knowledge base for restoration and identify and inform best practice.
4. Disseminate research and lessons learned from projects.
5. Build expertise and restoration capacity through collaborations between botanic gardens, large and small, as well as with partners in local communities, professional societies, academia, industry, government, NGOs, and international bodies.

The networking that is possible and the resources that are available through the ERA are invaluable, particularly to botanical gardens just beginning in ecological restoration.

It is important to understand that ecological recovery is not completed by anyone or any agency involved in ecological restoration; it is a matter of establishing the right conditions for the natural community to successfully recover. Thinking of it this way, restoration may be as straightforward as removing an invasive species or as complex as the wholesale alteration of a site involving topography, hydrology, and the entire biological community. In this process, there is often a great deal of debate about the historical context for the restoration; in other words, to what state or condition will the site be restored? This is the wrong concern and the wrong question. Degraded, damaged, and destroyed sites should be restored to a "historic trajectory" that will be governed by current conditions. An important factor in determining the historical trajectory is the impact of climate change. Once the restoration work has been done, it may still take a long period of time for that site, and the natural community it supports, to reach full recovery.[38]

The history of botanical garden involvement in ecological restoration could be said to have begun with the Curtis Prairie at the University of

Wisconsin Arboretum in 1935. Other noteworthy prairie restorations at botanical gardens are the Schulenberg Prairie at the Morton Arboretum and the Suzanne S. Dixon Prairie at the Chicago Botanic Garden. The extent to which each of these may be characterized as a "restoration" or a "creation" is debatable. They occupy a space that was once a small part of the great eastern tall grass prairie, an iconic landscape of the Midwest and eastern Great Plains at the time of European colonization. Through the research and applied practice that went into these projects, much has been learned about the restoration and management of prairies, including prescribed burns and soil inoculations, to name only two. At the Dixon Prairie, the appropriate substrate needed to be engineered on the site with raw materials.[39] All three of these prairies required painstaking plant translocations and reintroductions extending through multiple phases of transplanting and replacement. During these early phases of plant introduction, and continuing to this day, there are regular encroachments by invasive species that must be removed.

Ecological restoration is founded upon the following guiding principles adapted from the IUCN's *Biodiversity Guidelines for Forest Landscape Restoration Opportunities Assessments:*

- *Restore functionality*—Restore the functionality of a landscape, making it better able to provide a rich habitat, prevent erosion and flooding, and withstand the impacts of climate change and other disturbances.
- *Focus on landscapes*—Consider and restore entire landscapes [whenever possible] as opposed to individual sites. This typically entails balancing a mosaic of interdependent land uses, which include but are not limited to: agriculture, protected areas, agroforestry systems, well managed planted forests, ecological corridors, riparian plantings and areas set aside for natural regeneration.
- *Allow for multiple benefits*—Aim to generate a suite of ecosystem goods and services by intelligently and appropriately introducing trees, shrubs, and other plants within the landscape. This may involve planting trees on agricultural land to enhance food production, reduce erosion, provide shade and produce firewood, or trees may be planted to create a closed-canopy forest that sequesters large amounts of carbon, protects downstream water supplies and provides rich wildlife habitat.
- *Leverage a suite of strategies*—Consider the wide range of eligible technical strategies—from natural regeneration to planting—for restoring landscapes.
- *Involve stakeholders*—Actively engage local stakeholders in deciding restoration goals, implementation methods and trade-offs. Restoration

processes must respect their rights to land and resources, align with their land management practices and provide them with benefits.

- *Tailor strategies to local conditions*—Adapt restoration strategies to local social, economic and ecological contexts; there is no "one size fits all."
- *Avoid further reduction of natural cover or other natural ecosystems*—Address ongoing loss and aim to prevent further conversion of ecosystems.
- *Adaptively manage*—Be prepared to adjust a restoration strategy over time as environmental conditions, knowledge and societal values change. Leverage continuous monitoring and learning, and make adjustments as restoration progresses.[40]

As with any project, the first step is to formulate a plan, and ecological restoration, without exception, must be carefully planned before any work begins. First of all, those working on an ecosystem that is targeted for restoration should have located and identified a relevant, intact "reference ecosystem" to serve as a model for emulation for their project. For example, if you have targeted a damaged montane chaparral ecosystem in central California for restoration, you should be able to locate a nearby, reasonably intact, and fully functional montane chaparral ecosystem to serve as a useful reference. Such sites are commonly referred to as "reference ecosystems," or simply the "reference."[41]

> The problem with a simple reference is that it represents a single state or expression of ecosystem attributes. The reference that is selected could have been manifested as any one of many potential states that fall within the historic range of variation of that ecosystem. The reference reflects a particular combination of stochastic events that occurred during ecosystem development.[42]

Taking instruction from this quote, when you select a reference site at a particular stage of ecological development (trajectory), your restoration project may settle into a slightly different state (ideally) along that same trajectory. This is to be expected and isn't necessarily an indication of failure. It is unreasonable to expect your restoration project to fulfill a well-developed expression of, among other things, biological diversity when you've completed your work. What you should expect is to achieve a preliminary status along the trajectory of development that will ultimately attain the desired condition and functionality. Human impacts are also a concern to keep in mind; you will need to do your best to discern these in your reference and then compensate for them. Useful sources from the Society for Ecological Restoration for helping to describe and identify a reference ecosystem are

- ecological descriptions, species lists, and maps of the project site prior to damage;
- historical and recent aerial and ground-level photographs;
- remnants of the site to be restore, indicating previous physical conditions and biota;
- ecological descriptions and species lists of similar intact ecosystems;
- herbarium and museum specimens;
- historical accounts and oral histories by individuals familiar with the project site prior to damage; and
- paleoecological evidence (e.g., fossil pollen, charcoal, tree ring history, rodent middens).[43]

Identifying an appropriate reference site is a critically important part of ecological restoration and may require the help of someone with training in restoration ecology.

In addition to identifying a relevant ecosystem reference, the Society for Ecological Restoration recommends these other elements be included in a restoration plan:

- a clear rationale as to why restoration is needed;
- an ecological description of the site designated for restoration;
- a statement of the goals and objectives of the restoration project;
- a designation and description of the reference;
- an explanation of how the proposed restoration will integrate with the landscape and its flows of organisms and materials;
- explicit plans, schedules, and budgets for site preparation, installation, and post-installation activities, including a strategy for making prompt midcourse corrections;
- well-developed and explicitly stated performance standards, with monitoring protocols by which the project can be evaluated; and
- strategies for long-term protection and maintenance of the restored ecosystem.[44]

As you might expect, a key ingredient for successful ecological restorations, one that assists in the recovery of the ecological services that are the hallmark of fully functional ecosystems, is biological diversity. The impact of biological diversity on the success of restoration projects is the subject of ongoing research at some botanical gardens. Kayri Havens, the Medard and Elizabeth Welch director of plant science and conservation at the Chicago Botanic Garden, recommends several measures of biological diversity that are proving important for ecological restorations. At the level of species, restoring genetic diversity is a high priority, especially as it relates to local adaptability. In addition, age and growth stage diversity are important for all perennial plants. At the level of community, work toward restoring species diversity among the whole population, with an emphasis on functional, phylogenetic, and phenological diversity.[45]

The restoration principles from the IUCN listed at the beginning of this chapter include involving stakeholders. This is a critical element, especially concerning local and indigenous groups who, among several important issues, may have survival and other dependencies upon the habitat or landscape of concern. As first pointed out in chapter 9, local and indigenous populations must be approached as partners in any species recovery and ecological restoration projects. With this in mind, stakeholder dependencies will likely require project compromises toward the restoration of a mixed-use habitat different from the original. This may require moving forward without a clear reference ecosystem but with the capacity for research to create a hybrid ecosystem that will be functional and sustainable.

One example of such a scenario is an ecosystem restoration that takes into consideration managed grazing to accommodate the needs of traditional herdsmen. Prescribed and managed carefully, grazing has been incorporated as an asset rather than tolerated as a liability in ecological restoration.[46] Local community needs and habitat interfaces must be included in any restoration plans. An important benefit of involving the local community and accommodating their needs is that they will in turn assist in the necessary monitoring and management interventions that ecological restoration requires. Again, as partners in these projects, local and indigenous group participation must be fully documented and credited, and any benefits that accrue to or from the project must be shared with them.

Our understanding of and technical expertise in ecological restoration is still unfolding, and conditions that make this work necessary are intensifying. There is not only a great deal of work to be done but also a great need and many opportunities for botanical garden research in this area—both basic and applied.

Best Practice 11.2: Ecological Restoration

With "Best Practice," I try to provide a short synopsis of a practice or program that exemplifies what has been covered in the preceding section or chapter of the text. In this case and by way of providing more specific information on the subject, I include a longer account of an exceptional ecological restoration known as the Schulenberg Prairie. The details of this account were taken from the publication *The Schulenberg Prairie: A Benchmark in Ecological Restoration* published by the Morton Arboretum.[47]

Morton Arboretum, Schulenberg Prairie

The 41-acre Schulenberg Prairie at the Morton Arboretum was initiated in 1962 and is a restored example of the eastern tall grass prairie that was once the primary ecosystem of a large part of the eastern Great Plains and upper Midwest of North America. It once occupied 66 percent of the northeastern area of the state of Illinois where the city of Chicago and the Morton Arboretum are now located.

The work to restore tall grass prairie in the Morton Arboretum began with the project head, Ray Schulenberg, and three specialist colleagues studying the ecology, vegetation and floristics, and gene pool of prairie remnants within a 50-mile radius of the Arboretum. These remnants would now be considered reference ecosystems. Seeds were first collected and the initial soil preparation on the site of the restoration was implemented in 1962. The site was subdivided into developmental tracts—linear plots—that were plowed and disked that fall and then rototilled in spring just before planting. Seedlings were spring germinated in the Arboretum greenhouse and the seedlings planted out later that same spring on 1-foot-square centers in the prepared tract. Species were mixed according to observed associations in the wild on a 1:1 grass to forb ratio. *Rhizobium* bacteria was used to inoculate legume seeds. That spring, 20,000 seedlings representing 70 species were planted with the help of volunteers. These plants were then hand watered and weeded during the growing season with minimal losses. The following year another 64,000 seedlings representing 50 species were established in an adjoining plot also with hand watering and weeding. This process continued for several more years until 21 acres of diverse prairie vegetation containing 150 native species were established. During this time, all plots were burned on a biennial schedule.

In addition to the complete restoration of 21 acres of tall grass prairie from prepared soil, other large areas of existing mixed grassland were managed to help establish a new developmental trajectory toward tall grass prairie. A 15-acre tract was fire managed for succession to tall grass prairie while another 5-acre tract was mowed to achieve a similar effect. These sections were considered "successional" tracts. The developing Schulenberg Prairie was adjacent to a 25-acre restored woodland savanna, resulting in a diverse restored ecosystem containing more than 300 native plant species. As you can see, the Schulenberg Prairie is varied with three distinct developmental areas that serve broader restoration research purposes than any one area alone would provide. If small-scale species richness is a good indicator of restorative success, then the completely restored section established in the 1960s shows the most success with a species richness of 13.3 species per square meter at the time of this report (2012), while the other areas had less. Those areas being subjected to directed succession through burning and mowing also have a greater number of alien species per square meter. Several other types of vegetational and floristic analyses, some in comparison with undisturbed, late successional prairie remnants in the area, have been conducted on the Schulenberg Prairie revealing that restorations need a great deal of time to reach what may be construed as a stable or balanced dynamic.

The Schulenberg Prairie, a dynamic and evolving ecological restoration, is a rich laboratory for all kinds of restoration and ecological research. Some examples of the recent research focused on the Schulenberg Prairie are the success of high density out-planting versus seed sowing; the impact of controlled burns on species richness within several categories of prairie plants, such as plants with C4 versus C3 photosynthetic pathways; the importance of parasitic species that impact competition; successional dynamics and what it means for proportioning keystone species of certain successional stages in restoration plantings; the impact of various controlled burning schedules on the subterranean ecology of fire-adapted vegetation; and, of course, the predicted impacts of various climate change scenarios on the trajectory of prairie restorations.

Textbox 11.2. Recommendations for Ecological Restoration

Basic
- Partner with land management agencies and other conservation organizations experienced in ecological restoration and long-term management.
- Provide plant material for ecological restoration from ex situ conservation collections.
- Document those accessions used in restoration projects and their project histories.

Intermediate
- Restore a plant community or habitat within the property of the garden following a restoration plan based on a reference community or habitat.

- Focus on biological diversity, including plant species diversity and infraspecific diversity (functional, phylogenetic, and phenological).
- Include elements of experimental design in your on-site ecological restoration.
- Document all elements of the restoration project and its evolution.

Advanced

- Plan and conduct an independent ecological restoration of a local or regional ecosystem including recovery of threatened species.
- Partner with local and indigenous people to ensure their needs and habitat interfaces are preserved. Acknowledge and credit their involvement as well as share in any accrued benefit from the project.
- Involve local and indigenous partners in long-term project stewardship.
- Acquire all necessary permissions and permits; fully document the project.

Climate Change Modeling

A major challenge to ecological restoration, species recovery, and conservation in general is climate change. Climate change is the principal driver of so many research, preservation, and conservation actions now taken up by botanical gardens to fulfill their missions. So much of the conservation work described earlier is necessitated by climate change, and it will profoundly influence every aspect of botanical garden operation, to say nothing of our lives in general, going forward from here.

There is one specific aspect of climate change as it impacts and applies to curatorial practice that I want to touch on here: predicting the impacts of climate change. If we are to adapt our curatorial practices to climate change, we need to be able to better understand its likely impacts. Studies of climate change in ecological sciences have given rise to a number of predictive models. These models are employed to predict climate scenarios across a specified time frame. A particular kind of model—the climate envelope model—that may be used in botanical research on climate change uses wild species distribution data to build descriptions of an organism's climatic requirements.[48] This information can then be used to predict the climate suitability outside the species current range or in response to future climate changes. Chris Smart and Alan Elliott used a climate envelope modeling approach to gain insight into the climatic suitability of areas in Scotland for the cultivation of some common garden plants under climate conditions predicted for 2070.[49] For test subjects, they chose four common garden plants found in Scotland: *Clematis montana*, *Mecanopsis baileyi*, *Rhododendron luteum*, and *Skimmia japonica*.

I'm not going to go into all the details about how their climate envelope model was constructed; you may refer to the cited article in *Sibbaldia* for that information. I will tell you that they used WorldClim online for their climate data and the open-source software program Maxent for developing the climate envelopes in addition to the software program DIVA-GIS. At the end of collecting and processing the climate and plant distribution data, and then using Maxent to provide comparative analysis, they were able to show areas that, under both the 2015 and predicted 2070 conditions, were considered appropriate for their four species and those that were less so. For the inappropriate areas, Maxent could show the limiting factors involved, such as seasonal precipitation.[50] The climate envelope models for 2015 and 2070 were different for all four species; some showed suitable conditions going forward, and some were less suitable, similar to my palm example earlier in the section "Assisted Migration."

Climate envelope modeling can also be useful for biodiversity conservation. As part of the response to climate change, the conservation community is also starting to make decisions on longer time frames with a focus on "adaptation" strategies to help species and habitats adjust. One of the first steps in adaptation planning is to conduct vulnerability assessments to identify which species or systems are likely to be most affected by climate change and why, resembling the work of Smart and Elliott in Scotland. Climate envelope models are important to the conservation community for vulnerability assessments of how plants will respond to various climate change scenarios. Smart and Elliott compared the current climatic conditions with those predicted for 2070. Other researchers may choose more or different scenarios. The more climate data that is used in climate envelope modeling, the more reliable the models become. Highly credible models can provide information for natural resource planning by identifying species most at risk from climate change and highlighting areas of potential future conflict between human activities and conservation priorities.[51]

It should be obvious that climate envelope and other kinds of climate modeling could have interesting and important applications for collections development, modifications, preservation monitoring and potential trouble spots, conservation targets and programming, and other curatorial applications (see Longwood tree sucession planning, pgs. 177–78). Botanical gardens that don't have the staff or technical background to develop and run climate and other kinds of potentially useful models might be able to obtain these services from capable university, government, or NGO partners.

Notes

1. P. W. Jackson, *A Handbook for Botanic Gardens on the Reintroduction of Plants to the Wild* (London: Botanic Gardens Conservation International, 1995), 5.

2. V. H. Heywood, K. Shaw, Y. Harvey-Brown, and P. Smith, eds., *BGCI and IABG's Species Recovery Manual* (Richmond, Surrey, UK: Botanic Gardens Conservation International, 2018), 8, https://www.bgci.org/wp/wp-content/uploads/2019/04/Species_Recovery_Manual.pdf.

3. E. O. Guerrant, P. Fiedler, K. Havens, and M. Maunder, "Revised Genetic Sampling Guidelines for Conservation Collections of Rare and Endangered Plants: Supporting Species Survival in the Wild," in *Ex Situ Plant Conservation: Supporting Species Survival in the Wild*, ed. E. O. Guerrant, K. Havens, and M. Maunder (Washington, DC: Island Press. 2004), 419.

4. J. Gratzfeld, ed., *From Idea to Realisation: BGCI's Manual on Planning, Developing and Managing Botanic Gardens* (Richmond, Surrey, UK: Botanic Gardens Conservation International, 2016), 170.

5. See Heywood et al., *Species Recovery Manual.*

6. Jackson, *Reintroduction of Plants*, 19–21.

7. Heywood et al., *Species Recovery Manual*, 23.

8. See International Union for Conservation of Nature and Natural Resources, Species Survival Commission, *Strategic Planning for Species Conservation: A Handbook*, version 1.0 (Gland, Switzerland: IUCN/SSC, 2008).

9. J. C. Pierson, D. J. Coates, J. G. B. Oostermeijer, S. R. Beissinger, J. G. Bragg, P. Sunnucks, N. H. Schumaker, and A. G. Young, "Genetic Factors in Threatened Species Recovery Plans on Three Continents," *Frontiers in Ecology and the Environment* 14 (2016): 433–40.

10. Heywood et al., *Species Recovery Manual*, 82.

11. Heywood et al., *Species Recovery Manual*, 62.

12. Center for Plant Conservation, *CPC Best Plant Conservation Practices to Support Species Survival in the Wild* (Escondido, CA: CPC, 2019), 4-5.

13. Center for Plant Conservation, *CPC Best Plant Conservation Practices*, 4-9.

14. Center for Plant Conservation, *CPC Best Plant Conservation Practices*, 4-11.

15. Center for Plant Conservation, *CPC Best Plant Conservation Practices*, 4-23, 4-24.

16. Center for Plant Conservation, *CPC Best Plant Conservation Practices*, 4-29.

17. Hai Ren et al., "The Use of Grafted Seedlings Increases the Success of Conservation Translocations of *Manglietia longipedunculata* (*Magnoliaceae*), a Critically Endangered Tree," *Oryx* 50, no. 3 (2016): 437.

18. Heywood et al., *Species Recovery Manual*, 78.

19. Andrew R. Whiteley et al., "Genetic Rescue to the Rescue," *Trends in Ecology & Evolution* 30, no. 1 (2015): 1.

20. Gratzfeld, *From Idea to Realisation*, 171.

21. Heywood et al., *Species Recovery Manual*, 99.

22. S. Handler, C. Pike, and B. St. Clair, "Assisted Migration," USDA Forest Service Climate Change Resource Center, 2018, https://www.fs.usda.gov/ccrc/topics/assisted-migration.

23. Handler, Pike, and St. Clair, "Assisted Migration."

24. Christopher J. Butler, "Climate Change Winners and Losers: The Effects of Climate Change on Five Palm Species in the Southeastern United States," *Ecology and Evolution* 10, no. 19 (2020): 10408.

25. Adam B. Smith, Matthew A. Albrecht, and Abby Hird, "Chaperoned," *BG Journal* 11, no. 2 (2014): 19.

26. Handler, Pike, and St. Clair, "Assisted Migration."

27. Smith, Albrecht, and Hird, "Chaparoned," 20.

28. *Wikipedia*, "*Torreya taxifolia*," last updated April 30, 2021, https://en.wikipedia.org/wiki/Torreya_taxifolia.

29. Torreya Guardians, "About *Torreya taxifolia*," http://www.torreyaguardians.org/torreya.html.

30. Center for Plant Conservation, *CPC Best Plant Conservation Practices*, 4-39–4-41.

31. Jackson, *Reintroduction of Plants*, 17.

32. New England Wildflower Society, *State of the Plants* (Framingham, MA: NEWS, 2015), 17.

33. Atlanta Botanical Garden, "Cigar Orchid," https://atlantabg.org/plant-profile/cigar-orchid/.

34. Society for Ecological Restoration, "What Is Ecological Restoration?," https://www.ser-rrc.org/what-is-ecological-restoration/.

35. Rebecca S. Barak and Evelyn W. Williams, "Incorporating History to Improve Prairie Restorations," *BG Journal* 13, no. 2 (2016): 16.

36. K. Havens, "The Role of Botanic Gardens and Arboreta in Restoring Plants from Populations to Ecosystems," in *Plant Conservation Science and Practice: The Role of Botanic Gardens*, ed. S. Blackmore and S. Oldfield (New York: Cambridge University Press, 2017), 140–42.

37. Botanic Gardens Conservation International, "Ecological Restoration Alliance of Botanic Gardens," https://www.bgci.org/our-work/projects-and-case-studies/ecological-restoration-alliance-of-botanic-gardens/.

38. Botanic Gardens Conservation International, "Ecological Restoration Alliance of Botanic Gardens."

39. Barak and Williams, "Incorporating History," 17.

40. C. R. Beatty, N. A. Cox, and M. E. Kuzee, *Biodiversity Guidelines for Forest Landscape Restoration Opportunities Assessments* (Gland, Switzerland: IUCN, 2018), 2.

41. Society for Ecological Restoration International Science & Policy Working Group, *The SER International Primer on Ecological Restoration* (Tucson, AZ: SERI, 2004), 8, https://cdn.ymaws.com/www.ser.org/resource/resmgr/custompages/publications/ser_publications/ser_primer.pdf

42. Society for Ecological Restoration International Science & Policy Working Group, *SER International Primer*, 8.

43. Society for Ecological Restoration International Science & Policy Working Group, *SER International Primer*, 8.

44. Society for Ecological Restoration International Science & Policy Working Group, *SER International Primer*, 11.

45. Havens, "The Role of Botanic Gardens," 147–48.

46. Havens, "The Role of Botanic Gardens," 154.

47. Marlin Bowles et al., *The Schulenberg Prairie: A Benchmark in Ecological Restoration* (Chicago: Morton Arboretum, 2012).

48. I. Ibanez et al., "Predicting Biodiversity Change: Outside the Climate Envelope, beyond the Species Area Curve," *Ecology* 87, no. 8 (2006): 1896–1906.

49. Chris Smart and Alan Elliott, "Forward Planning for Scottish Gardens in the Face of Climate Change," *Sibbaldia* 13 (2015): 132.

50. Smart and Elliott, "Forward Planning," 133.

51. The Croc Docs, University of Florida, "Climate Envelope Modeling for Threatened and Endangered Species," https://crocdoc.ifas.ufl.edu/projects/climateenvelopemodeling/.

CHAPTER TWELVE

~

New Plant Introduction Programs and Part II Epilogue

"Plant introduction is the process by which an institution, such as a botanical garden, systematically develops a procedure for the dissemination and release of new plants to the public through either profit or not-for-profit programs."[1]

"The use of the living collections of a botanical garden or arboretum for the development of better plants for agricultural production, for landscape use in suburban and urban settings, and for improving basic mechanisms of stress tolerance and pest resistance has been closely tied to the land grant university system with its long history of support from the Department of Agriculture and related commercial sources."[2]

Plant introductions are one of the primary ways in which botanical gardens can make their plant collections directly accessible to the public. These programs offer a wide range of research opportunities within all facets of a garden's collections management program. They also provide a context for program integration and help galvanize garden staff in fulfilling an interdisciplinary goal of great public benefit. However, we cannot take such programs lightly for they draw resources from nearly every segment of the botanical garden operation. Consequently, plant introduction programs must be focused, governed, and supported by the institutional mission and collections management policy.

Before getting into the details and principal types of plant introduction programs, I want to emphasize something that has been written about at length elsewhere in this text: the Convention on Biological Diversity (CBD) and the Nagoya Protocol (NP). I hope that any thoughts of plant introduction, especially commercial plant introduction, will bring these two very important multilateral agreements to mind. As a brief reminder, these pertain to benefit sharing from the use or marketing of germplasm back to the country of origin. I'm certain that some breeders are unaware of such agreements or may not fully appreciate their implications. Taking the necessary steps to arrange for benefit-sharing agreements, memoranda of understanding, or contracts takes time and may be construed as an unnecessarily bureaucratic nuisance to plant introduction plans and processes. Since the United States is one of only two countries that have declined to ratify the CBD, breeders, botanical gardens, and others wishing to commercialize plant germplasm in the United States may decline to participate. Fortunately, this is not the case, and botanical gardens in the United States that release plants have adopted the CBD and NP and operate within their frameworks.[3]

Plant Introduction Programs

The principal elements of a plant introduction program from a collections management standpoint are acquisitions, propagation and production, documentation, preservation, breeding, evaluation, and selection. Here are some useful guidelines for developing a woody plant introduction program adapted from "Woody Plant Introduction Programs," a master's thesis by Carla Pastore:

- Examine the mission statement: consistent with the majority of guidelines provided in this manual, ensure continuity by building on established institutional goals.
- Analyze resources: plant introduction programs must maximize resources without overburdening staffs and budgets. These programs are not cheap.
- Design the program before implementing it: consider focus, sources of plants, application and suitability of introduced plants, budget, collaborators, and/or recipients of introduced plants.
- Plants: use documented, true-to-name stocks.
- Test: monitor and evaluate potential introductions in a continuous fashion; look for invasive qualities.
- Cultivate cooperators: articulate criteria for selecting cooperators, consider binding agreements, simplify cooperator paperwork, and follow through.
- Develop diverse promotional strategies.[4]

Pastore has categorized plant introduction programs into four groups based on some of these elements.

Group 1: Complete Programs

The most sophisticated, formalized programs involve cooperative efforts between public gardens, commercial nurseries, and farms. These complete programs include plant breeding, evaluation, selection, introduction, and marketing. Plant introductions may be trademarked and patented, and royalties may be collected.

The following are examples:

- Minnesota Hardy, University of Minnesota, St. Paul, Minnesota
- Chicagoland Grows®, Chicago Botanic Garden and the Morton Arboretum, Chicago, Illinois
- Plant Select®, a nonprofit collaborative between the Denver Botanic Garden and Colorado State University
- Sego Supreme™, Utah State University Botanical Center, Kaysville, Utah

This group of plant introduction programs is primarily based on clonal selections of plants but may also be the result of plant breeding. The plant collections of the parent institutions most often serve as the source of introductions. The cooperative nature of these programs is used at several operational levels including the selection of prospective introductions, development of propagation and production protocols, cooperative evaluation, and determination of the final selections. These programs require plant propagation and production facilities as well as growing space.

Typically, prospective introductions are monitored and evaluated on a regular basis using a rigorous set of criteria. Plants considered worthy of continued evaluation are distributed to cooperating stations for further, more rigorous evaluation under a varying set of environmental conditions. Chicagoland Grows®, Minnesota Hardy, and Sego Supreme engage in plant breeding as well as selection.

Group 2: Plant Breeding Programs

This group of programs has evolved as the result of plant breeding efforts. Plant testing is done with a large number of cooperators throughout the region. These programs include evaluation, selection, and introduction, and they rely on producers for propagation, marketing, and sales. Plant introductions may or may not be patented.

The following are examples:

- Oregon State University Ornamental Plant Breeding Program in cooperation with the Hoyt Arboretum and Atlanta Botanical Garden
- U.S. National Arboretum, Washington, DC
- Agriculture Canada Research Station, Prairie Regional Zonation Trials for Woody Ornamentals, Morden, Manitoba, Canada[5]

Because breeding programs require a long-term commitment, gardens should examine their mission statements carefully for justification and support of such a commitment. In addition, breeding programs need careful planning, adequate financial resources, appropriate infrastructure, and competent staff. Still, botanical gardens are well suited to this work as repositories for germplasm and a working environment exempt from the "publish or perish" pressure of academia.

From a plant selection standpoint, program staff must be sensitive to several factors in pursuing a breeding strategy. According to Pastore:

1) There must be a source of potential germplasm. 2) The plant group needs to have a major problem [or tremendous undeveloped potential] that could be improved through plant breeding, such as cold hardiness, disease or insect resistance. 3) Plant groups selected must have consumer appeal, such as showy flowers or good fall color. 4) It must be economically feasible for the wholesale nurseryman to produce the plant.[6]

Because of the time frame required for successful breeding, gardens should plan to pursue several projects simultaneously. If thoughtfully planned, these projects will come to fruition at different times. As an example, the U.S. National Arboretum may be working on as many as 24 genera of trees and shrubs at once.

These programs require adequate plant propagation, production, storage, and field as well as laboratory space. Often, germplasm is selected from wild populations, plant collections, or advanced generations in controlled crosses. On-site breeding, monitoring, and evaluation may continue for 10 to 20 years before seedlings are selected for distribution to cooperative evaluators. Cooperative evaluators are selectively chosen based on interest, capability, and location.

Cooperators may evaluate plants for two to five years before a final selection is made. It may then take an additional two to five years for select cooperating wholesale nurseries to bulk up the stock for general sale to other wholesalers. Before this general distribution, clones are formally registered and described for publication.

Group 3: Non-breeding Programs

The focus of these programs is on the testing and evaluation of clonal selections of wild-collected plants. They are not involved in plant breeding and have no system for marketing plants. A number of these programs are administered through the U.S. Department of Agriculture–Agricultural Research Service. Plants are not patented.

The following are examples:

- NC-7 Regional Ornamental Plant Trials, North Central Regional Plant Introduction Station, Ames, Iowa
- NE-9 Regional Ornamental Plant Trials, Northeastern Regional Plant Introduction Station, Geneva, New York[7]
- Choice Plants, JC Raulston Arboretum, North Carolina State University, Raleigh, North Carolina

The Ames, Iowa, program focuses on herbaceous perennials native to the Midwest prairie and woody plants with the same adaptability. The Choice Plants program at the JC Raulston Arboretum works in collaboration with the Johnson County Nursery Marketing Association.

Group 4: Plant Advocate Programs

These types of programs are a result of the energies and interests of one or several individuals. Plants are promoted through word of mouth, giveaways, talks, classes, exhibits, and displays. These programs work because the plant advocate has established credibility and respect in the field.

The following are examples:

- Arnold Arboretum of Harvard University, Jamaica Plains, Massachusetts
- International Succulent Introductions (ISI), The Huntington Botanical Gardens, Pasadena California

Plant advocate programs are common among botanical gardens and arboreta and are often the genesis of what may become more elaborate programs because they have plant advocates on staff. From the above description, you can see that these are informal programs that rely on clonal selection, testing, and promotion. The key to success, albeit a tenuous one, is the drive and enthusiasm of the plant advocate. In the case of the ISI program at The

Huntington Botanical Garden, the principal advocate is John Trager, long-time curator and notable authority on succulent and desert plants.

The facilities for such a program may be minimal as long as there is some space to grow and evaluate plants. Plants of potential interest may be newly acquired, rare, or underused. They may be chosen because they represent the specific collecting interests of the garden or serve the purposes of the advocate program.

There are usually no cooperators for these types of programs although advocate networks may provide for useful testing sites and constructive feedback. Plant selections are usually introduced and promoted at professional meetings, garden plant sales, and other venues attended and/or organized by the advocate.

Plant introduction programs of any type, particularly the "complete" and "plant breeding" programs, that plan to reach out to partners with expertise in plant development and marketing should consider Matthew Taylor's following lists of dos and don'ts from his article "Considerations for Commercial Plant Introduction from Public Gardens" in *Sibbaldia*:

- Do take lots of good photographs of the plant in a garden setting. Remember, you are also marketing your plant to these companies.
- Do record cultural and phenological data so that you can provide as much information as possible to a potential partner company.
- If the plant was created through an active breeding programme, do record pedigree information as it may be needed in a patent application or to direct future crosses.
- If the plant is given to a company (licensor) for trial by your organisation (licensee), do be sure to sign a trial/testing agreement that indicates limited ownership and use by the licensor. The agreement should also state that any sports or mutations that occur during the trial must be reported immediately and are owned by the licensee.
- Do test for viruses and propagate in tissue culture. This can be expensive, but can pay off in the long run. The presence of microorganisms within plant tissue can be a major setback for a plant introduction. Ridding plants of virus or bacteria costs time and money. If the selected plant is tested, determined to be clean and then put into tissue culture, it is then protected from viral and bacterial vectors for the future.

The following are a few important don'ts from Taylor that may be more crucial to success than the dos:

- Do not share the plant with any other organisations unless a trial agreement has been signed and states what the organisation can and cannot do with the plant.
- Do not name the plant and publicise it in any way.
- Do not sell the plant to anyone. Once a cultivar name has been publicly announced or the plant has been sold, this sale can start the clock on the window to patent.

When applying for a plant patent, the United States Patents and Trademark Office will do a background check on the plant material and reject any application for a plant that has been sold, named or publicised for any time period greater than one year of the application. Simply naming a plant and putting it into a garden newsletter can start this one-year patent clock.[8]

Textbox 12.1. Recommendations for Commercial Plant Introductions

Basic
- The garden promotes commercial plant introductions informally through word of mouth, giveaways, talks, classes, exhibits, and displays following a plant advocacy approach.

Intermediate
- The garden conducts a commercial plant introduction program through testing and evaluation of clonal selections of wild-collected plants and/or clones of garden origin.
- The garden partners with a professional organization of commercial growers for the evaluation, multiplication, and commercial distribution of plant introductions.

Advanced
- The garden conducts a program of plant breeding and/or selection with the participation of a group of cooperators to develop plants of improved hardiness, productivity, and/or aesthetic appeal that are made available for commercial introduction.
- The garden actively works with cooperators to multiply and market plant introductions, some of which may be trademarked and patented.

Epilogue to Part II

By way of concluding part II on research and collections, there are a few points I think worth reiterating. The subject of accessibility has come up previously, but I don't think it can be overstated. A collection's research value depends not only upon its quality and organization but also upon its accessibility to researchers.[9] Collections frequently contain eclectic groups of specimens and rare curiosities that have value apparent only to specialists.[10] Needless to say, these specialists need to know what's available, and the garden curator needs to know who's out there and what they need to work with. There's clearly a marketing obligation here for curators.

Online information is probably the best way to alert researchers to the resources available to them at various gardens. This situation has drastically improved since the first edition of this book with most gardens having complete inventories available online as well as webpages dedicated to their research programs. The utility of online databases becomes even more useful through networking, and powerful networking opportunities are available through Botanic Gardens Conservation International (BGCI), the American Public Gardens Association (APGA), and the Center for Plant Conservation (CPC), professional organizations whose programs have been profiled in several places throughout this text. Something else about data sharing should be of interest to curators: it can provide useful curatorial feedback akin to an audit. Here's an opportunity for additional assistance in revising the collection and for validating its uniqueness. With increased use and exchange, it's more likely that errors and problems in the documentation will be detected but also more likely that the rare and unique attributes of specimens in the collections will be acknowledged. It's a double-edged sword, yes, but helpful in collections management nevertheless.[11]

A spin-off of collaborative research work is the prospect for new accessions. The scientists you work with may have assembled research collections of their own and may be willing to donate them at the conclusion of their work. Consider that these collections, depending upon the kind of research involved, may be the most comprehensive of their kind.

No discussion of research would be complete without reiterating the important role volunteers can play in helping carry out research work. So much of the research work and many of the programs I've outlined in this chapter present opportunities for engaging citizen science projects. Of course, botanical gardens are no strangers to the use of volunteers, and they may be successfully employed to further the goals of research as pointed out in several places throughout part II of this text and in part I, chapter 6, under "Citizen Science."

I would like to offer some final words about the importance of collections research. Many curators complain that garden administrators and board members undervalue collections programs, and research in particular, in favor of more peripheral but engaging public programs. Tom Elias challenges that approach: "Never compromise the quality of our collections, the importance of educational opportunities and the value of research and conservation efforts, even if they are not as cost effective as more popular but non-botanical and peripheral activities designed to increase attendance and admission fees."[12]

I hope the various programmatic examples and cases described in this section will provide some instruction and inspiration for botanical garden curators on the research values and possibilities for the living collections in their charge. Botanical gardens and arboreta must maintain a preeminence in research and education in order to continue their complementary function as a unique public amenity and authoritative source of information on plants. "When research ceases, the facility, of whatever kind, retains only historical and curiosity value, and all concerned tend to look backward only. . . . [W]ithout the ability to look and move forward through active research, we make no progress."[13]

Notes

1. Roy L. Taylor, "Is a Plant Introduction Program Right for Your Garden?," *Public Garden* 2, no. 4 (1987): 14.

2. R. E. Cook, "Botanical Collections as a Resource for Research," *Public Garden* 21, no. 1 (2006): 19.

3. Matthew Taylor, "Considerations for Commercial Plant Introduction from Public Gardens," *Sibbaldia* 14 (2016): 176–77.

4. Carla Pastore, "Plant Introduction Programs in the United States and Canada," *Public Garden* 2, no. 4 (1987): 16.

5. Carla Pastore, "Woody Plant Introduction Programs" (master's thesis, University of Delaware, 1988), 34.

6. Pastore, "Woody Plant Introduction," 34.

7. Pastore, "Plant Introduction Programs," 16.

8. Taylor, "Considerations for Commercial Plant Introduction," 176.

9. A. J. Pekarik, "Long-Term Thinking: What about the Stuff?," *Curator* 46, no. 4 (2003): 367.

10. J. M. Bryant, "Biological Collections: Legacy or Liability?," *Curator* 26, no. 3 (1983): 204.

11. Michael S. Dosmann, "Research in the Garden: Averting the Collections Crisis," *Botanical Review* 72, no. 3 (2006): 228.

12. T. S. Elias, "About This Issue," *Public Garden* 6, no. 3 (1991): 6.

13. Alden H. Miller, "The Curator as a Research Worker," *Curator* 6, no. 4 (1963): 286.

Bibliography

Abbel-Seddon, B. "Reforming Collection Documentation: A New Approach." *International Journal of Museum Management and Curatorship* 8, no. 1 (1989).

Adams, John. "In Support of Collections Policy." Opening address at the Collections Policy Panel of the B.C. Museums Association 1979 Seminar, British Columbia Museum Association. *Museum Roundup*, no. 76 (1979).

Aguilar, Kristina. *Assigning Disaster Priorities to Your Collection*. Kennett Square, PA: Longwood Gardens, 2010.

Aiello, Anthony S., A. Gapinski, and K. Wang. "Collaboration across Continents and Cultures." *BG Journal* 16, no. 2 (2019).

Alberta Museums Association. *Standard Practices Handbook for Museums*. Edmonton, AB: AMA, 1990.

Albrecht, M. A., E. O. Guerrant Jr., K. Kennedy, and J. Maschinski. "A Long-Term View of Rare Plant Reintroduction." *Biological Conservation* 144 (2011).

Albrecht, M. A., and J. Maschinski. "Influence of Founder Population Size, Propagule Stages, and Life History on the Survival of Reintroduced Plant Populations." In *Plant Reintroduction in a Changing Climate: Promises and Perils*, edited by J. Maschinski and K. E. Haskins. Washington, DC: Island Press, 2012.

Albrecht, M. A., et al. "Effects of Life History and Reproduction on Recruitment Time Lags in Reintroductions of Rare Plants." *Conservation Biology* 33, no. 3 (2019).

Alexander, E. P. *Museums in Motion*. Nashville, TN: American Association for State and Local History, 1989.

———. "What Is Interpretation?" *Longwood Graduate Program Seminars* 9 (1977).

Alliance for Public Gardens GIS. "Alliance for Public Gardens GIS." https://public gardensgis.ucdavis.edu.

———. "Guide to GIS for Public Gardens: Botanical Gardens, Zoos, and Parks." March 2013. https://publicgardensgis.ucdavis.edu/sites/g/files/dgvnsk6621/files/inline-files/Guide-to-GIS-for-Public-Gardens-March-2013.pdf.

Ambrose, J. "Conservation Strategies for Natural Areas." *Public Garden* 3, no. 2 (1988).

Ambrose, T. M. *New Museums: A Start-Up Guide.* Edinburgh: HMSO, 1987.

American Alliance of Museums. *A Code of Ethics for Curators.* Washington, DC: AAM, 2009.

American Association of Museums. *Caring for Collections: Strategies for Conservation, Maintenance and Documentation.* Washington, DC: AAM, 1984.

———. *Code of Ethics for Museums.* Washington, DC: AAM, 2000.

———. *Gifts of Property: A Guide for Donors and Museums.* Washington, DC: AAM, 1985.

———. *Museum Ethics.* Washington, DC: AAM, 1978.

———. *Museums for a New Century.* Washington, DC: AAM, 1984.

———. "Stewards of a Common Wealth." In *Museums for a New Century.* Washington, DC: AAM, 1984.

American Association of Museums, Museum Studies Committee. "Museum Positions, Duties and Responsibilities." *Museum News* 57, no. 7 (1978).

American Association of Museums, Standing Professional Committee on Education. "Standards: A Hallmark in the Evolution of Museum Education." *Museum News* 69, no. 1 (1990).

American Heritage Dictionary. 2nd college ed. Boston: Houghton Mifflin, 1982.

American Public Gardens Association. "About the Plant Collections Network." https://www.publicgardens.org/programs/about-plant-collections-network.

———. "Invasive Plant Species Voluntary Codes of Conduct for Botanic Gardens and Arboreta." https://www.publicgardens.org/resources/invasive-plant-species-voluntary-codes-conduct-botanic-gardens-arboreta.

———. "*Metasequoia glyptostroboides.*" https://www.publicgardens.org/programs/plant-collections-network/collections-showcase/metasequoia-glyptostroboides.

———. "Sentinel Plant Network." https://www.publicgardens.org/programs/sentinel-plant-network/about-spn.

———. "Sustainability Index." https://www.publicgardens.org/sustainability-index.

Ames, Peter J. "Guiding Museum Values: Trustees, Missions and Plans." *Museum News* 63, no. 6 (1985).

Anderson, N. O., S. M. Galatowitsch, and N. Gomez. "Selection Strategies to Reduce Invasive Potential in Introduced Plants." *Euphytica* 148 (2006).

Andrianandrasana, H. T. "Testing the Effectiveness of Community-Based Conservation in Conserving Biodiversity, Protecting Ecosystem Services, and Improving Human Well-Being in Madagascar." PhD thesis, University of Oxford, 2016.

Angiosperm Phylogeny Group. "An Update of the Angiosperm Phylogeny Group Classification for the Orders and Families of Flowering Plants: APG IV." *Botanical Journal of the Linnean Society* 181, no. 1 (2016).

Aplin, D. M. "Assets and Liabilities: The Role of Evaluation in the Curation of Living Collections." *Sibbaldia* 11 (2013).

———. "How Useful Are Botanic Gardens for Conservation?" *Plantsman* 7, no. 3 (2008).

Applebaum, B., and P. Hammerstein. "Planning for a Conservation Survey." *Museum News* 64, no. 3 (1986).

Arnold Arboretum. *The Arnold Arboretum Expedition Toolkit.* Jamaica Plain, MA: Arnold Arboretum, 2018.

———. *Landscape Management Plan.* Jamaica Plain, MA: Arnold Arboretum, 2011.

———. *Plant Inventory Operations Manual.* 2nd ed. Jamaica Plain, MA: Arnold Arboretum, 2011.

Aronson, J., et al. "An Experimental Technique for Long-Distance Transport of Evergreen and Deciduous Cuttings under Tropical Conditions." *FAO/IBPGR Plant Genetic Resources Newsletter* 81/82 (1990).

Ashton, Peter. "Biological Considerations in *In Situ* vs. *Ex Situ* Plant Conservation." In *Botanic Gardens and the World Conservation Strategy*, edited by D. Bramwell et al. London: Academic Press, 1987.

———. "Botanic Gardens and Experimental Grounds." In *Current Concepts in Plant Taxonomy*, edited by V. H. Heywood and D. M. Moore. London: Academic Press, 1984.

———. "Museums and Botanical Gardens: Common Goals?" In *Museum Collections: Their Roles and Futures in Biological Research*, edited by E. H. Miller. Occasional Papers 25. Victoria, BC: British Provincial Museum, 1985.

———. "Tropical Botanical Gardens: Meeting the Challenge of Declining Resources." *Longwood Graduate Program Seminars* 13 (1981).

Association of British Columbia Archivists. *A Manual for Small Archives.* Vancouver: ABCA, 1988.

Atkins, P. "Outsourcing: A User's Perspective." *Public Garden* 10, no. 2 (1995).

Atlanta Botanical Garden. "Cigar Orchid." https://atlantabg.org/plant-profile/cigar-orchid/.

Atwood, J. "Spirit Collections." *Public Garden* 12, no. 1 (1997).

Audubon. "Audubon Christmas Bird Count." https://www.audubon.org/conservation/science/christmas-bird-count.

Auerbach, N. A., K. A. Wilson, A. I. T. Tulloch, R. Rhodes, J. O. Hanson, and H. P. Possingham. "Effects of Threat Management Interactions on Conservation Priorities." *Conservation Biology* 29 (2015).

Aughanbaugh, J. E., and O. Dillard. *Performance Records of Woody Plants in the Secrest Arboretum.* Research Circular 139. Wooster, OH: Agriculture Research and Development Center, 1965.

Avery, G. S., Jr. "Botanic Gardens—What Role Today?" *American Journal of Botany* 44, no. 3 (1957).

Bachman S., et al. "Extinction Risk and Threats to Plants." In *The State of the World's Plants Report.* Richmond, Surrey, UK: Royal Botanic Gardens, Kew, 2016.

Baker, Charles. "Planning Exhibits: From Concept to Opening." Technical Leaflet 137. *History News* 36, no. 4 (1981).

Barker P. *A Technical Manual for Vegetation Monitoring*. Hobart, Australia: Resource Management and Conservation, Department of Primary Industries, Water and Environment, 2001.

Ballard, Jennifer Schwarz, Amy Padolf, Jessica A. Schuler, and Karen Oberhauser. "Engaging Diverse Audiences in Conservation and Research through Citizen Science." Presented to the American Public Garden Association Conference, Washington, DC, June 17–21, 2019.

Ballesteros, D., and V. C. Pence. "Survival and Death of Seeds during Liquid Nitrogen Storage: A Case Study on Seeds with Short Lifespans." *CryoLetters* 38 (2017).

Barak, Rebecca S., and Evelyn W. Williams. "Incorporating History to Improve Prairie Restorations." *BG Journal* 13, no. 2 (2016).

Barber, S., and L. Galloway. *Guide to Collecting Living Plants in the Field*. Edinburgh: Royal Botanic Garden Edinburgh, 2014.

Barber, Sadie, and S. Scott. "Botanical Envelopes." *Sibbaldia* 7 (2009).

Barr, David W. "Top Down or Bottom Up? Which Is the Most Useful Way to Develop Our 'First Principles' of Collecting?" *Museum Quarterly* 17, no. 3 (1989).

Barzun, J., and H. Graff. *The Modern Researcher*. Rev. ed. New York: Harcourt, Brace and World, 1970.

Basey, A. C., J. B. Fant, and A. T. Kramer. "Producing Native Plant Materials for Restoration: 10 Rules to Collect and Maintain Genetic Diversity." *Native Plants Journal* 16 (2015).

Baskin, C. C., and J. M. Baskin. *Seeds: Ecology, Biogeography, and Evolution of Dormancy and Germination*. Waltham, MA: Academic Press, 2014.

Bearman, D., comp. *Directory of Software for Archives & Museums*. Archival Informatics Technical Report. Pittsburgh: Archives & Museum Informatics, 1988.

———. "Functional Requirements for Collections Management Systems." *Archival Informatics Technical Report* 1, no. 3 (1987).

Beatty, C. R., N. A. Cox, and M. E. Kuzee. *Biodiversity Guidelines for Forest Landscape Restoration Opportunities Assessments*. Gland, Switzerland: IUCN, 2018.

Bechtol, N. "Integrated Pest Management: A Viable Alternative for Public Gardens." *Longwood Graduate Program Seminars* 16 (1984).

Beckman, E., A. Meyer, A. Denvir, D. Gill, G. Man, D. Pivorunas, K. Shaw, and M. Westwood. *Conservation Gap Analysis of Native U.S. Oaks*. Lisle, IL: Morton Arboretum, 2019.

Beer, Valorie. "The Problem and Promise of Museum Goals." *Curator* 33, no. 1 (1990).

Beever E. A. "Monitoring Biological Diversity: Strategies, Tools, Limitations, and Challenges." *Northwestern Naturalist* 87 (2006).

Benson, M. H. "Intelligent Tinkering: The Endangered Species Act and Resilience." *Ecology and Society* 17 (2012).

Bergmann, Eugene. "Exhibits: A Proposal for Guidelines." *Curator* 19, no. 2 (1976).

Berjak P., and N. W. Pammenter. "Cryostorage of Germplasm of Tropical Recalcitrant-Seeded Species: Approaches and Problems." *International Journal of Plant Sciences* 175 (2014).

Blackaby, J. *Managing Historical Data: The Report of the Common Agenda Task Force.* AASLH Special Report 3. Nashville, TN: American Association for State and Local History, 1989.

Blackaby, J., Patricia Greeno, and the Nomenclature Committee, eds. *The Revised Nomenclature for Museum Cataloguing.* Nashville, TN: American Association for State and Local History, 1988.

Booth, J. H., et al. *Creative Museum Methods and Educational Techniques.* Springfield, IL: Charles C. Thomas, 1982.

Bossler, N. "Interpretation of Historical Gardens." *Longwood Graduate Program Seminars* 11 (1979).

Botanic Gardens Conservation International. "BGCI ABS Learning Tool." https://www.bgci.org/policy/abs_learning.

———. "BGCI Databases." https://www.bgci.org/resources/bgci-databases/.

———. *Building Living Plant Collections to Support Conservation: A Guide for Public Gardens.* Richmond, Surrey, UK: BGCI, 2014.

———. "CITES Learning Tool." https://www.bgci.org/resources/links.

———. *Cycads: A Model Group for Ex Situ Plant Conservation.* Richmond, Surrey, UK: BGCI, 2015.

———. *Developing Botanic Garden Policies and Practices for Environmental Sustainability.* Richmond, Surrey, UK: BGCI, 2009.

———. "Ecological Restoration Alliance of Botanic Gardens." https://www.bgci.org/our-work/projects-and-case-studies/ecological-restoration-alliance-of-botanic-gardens/.

———. GardenSearch. Database. https://www.bgci.org/resources/bgci-databases/gardensearch/.

———. "Global Seed Conservation Challenge." https://www.bgci.org/our-work/plant-conservation/seed-conservation/global-seed-conservation-challenge/.

———. GlobalTreeSearch. Database. https://tools.bgci.org/global_tree_search.php.

———. *A Handbook for Botanic Gardens on the Reintroduction of Plants to the Wild.* London: BGCI, 1995.

———. *International Agenda for Botanic Gardens in Conservation.* 2nd ed. Richmond, Surrey, UK: BGCI, 2012.

———. *IPEN Code of Conduct.* Richmond, Surrey, UK: BGCI, 2018.

———. PlantSearch. Database. https://www.bgci.org/resources/bgci-databases/plantsearch/.

———. ThreatSearch. Database. https://www.bgci.org/threat_search.php.

Bowles, Marlin, et al. *The Schulenberg Prairie: A Benchmark in Ecological Restoration.* Chicago: Morton Arboretum, 2012.

Bradley, S. "Conservation Recording in the British Museum." *Conservator* 7 (1985).

Brady, T. S. "Six-Step Method to Long-Range Planning for Nonprofit Organizations." *Managerial Planning* (1984).

Brandes, G. "Plants or People: A Fallacious Choice?" *Longwood Graduate Program Seminars* 7 (1975).

Breed, M. F., et al. "Priority Actions to Improve Provenance Decision-Making." *Bioscience* 68, no. 7 (2018).

Brennan, D. "Botanical Gardens and Plant Evaluation." *Public Garden* 2, no. 3 (1987).

Bridson, D., and L. Forman. *The Herbarium Handbook*. London: Royal Botanic Gardens, Kew, 1992.

———. *The Herbarium Handbook*. 3rd ed. Richmond, Surrey, UK: Royal Botanic Gardens, Kew, 2004.

Bristol, P. "Collectors, Start Your Engines." In *Plant Exploration: Protocols for the Present, Concerns for the Future*, edited by James R. Ault (Glencoe, IL: Chicago Botanical Garden, 1999), 41.

Brown, R. A. *A Guide to the Computerization of Plant Records*. Swarthmore, PA: American Association of Botanical Gardens and Arboreta, 1988.

Brumback, W. "Endangered Plant Species and Botanic Gardens." *Longwood Graduate Program Seminars* 12 (1980).

Brunel, S., et al. "Invasive Alien Species: A Growing but Neglected Threat?" In *Late Lessons from Early Warnings: Science, Precaution, Innovation*. EEA Report 1/2013. Copenhagen: European Environment Agency, 2013.

Brunton, Howard. "Conflict or Co-operation? Natural Sciences: Collecting Policies." *Museums Journal* 87, no. 2 (1987).

Bryan, Charles F. "Put It in Writing: Developing an Effective Staff and Policy Manual." *History News* 42, no. 6 (1987).

Bryant, J. M. "Biological Collections: Legacy or Liability?" *Curator* 26, no. 3 (1983).

Bryant, P. "Progress in Documentation: The Catalogue." *Journal of Documentation* 36, no. 2 (1980).

Buma, D. "Effective Garden Interpretation." *Longwood Graduate Program Seminars* 9 (1977).

Bunce, F., et al. *Arboreta, Botanical Gardens, Special Gardens*. No. 90. Washington, DC: National Recreation and Park Association, 1971.

Bunn, Eric, and Kingsley Dixon. "Botanic Gardens: Meeting the Restoration Challenge with Critically Endangered Plants: A Case History, *Symonanthus bancroftii* (*Solanaceae*)." *BG Journal* 6, no. 1 (2009).

Burcaw, G. E. *Introduction to Museum Work*. Nashville, TN: American Association for State and Local History, 1975.

———. *Introduction to Museum Work*. 3rd ed. Lanham, MD: AltaMira Press, 1997.

Bureau of Land Management. *Technical Protocol for the Collection, Study, and Conservation of Seeds from Native Plant Species for Seeds of Success*. Denver, CO: BLM, 2016.

Butler, Cristopher J. "Climate Change Winners and Losers: The Effects of Climate Change on Five Palm Species in the Southeastern United States." *Ecology and Evolution* 10, no. 19 (2020).

Camacho, A. E., E. M. Taylor, and M. L. Kelly. *Lessons from Area-Wide, Multiagency Habitat Conservation Plans in California*. Washington, DC: Environmental Law Institute, 2016.

Campbell, F. "What Every Public Garden Should Know about CITES." *Public Garden* 6, no. 4 (1991).

Canadian Museums Association, Professional Development and Standards Committee. "Guidelines for Writing a Professional Development Policy." *Muse* 2, no. 3 (1984).

Cannon-Brookes, P., and C. Cannon-Brookes. "Control of Terminology but Not Its Limitation." *International Journal of Museum Management and Curatorship* 7 (1988).

Capelle, A. "Interpretive Planning . . . New Trends." *Interpreter* 15, no. 3 (1984).

Carter, John. "Developing an Interpretation and Education Policy for the Community Museum." *Journal of Museum Education: Roundtable Reports* 10, no. 1 (1985).

Case, Mary, ed. *Registrars on Record: Essays on Museum Collections Management*. Washington, DC: American Association of Museums, 1988.

Center for Plant Conservation. *CPC Best Plant Conservation Practices to Support Species Survival in the Wild*. Escondido, CA: CPC, 2019.

———. "National Collection." https://saveplants.org/national-collection/.

———. "Rare Plants." https://saveplants.org/search/?category=plants/.

Chen, J., R. T. Corlett, and C. Cannon. "The Role of Botanic Gardens in In Situ Conservation." In *Plant Conservation Science and Practice*, edited by S. Blackmore and S. Oldfield. New York: Cambridge University Press, 2017.

Chenhall, R. *Museum Cataloguing in the Computer Age*. Nashville, TN: American Association for State and Local History, 1975.

Chenhall, R., and Peter Homulos. "Museum Data Standards." *Gazette* 11, no. 3 (1978).

Chicago Botanic Garden. "Plant Evaluation." https://www.chicagobotanic.org/collections/ornamental_plant_research/plant_evaluation.

Chiou, C.-R., T.-Y. Hsieh, and C.-C. Chien. "Plant Bioclimatic Models in Climate Change Research." *Botanical Studies* 56 (2015).

CitizenScience.gov. "About CitizenScience.gov." https://www.citizenscience.gov/about/#.

———. "Project BudBurst." https://www.citizenscience.gov/project-budburst/#.

CitizenScience.org. "Citizen Science: Activating STEM Learning out of School." https://www.citizenscience.org/wpcontent/uploads/2018/08/AfterSchoolSTEM-170510.pdf.

Climate Toolkit. "The Climate Toolkit for Museums, Gardens and Zoos." https://climatetoolkit.org.

Clugston, Richard M., and Wynn Calder. "Critical Dimensions of Sustainability in Higher Education." In *Sustainability and University Life*, edited by W. L. Filho. New York: Peter Lang, 1999.

Coffman, L. "Visual Literacy in Museum Education." *Dawson & Hind* 15, no. 1 (1988/1989).

Colas, B., et al. "Restoration Demography: A 10-Year Demographic Comparison between Introduced and Natural Populations of Endemic *Centaurea Corymbosa* (Asteraceae)." *Journal of Applied Ecology* 45 (2008).

Colbert, E. H. "On Being a Curator." *Curator*, no. 1 (1958).

Committee on Earth Observation Satellites. "International Directory Network." https://idn.ceos.org.

Communications Canada. *Challenges and Choices: Federal Policy and Program Proposals for Canadian Museums*. Ottawa: Minister of Supply and Services Canada, 1988.

Concise Oxford Dictionary. 7th ed. Oxford: Clarendon Press, 1982.

Connor, Sheila. "Public Garden Archives." In *Public Garden Management*, edited by D. Rakow and S. Lee. Hoboken, NJ: Wiley, 2011.

Convention on Biological Diversity. "Global Strategy for Plant Conservation." https://www.cbd.int/gspc.

Cook, R. E. "Botanical Collections as a Resource for Research." *Public Garden* 21, no. 1 (2006).

———. "Research at Cornell." *Public Garden* 1, no. 1 (1986).

Corcoran, M., M. A. Hamilton, and C. Clubbe. "Developing Horticultural Protocols for Threatened Plants from the UK Overseas Territories." *Sibbaldia* 12 (2014).

Corlett, R. T. "Restoration, Reintroduction and Rewilding in a Changing World." *Trends in Ecology and Evolution* 31, no. 6 (2016).

Correll, Phillip G. *Botanical Gardens and Arboreta of North America: An Organizational Survey*. Los Angeles: American Association of Botanical Gardens and Arboreta, 1980.

Cowen, David J. "GIS versus CAD versus DBMS: What Are the Differences?" *Photogrammetric Engineering and Remote Sensing* 54, no. 11 (1988).

Crammond, A. "Starting a Botanical Garden." *Longwood Graduate Program Seminars* 20 (1988).

Croat, T. B. "Research under Glass." *Public Garden* 3, no. 4 (1988).

The Croc Docs, University of Florida. "Climate Envelope Modeling for Threatened and Endangered Species." https://crocdoc.ifas.ufl.edu/projects/climateenvelopemodeling/.

Cromarty, A. S., et al. *The Design of Seed Storage Facilities for Genetic Conservation*. Rome, Italy: IBGR, 1982.

Cronk, Q. C. B., and J. L. Fuller. *Plant Invaders: The Threat to Natural Ecosystems*. Abingdon, UK: Routledge, 2014.

Cubey, R., and Martin F. Gardner. "A New Approach to Targeting Verifications at the Royal Botanic Garden Edinburgh." *Sibbaldia* 1 (2003): 22–23.

Cubey, R., and D. Rae. *PlantNet Directory of Botanical Collections in Britain and Ireland*. Cambridge: PlantNet, 1999.

Cullen, J. "Use of Records Systems in the Planning of Botanic Garden Collections." In *Conservation of Threatened Plants*, edited by John Simonds. New York: Plenum Press, 1976.

Dansi, A. "Chapter 21: Collecting Vegetatively Propagated Crops (Especially Roots and Tubers)." In *Collecting Plant Genetic Diversity: Technical Guidelines—2011 Update*, edited by L. Guarino et al. Rome, Italy: Bioversity International, 2011.

Darke, Rick, ed. *Plant Records Procedures Manual for Longwood Gardens Employees*. Kennett Square, PA: Longwood Gardens, 1991.

———. "Trademarks, Patents and Cultivars." *Public Garden* 6, no. 1 (1991).

Davis, K. *A CBD Manual for Botanic Gardens*. Richmond, Surrey, UK: Botanic Gardens Conservation International, 2008.

Daws, M. I., et al. "Developmental Heat Sum Influences Recalcitrant Seed Traits in *Aesculus hippocastanum* across Europe." *New Phytologist* 162, no. 1 (2004).

Debergh, P. C., and R. H. Zimmerman, eds. *Micropropagation Technology and Application*. Dordrecht: Kluwer Academic, 1991.

Deblinger, R. "Deer and Open Spaces." *Public Garden* 3, no. 2 (1988).

De Groot, S. J. "Collecting and Processing Cacti into Herbarium Specimens, Using Ethanol and Other Methods." *Systematic Botany* 36, no. 4 (2011).

deJong, P. C. "The Establishment of a Complete Reference Collection of Cultivated Betula." *Acta Horticulturae* 182 (1986).

DeMarie, E. T., III. "The Value of Plant Collections." *Public Garden* 11, no. 2 (1996).

Denver Botanic Gardens. "Citizen Science." https://www.botanicgardens.org/citizen -science.

———. *Living Collections Management Plan*. Denver, CO: DBG, 2017.

de von Flynt, M., ed. *The ABCs of Collections Care*. Winnipeg: Association of Manitoba Museums, 1989.

de Vere, Natasha. "Barcode Wales." *BG Journal* 9, no. 1 (2012).

de Vicente, M. C. "Collecting DNA for Conservation." In *Collecting Plant Genetic Diversity: Technical Guidelines—2011 Update*, edited by L. Guarino et al. Rome, Italy: Bioversity International, 2011.

DeWolf, G. "Educational Systems in Public Gardens." *Longwood Graduate Program Seminars* 5 (1973).

Dickie, J. B., et al., eds. *Seed Management Techniques for Gene Banks*. Rome, Italy: IBGR, 1984.

DIVA-GIS. "DIVA-GIS." http://www.diva-gis.org.

Dobbs, R. C., et al. *Guide to Collecting Cones of B.C. Conifers*. Joint Report 3. Victoria: British Columbia Forest Service / Canadian Forest Service, 1976.

Dodge, A. F. "North Central Regional Plantings of Woody Ornamental and Shelter Plant Introductions." *Proceedings of the Plant Propagators Society* 12 (1962).

Donaldson, J. S. "Botanic Gardens Science for Conservation and Global Change." *Trends in Plant Science* 14 (2009).

Dondis, D. *A Primer on Visual Literacy.* Cambridge, MA: Massachusetts Institute of Technology, 1973.

Donnelly, Gerard T., and William R. Feldman. "How to Write a Plant Collections Policy." *Public Garden* 5, no. 1 (1990).

Dosmann, Michael S. "Research in the Garden: Averting the Collections Crisis." *Botanical Review* 72, no. 3 (2006).

Dosmann, Michael S., and P. Del Tredici. "Plant Introduction, Distribution, and Survival: A Case Study of the 1980 Sino-American Botanical Expedition." *BioScience* 53, no. 6 (2003).

———. "The Sino-American Botanical Expedition of 1980: A Retrospective Analysis of Success." *HortScience* 40, no. 2 (2005).

Dosmann, Michael S., and K. Port. "The Art and Act of Acquisition." *Arnoldia* 73, no. 4 (2016).

Dougherty, Carissa K. "Disaster Recovery Planning: An IT Perspective." *Public Garden* 31, no. 3 (2016).

DuBois, J. "Care of Natural History Specimens." *Dawson & Hind* 10, no. 4 (1982).

Dudley, D., and I. Wilkinson. *Museum Registration Methods.* 3rd ed. Washington, DC: American Association of Museums, 1979.

Dulloo, M. E., J. Labokas, J. M. Iriondo, N. Maxted, A. Lane, E. Laguna, A. Jarvis, and S. P. Kell. "Genetic Reserve Location and Design." In *Conserving Plant Genetic Diversity in Protected Areas*, edited by J. M. Iriondo, N. Maxted, and M. E. Dulloo. Wallingford, UK: CAB International, 2008.

Dunn, John R. "Museum Interpretation/Education: The Need for Definition." *Gazette* 10, no. 1 (1977).

Ebbels, D. L. *Principles of Plant Health and Quarantine.* Wallingford, UK: CAB International, 2003.

Eberhart, S. A., et al. "Strategies for Long-Term Management of Germplasm Collections." In *Genetics and Conservation of Rare Plants*, edited by Donald Falk and T. Holzinger. New York: Oxford University Press, 1991.

Ebert, A.W., J. L. Karihaloo, and A. Ferraira. "Opportunities, Limitations and Needs for DNA Banks." In *DNA Banks—Providing Novel Option for Genebanks? Topical Reviews in Agricultural Biodiversity*, edited by M. C. Vicente and M. S. Andersson. Rome, Italy: International Plant Genetic Resources Institute, 2006.

EC and TRAFFIC. *Reference Guide to the European Union Wildlife Trade Regulations.* Brussels: European Commission and TRAFFIC, 2015. https://ec.europa.eu/environment/cites/legis_refguide_en.htm.

Edwards, R. Y. "Research: A Museum Cornerstone." In *Museum Collections: Their Roles in Biological Research*, edited by E. H. Miller. Victoria: British Columbia Provincial Museum, 1985.

———. "Research and Education: Museums Need Both." *Muse* 3, no. 1 (1985).

Elias, T. S. "About This Issue." *Public Garden* 6, no. 3 (1991).

Ellegren, H., and N. Galtier. "Determinants of Genetic Diversity." *Nature Reviews* 17 (2016).

Elliot, S., D. Blakesley, and K. Hardwick. *Restoring Tropical Forests: A Practical Guide.* Richmond, Surrey, UK: Royal Botanic Gardens, Kew, 2013.

Ellis, R. H. "Revised Table of Seed Storage Characteristics." *Plant Genetic Resources Newsletter* 58 (1984).

Ellis, R. H., and E. H. Roberts. *Use of Deep-Freeze Chests for Medium and Long-Term Storage of Small Seed Collections.* Rome, Italy: IBGR, 1982.

Ellis, R. H., et al. *Handbook of Seed Technology for Gene Banks.* Vol. 1, *Principles and Methodology.* Rome, Italy: IBGR, 1985.

———. "Procedures for Monitoring the Viability of Accessions during Storage." In *Crop Genetic Resources: Conservation and Evaluation,* edited by J. H. W. Holden and J. T. Williams. London: Allen and Unwin, 1984.

Elsik, S. "From Each a Voucher: Collecting in the Living Collections." *Arnoldia* 49, no. 1 (1989).

Elzinga, C. L., D. W. Salzer, and J. W. Willoughby. *Measuring & Monitoring Plant Populations.* Denver, CO: U.S. Department of the Interior, Bureau of Land Management, 1998.

Engels, J. M. M., and L. Visser, eds. *A Guide to Effective Management of Germplasm Collections.* IPGRI Handbooks for Genebanks 6. Rome, Italy: International Plant Genetic Resources Institute, 2003.

Ensslin, Andreas, and Sandrine Godefroid. "How the Cultivation of Wild Plants in Botanic Gardens Can Change Their Genetic and Phenotypic Status and What This Means for Their Conservation Value." *Sibbaldia* 17 (2019).

Ensslin, A., O. Tschöpe, M. Burkart, and J. Joshi. "Fitness Decline and Adaptation to Novel Environments in *Ex Situ* Plant Collections: Current Knowledge and Future Perspectives." *Biological Conservation* 192 (2015).

European Native Seed Conservation Network. *ENSCONET Seed Collecting Manual for Wild Species.* Richmond, Surrey, UK: Royal Botanic Gardens, Kew, 2009.

———. *Curation Protocols and Recommendations.* Richmond, Surrey, UK: Royal Botanic Gardens, Kew, 2009.

Fairchild Tropical Garden. *Fairchild Tropical Garden Bulletin.* Miami, FL: FTG, July 1961.

Falk, D. A., and T. Holzinger. "Sampling Guidelines for Conservation of Endangered Plants." In *Genetics and Conservation of Rare Plants,* edited by D. A. Falk and T. Holzinger. New York: Oxford University Press, 1991.

Falk, D. A., C. I. Millar, and M. Olwell, eds. *Restoring Diversity: Strategies for Reintroduction of Endangered Plants.* Washington, DC: Island Press, 1996.

Falk, D. A., M. A. Palmer, and J. B. Zedler, eds. *Restoration Ecology.* Washington, DC: Island Press, 2006.

Fant, Jeremie B., et al. "What to Do When We Can't Bank on Seeds." *American Journal of Botany* 103, no. 9 (2016).

Fay, Neville. "Natural Fracture Pruning Techniques and Coronet Cuts." April 2003. https://www.semanticscholar.org/paper/Natural-Fracture-Pruning-Techniques-and-Coronet-Fay/6da7d4f3c302240af3cacc96e25a433406f31f88.

Fleming, D., and R. Higginson. "Collections Management: An Independent Approach." *Museums Journal* 84, no. 2 (1984).

Fleming, David. "Sense or Suicide? The Disposal of Museum Collections." *Museums Journal* 87, no. 2 (1987).

Flower, Charles E., et al. "Optimizing Conservation Strategies for a Threatened Tree Species: In Situ Conservation of White Ash (*Fraxinus americana* L.) Genetic Diversity through Insecticide Treatment." *Forests* 9 (2018).

Folsom, J. "The Issues and Ethics of Plant Collections." *Public Garden* 11, no. 4 (1996).

Fong, H. "The Art and Science of Management." *Public Garden* 4, no. 3 (1989).

Food and Agriculture Organization. *Genebank Standards for Plant Genetic Resources for Food and Agriculture*. Rev. ed. Rome, Italy: FAO, 2014.

Foulger, N. "Interpretation Programs for Botanic Gardens and Arboretums." *Longwood Graduate Program Seminars* 11 (1979).

Frachon, N. "Plant Health Protocols for the Reintroduction of Native Plants." *Sibbaldia* 11 (2013).

Franchon, Natasha, Martin Gardner, and David Rae. "The Data Capture Project at the Royal Botanic Garden Edinburgh." *Sibbaldia* 7 (2009).

Frankham, R., C. J. A. Bradshaw, and B. W. Brook. "Genetics in Conservation Management: Revised Recommendations for the 60/600 Rules, Red List Criteria and Population Viability Analyses." *Biological Conservation* 170 (2014).

Freedman, G. "The Changing Nature of Museums." *Curator* 43, no. 4 (2000).

Freeman, R. *A Guide to Collections Records Management*. Toronto: Ontario Ministry of Citizenship and Culture, 1985.

Friedmann, H. "The Curator: Introduction." *Curator* 6, no. 4 (1963).

Funk, Vicki A., et al. "Guidelines for Collecting Vouchers and Tissues Intended for Genomic Work (Smithsonian Institution): Botany Best Practices." *Biodiversity Data Journal* 5 (2017).

Galbraith, D. A. "Natural Areas at Public Gardens: Creative Tensions and Conservation Opportunities." *Public Garden* 18, no. 3 (2003).

Galbraith, David. "Suggestions for Field Preservation of Plant Tissue for Subsequent DNA Studies." *CBCN Newsletter* 5, no. 3 (November 2000).

Galbraith, M., B. Bollard-Breen, and D. R. Towns. "The Community-Conservation Conundrum: Is Citizen Science the Answer?" *Land* 5 (2016).

Gates, G. "Characteristics of an Exemplary Plant Collection." *Public Garden* 21, no. 1 (2006).

Geldman, J., M. Barnes, L. Coad, I. Craigie, M. Hockings, and N. Burgess. "Effectiveness of Terrestrial Protected Areas in Reducing Biodiversity and Habitat Loss." CEE 10-007 (2013).

Gemeinholzer, B., et al. "Organizing Specimen and Tissue Preservation in the Field for Subsequent Molecular Analyses." In *Manual on Field Recording Techniques and Protocols for All Taxa Biodiversity Inventories*, edited by J. Eymann et al. ABCTaxa 8. N.p.: Belgian Development Cooperation, 2010.

George, E. F. *Plant Propagation by Tissue Culture: Part 1*. Westbury, UK: Exergetics, 1993.

———. *Plant Propagation by Tissue Culture: Part 2*. Westbury, UK: Exergetics, 1996.

George, G., and Cindy Sherrell-Leo. *Starting Right: A Basic Guide to Museum Planning*. Nashville, TN: American Association for State and Local History, 1986.

Gilbert, Stephen P. *Planning for the Future: Long-Range Planning for Associations*. Washington, DC: Foundation of the American Society of Association Executives, 1973.

Gimenez Ferrer, R. M., and V. B. Steward. "IPM at Public Gardens." *Public Garden* 10, no. 3 (1995).

Given, David R. *Principles and Practice of Plant Conservation*. Portland, OR: Timber Press, 1995.

Glaser, Jane R., and Artemis A. Zenetou. *Museums: A Place to Work*. Ottawa: Canadian Museums Association, 1979.

Global Genomic Biodiversity Network. "About GGBN." https://wiki.ggbn.org/ggbn/About_GGBN.

———. "The GGBN Data Portal." http://www.ggbn.org/ggbn_portal/.

Godefroid, S., et al. "How Successful Are Plant Species Reintroductions?" *Biological Conservation* 144, no. 2 (2011).

Goff, P. "Using Temporary Exhibits as Valuable Interpretive Tools." *Longwood Graduate Program Seminars* 13 (1981).

Goodman, J. "Horticultural Potential in Cultural Properties." *Longwood Graduate Program Seminars* 11 (1979).

Gordon, D., and C. Gantz. "Screening New Plant Introductions for Potential Invasiveness: A Test of Impacts for the United States." *Conservation Letters* 1 (2008).

Gorzelak, Monika A., Amanda K. Asay, Brian J. Pickles, and Suzanne W. Simard. "Inter-plant Communication through Mycorrhizal Networks Mediates Complex Adaptive Behaviour in Plant Communities." *AoB Plants* 7 (2015).

Government of Canada. "Recovery Strategies." https://www.canada.ca/en/environment-climate-change/services/species-risk-public-registry/recovery-strategies.html.

Graham, W. "Tree Conservation." *Public Garden* 4, no. 3 (1989).

Graham-Bell, Maggie. *Preventive Conservation: A Manual*. 2nd ed. Victoria: British Columbia Museums Association, 1986.

Gratzfeld, J., ed. *From Idea to Realisation: BGCI's Manual on Planning, Developing and Managing Botanic Gardens*. Richmond, Surrey, UK: Botanic Gardens Conservation International, 2016.

Greenwood, O., et al. "Using In Situ Management to Conserve Biodiversity under Climate Change." *Journal of Applied Ecology* 53 (2016).

Greiber, T., et al. *An Explanatory Guide to the Nagoya Protocol on Access and Benefit-Sharing*. IUCN Environmental Policy and Law Paper 83. Gland, Switzerland and Cambridge: IUCN, 2012.

Griffith, M. P., et al. "Can a Botanic Garden Cycad Collection Capture the Genetic Diversity in a Wild Population?" *International Journal of Plant Sciences* 176, no. 1 (2015).

Griffith, M. Patrick, et al. *Toward the Metacollection: Safeguarding Plant Diversity and Coordinating Conservation Collections.* San Marino, CA: Botanic Gardens Conservation International, 2019.

Grogg, C. "How to Establish a Coordinate Locating System." *Public Garden* 4, no. 3 (1989).

Gropp, R. E. "Are University Natural Science Collections Going Extinct?" *BioScience* 53, no. 6 (2003).

Gross, Michael. "Can Botanic Gardens Save All Plants?" *Current Biology* 28 (2018).

Grubb, P. J. "The Maintenance of Species-Richness in Plant Communities: The Importance of the Regeneration Niche." *Biological Reviews* 52 (1977).

Guarino, L. "Secondary Sources on Cultures and Indigenous Knowledge Systems." In *Collecting Plant Genetic Diversity*, edited by L. Guarino et al. Wallingford, Oxon, UK: CAB International, 1995.

Guarino, L., et al., eds. *Collecting Plant Genetic Diversity.* Wallingford, Oxon, UK: CAB International, 1995.

———. *Collecting Plant Genetic Diversity: Technical Guidelines—2011 Update.* Rome, Italy: Bioversity International, 2011.

Guerrant, E., and Linda McMahan. "Practical Pointers for Conserving Genetic Diversity in Botanic Gardens." *Public Garden* 6, no. 3 (1991).

Guerrant, E. O. "Designing Populations: Demographic, Genetic, and Horticultural Dimensions." In *Restoring Diversity: Strategies for Reintroduction of Endangered Species*, edited by D. A. Falk, C. I. Miller, and M. Olwell. Washington, DC: Island Press, 1996.

Guerrant, E. O., P. Fiedler, K. Havens, and M. Maunder. "Revised Genetic Sampling Guidelines for Conservation Collections of Rare and Endangered Plants: Supporting Species Survival in the Wild." In *Ex Situ Plant Conservation: Supporting Species Survival in the Wild*, edited by E. O. Guerrant, K. Havens, and M. Maunder. Washington, DC: Island Press, 2004.

Guerrant, E. O., K. Havens, and P. Vitt. "Sampling for Effective *Ex Situ* Plant Conservation." *International Journal of Plant Sciences* 175, no. 1 (2014).

Guerrant, E. O., and T. N. Kaye. "Reintroduction of Rare and Endangered Plants: Common Factors, Questions, and Approaches." *Australian Journal of Botany* 55 (2007).

Gunderson, L. "How the Endangered Species Act Promotes Unintelligent, Misplaced Tinkering." *Ecology and Society* 18 (2013).

Guthe, Carl. *Documenting Collections: Museum Registration & Records.* American Association for State and Local History Technical Leaflet 11. Nashville, TN: American Association for State and Local History, 1970.

———. *The Management of Small History Museums.* 2nd ed. Nashville, TN: American Association for State and Local History, 1964.

Gutowski, R. "Documenting Humphry Marshall's Botanic Garden Plantings." *Longwood Graduate Program Seminars* 17 (1985).

Haggerty, B. P., A. A. Hove, and S. A. Mazer. *Primer on Herbarium-Based Phenological Research*. Santa Barbara: University of California, 2012.

Halbrooks, Mary C. "Decision Making in the Restoration of a Historic Landscape." *Public Garden* 20, no. 1 (2005).

Hall, A. V. "Pest Control in Herbaria." *Taxon* 37 (1988).

Hammer, K., and Y. Morimoto. "Chapter 7: Classifications of Infraspecific Variation in Crop Plants." In *Collecting Plant Genetic Diversity: Technical Guidelines—2011 Update*, edited by L. Guarino et al. Rome, Italy: Bioversity International, 2011.

Hamrick, J. L. "The Distribution of Genetic Variation within and among Plant Populations." In *Genetics and Conservation: A Reference for Managing Wild Animal and Plant Populations*, edited by S. M. Schonewald-Cox et al. Menlo Park, CA: Benjamin/Cummings, 1983.

Handler, S., C. Pike, and B. St. Clair. "Assisted Migration." USDA Forest Service Climate Change Resource Center, 2018. https://www.fs.usda.gov/ccrc/topics/assisted-migration.

Hangay, G., and M. Dingley. *Biological Museum Methods*. Vol. 2, *Plants and Invertebrates*. Sydney: Academic Press, 1985.

Harding, C., and M. Williams. *Designing a Monitoring Project for Significant Native Flora*. Version Number 1.0, January 2010. Prepared for Resource Condition Monitoring Project: Significant Native Species and Ecological Communities. Department of Environment and Conservation, 2010.

Hardwick, K. A., et al. "The Role of Botanic Gardens in the Science and Practice of Ecological Restoration." *Conservation Biology* 25, no. 2 (2011).

Harris, R. W. "A Management Approach to Maintenance." Presented at the Golf Course Superintendent's Institute, Minneapolis, MN, March 1976.

Hartman, H., and Suzanne B. Schell. "Institutional Master Planning for Historical Organizations and Museums." AASLH Technical Report 11. *History News* 41, no. 6 (1986).

Hartmann, H. T., D. E. Kester, F. T. Davies, and R. L. Geneve. *Plant Propagation: Principles and Practices*. 9th ed. Harlow, UK: Pearson Education, 2017.

Harty, M. C., et al. "Cataloguing in the Metropolitan Museum of Art, with a Note on Adaptations for Small Museums." In *Museum Registration Methods*, edited by D. Dudley and I. Wilkinson. Washington, DC: American Association of Museums, 1979.

Haskins, K. E., and B. Keel. "Managed Relocation: Panacea or Pandemonium?" In *Plant Reintroduction in a Changing Climate: Promises and Perils*, edited by J. Maschinski and K. E. Haskins. Washington, DC: Island Press, 2012.

Haskins, K. E., and V. Pence. "Transitioning Plants to New Environments: Beneficial Applications of Soil Microbes." In *Plant Reintroduction in a Changing Climate: Promises and Perils*, edited by J. Maschinski and K. E. Haskins, 89–108. Washington, DC: Island Press, 2012.

Hasselkus, E. R. "Maximizing the Use of Collections." *Longwood Graduate Program Seminars* 17 (1985).

———. "Whitespire Birch: From Collection to Introduction." *Public Garden* 2, no. 3 (1987).

Hatcher, Thomas, and Rosemary Hatcher. *The Definitive Guide to Long-Range Planning: Designed Especially for Non-profit and Volunteer Organizations.* Minneapolis, MN: Futures Unlimited, 1981.

Havens, K. "Developing an Invasive Plant Policy: The Chicago Botanic Garden's Experience." *Public Garden* 17, no. 4 (2002).

———. "The Role of Botanical Gardens and Arboreta in Restoring Plants from Populations to Ecosystems." In *Plant Conservation Science and Practice: The Role of Botanic Gardens,* edited by S. Blackmore and S. Oldfield. New York: Cambridge University Press, 2017.

Havens, K., E. O. Guerrant Jr., P. Vitt, and M. Maunder. "Conservation Research and Public Gardens." *Public Garden* 19, no. 3 (2004).

Havens, K., P. Vitt, M. Maunder, E. O. Guerrant Jr., and K. Dixon. "Ex Situ Plant Conservation and Beyond." *BioScience* 56 (2006): 525–31.

Havinga, Reinout, and Havard Ostgaard. "Barcodes Are Dead, Long Live Barcodes! Improving the Inventory of Living Plant Collections Using Optical Technology." *Sibbaldia* 14 (2016).

Hawkes, J. G. *Crop Genetic Resources Field Collection Manual.* Rome, Italy: IBPGR and EUCARPIA, 1980.

———. "A Strategy for Seed Banking in Botanic Gardens." In *Botanic Gardens and the World Conservation Strategy,* edited by D. Bramwell et al. London: Academic Press, 1987.

———. "Theory and Practice of Collecting Germplasm in a Centre of Diversity." In *Plant Genetic Resources of Ethiopia,* edited by J. M. Engels et al. Cambridge, U.K.: Cambridge University Press, 1991.

Heckon, D. "Free of Guilt and Full of Care: Some Aspects of Legal Requirements and Problems in Collecting." *Gazette* 12, no. 2 (1979).

Heeps, A. "The Arboretum from the Horticulturist's Viewpoint." *Longwood Graduate Program Seminars* 4 (1972).

Hensley, D. L. "Evaluation of Ten Landscape Trees for the Midwest." *Journal of Environmental Horticulture* 9, no. 3 (1991).

Herrmann, L., et al. "Chapter 26: Collecting Symbiotic Bacteria and Fungi." In *Collecting Plant Genetic Diversity: Technical Guidelines—2011 Update,* edited by L. Guarino et al. Rome, Italy: Bioversity International, 2011.

Heyning, J. E. "The Future of Natural History Collections." *Collections* 1, no. 1 (2004).

Heywood, V. H. "The Future of Plant Conservation and the Role of Botanic Gardens." *Plant Diversity* 39 (2017).

———. "A Global Strategy for the Conservation of Plant Diversity." *Grana* 34, no. 6 (1995).

———. "*In Situ* Conservation of Plant Species—an Unattainable Goal?" *Israel Journal of Plant Sciences* 63 (2015).

———. "Is the Conservation of Vegetation Fragments and Their Biodiversity Worth the Effort?" In *Ecosystem Management: Questions for Science and Society*, edited by E. Maltby, M. Holdgate, M. Acreman, and A. G. Weir. London: Royal Holloway Institute for Environmental Research, University of London, 1999.

———. "An Overview of *In Situ* Conservation of Plant Species in the Mediterranean." *Flora Mediterranea* 24 (2014): 5–24.

———. "Role of Seed Lists in Botanic Gardens Today." In *Conservation of Threatened Plants*, edited by John Simonds. New York: Plenum Press, 1976.

Heywood, V. H., and M. E. Dulloo. In Situ *Conservation of Wild Plant Species. A Critical Global Review of Good Practices*. IPGRI Technical Bulletin 11. Rome, Italy: FAO and IPGRI, 2005.

Heywood, V. H., and S. Sharrock. *European Code of Conduct for Botanic Gardens on Invasive Alien Species*. Richmond, Surrey, UK: Council of Europe, Strasbourg, and Botanic Gardens Conservation International, 2013.

Heywood, V. H., K. Shaw, Y. Harvey-Brown, and P. Smith, eds. *BGCI and IABG's Species Recovery Manual*. Richmond, Surrey, UK: Botanic Gardens Conservation International, 2018. https://www.bgci.org/wp/wp-content/uploads/2019/04/Species_Recovery_Manual.pdf.

Hijmans, R. J., M. Schreuder, J. De la Cruz, and L. Guarino. "Using GIS to Check Co-ordinates of Genebank Accessions." *Genetic Resources and Crop Evolution* 46, no. 3 (1999).

Hill, D., M. Fasham, G. Tucker, M. Shewry, and P. Shaw. *Handbook of Biodiversity Methods: Survey, Evaluation and Monitoring*. Cambridge: Cambridge University Press, 2005.

Hitchcock, Ann. "Collections Policy: An Example from the Manitoba Museum of Man and Nature." *Gazette* 13, no. 2 (1980).

Hoachlander, M. *Profile of a Museum Registrar*. Washington, DC: Academy for Educational Development, 1979.

Hoban, S., and A. Strand. "*Ex Situ* Seed Collections Will Benefit from Considering Spatial Sampling Design and Species' Reproductive Biology." *Biological Conservation* 187 (2015).

Holt, Daniel D. "An Essay on Deaccessioning: A Professional Point of View." *Midwest Museums Quarterly* 41, nos. 1 and 2 (1981).

Homulos, P., and C. Sutyla. *Information Management in Canadian Museums*. Background Paper for Museum Policy Working Group. Ottawa: Communications Canada, 1988.

Horie, C. V. "Who Is a Curator?" *International Journal of Museum Management and Curatorship*, no. 5 (1986).

Horwich, R. H., and J. Lyon. "Community Conservation: Practitioners' Answer to Critics." *Oryx* 41 (2017).

Howard, R. A. "Comments on 'Seed Lists.'" *Taxon* 13 (1964).

Hoyt, Sean, et al. "A Tree Tour with Radio Frequency Identification (RFID) and a Personal Digital Assistant (PDA)." Paper presented at IEEE IECon '03, Roanoke, VA, November 2–6, 2003.

Huckins, C. "Collections: Forecasting the Future." *Longwood Graduate Program Seminars* 13 (1981).

Hughes, K., K. Price, and I. Lawrie. "Developing a Carbon Management Plan for the Royal Botanic Gardens Edinburgh." *Sibbaldia* 9 (2013).

Hummer, Kim E., ed. *Operations Manual.* Corvallis, OR: National Clonal Germplasm Repository, 1995.

Hunter, D., and V. Heywood, eds. *Crop Wild Relatives: A Manual of In Situ Conservation.* London: Earthscan, 2011.

Hyland, R. "Interpreting Landscapes and Gardens around Historic Buildings." *Longwood Graduate Program Seminars* 14 (1982).

Ibanez, I., et al. "Predicting Biodiversity Change: Outside the Climate Envelope, beyond the Species Area Curve." *Ecology* 87, no. 8 (2006).

Illg, Paul L. "The Recruitment and Training of Curators for Natural Science Museums." *Curator* 6, no. 4 (1963).

International Board for Plant Genetic Resources. *Conservation and Movement of Vegetatively Propagated Crops: In Vitro Culture and Disease Aspects.* Rome, Italy: IBPGR, 1988.

International Council of Museums. "Ethics of Acquisition." *ICOM News* 23, no. 2 (1970).

International Plant Sentinel Network. "Home." https://www.plantsentinel.org/index/.

International Union for the Conservation of Nature and Natural Resources. *Guidelines for Reintroductions and Other Conservation Translocations.* Ver. 1.0. Gland, Switzerland: IUCN, 2013.

———. *The IUCN Red List of Threatened Species.* Ver. 2020-3. Gland, Switzerland: IUCN, 2020.

———. *WCPA Best Practice Guidance on Recognising Protected Areas and Assigning Management Categories and Governance Types.* Best Practice Protected Area Guidelines Series 21. Gland, Switzerland: IUCN, 2013.

International Union for the Conservation of Nature and Natural Resources, Botanic Gardens Secretariat. *The Botanic Gardens Conservation Strategy.* London: IUCN, 1989.

International Union for Conservation of Nature and Natural Resources, Species Survival Commission. *Strategic Planning for Species Conservation: A Handbook.* Version 1.0. Gland, Switzerland: IUCN/SSC, 2008.

Iriondo, J. M., et al. "Plant Population Monitoring Methodologies for the *In Situ* Genetic Conservation of CWR." In *Conserving Plant Diversity in Protected Areas,* edited by J. M. Iriondo, N. Maxted, and M. E. Dulloo. Wallingford, Oxon, UK: CAB International, 2008.

Jackson, Peter Wyse. "Convention on Biological Diversity." *Public Garden* 12, no. 2 (1997).

———. "Developing Botanic Garden Policies and Practices for Environmental Sustainability." *BG Journal* 6, no. 2 (2009).

———. *A Handbook for Botanic Gardens on the Reintroduction of Plants to the Wild.* London: Botanic Gardens Conservation International, 1995.

Jefferson, L., K. Havens, and J. Ault. "Implementing Invasive Screening Procedures: The Chicago Botanic Garden Model." *Weed Technology* 18 (2004).

Jenderek, Maria M., and Barbara M. Reed. "Cryopreserved Storage of Clonal Germplasm in the USDA National Plant Germplasm System." *In Vitro Cellular & Developmental Biology—Plant* 4 (2017).

Jones, L., et al. "Collections Policy Development: A Case Study." *Longwood Graduate Program Seminars* 16 (1984).

———. "Collections Policy: The Basics." *Public Garden* 1, no. 2 (1986).

Jones-Roe, C., and J. Shaw. "Conservation in North American Gardens." *Public Garden* 3, no. 1 (1988).

Junak, S. "The Herbarium: A Botanical Encyclopedia." *Public Garden* 6, no. 3 (1991).

Kadis, C., C. A. Thanos, and E. Laguna. "The Future of PMRs: Towards a European PMR Network." In *Plant Micro-Reserves: From Theory to Practice: Experiences Gained from 168 169 EU LIFE and Other Related Projects*, edited by C. Kadis et al. Athens, Greece: PlantNet CY Project Beneficiaries, Utopia, 2013.

Kahn, R. P. *Plant Protection and Quarantine.* Vols. 1–3. Boca Raton, FL: CRC Press, 1989.

Kallow, S. *UK National Tree Seed Project: Seed Collecting Manual.* London: Royal Botanic Gardens, Kew, 2014.

Kamisher, L. M. "A Model for Computerization of Museum Collections." *International Journal of Museum Management and Curatorship* 8, no. 1 (1989).

Kell, S. P., E. Laguna, J. M. Iriondo, and M. E. Dulloo. "Population and Habitat Recovery Techniques for the *In Situ* Conservation of Plant Genetic Diversity." In *Conserving Plant Genetic Diversity in Protected Areas*, edited by J. M. Iriondo, N. Maxted, and M. E. Dulloo. Wallingford, Oxon, UK: CAB International, 2008.

Keller, E. R. J., A. Kaczmarczyk, and A. Senula. "Cryopreservation for Plant Genebanks: A Matter between High Expectations and Cautious Reservation." *Cryoletters* 29, no. 1 (2008).

Kenis, M., et al. *Guidelines on Legislation, Import Practices and Plant Quarantine for Botanic Gardens and Kindred Institutions.* Cambridge: PlantNetwork, 2006.

Kennedy, K. "Twenty Years of Ex Situ Plant Conservation." *Public Garden* 19, no. 3 (2004).

Kenworthy, M. A., et al. *Preserving Field Records: Archival Techniques for Archaeologists and Anthropologists.* Philadelphia: University Museum of Pennsylvania, 1985.

Kister, Shawn. "Tree Management and Climate Change." *Public Garden* 31, no. 2 (2016).

Knudson, J. W. *Biological Techniques: Collecting, Preserving, and Illustrating Plants and Animals*. New York: Harper & Row, 1966.

————. *Collecting and Preserving Plants and Animals*. New York: Harper & Row, 1972.

Koller, G. "Landscape Curation: Maintaining the Living Collections." *Arnoldia* 49, no. 1 (1989).

Krahn, A. H. "Access and the Public Trust: A Conservation Perspective." *Muse* 1, no. 4 (1984).

Kramer, A. T., and K. Havens. "Plant Conservation Genetics in a Changing World." *Trends in Plant Science* 14 (2009).

Krishnamurthy, K. V. *Textbook of Biodiversity*. Enfield, NH: Science, 2003.

Kunz, Michael, et al. "Germination and Propagation of *Astragalus michauxii*, a Rare Southeastern US Endemic Legume." *Native Plants Journal* 17, no. 1 (2016).

Kyte, L. A., and J. Kleyn. *Plants from Test Tubes. An Introduction to Micropropagation*. 3rd ed. Portland, OR: Timber Press, 2001.

Labriola, L. "Outsourcing: A Maintenance Alternative." *Public Garden* 10, no. 2 (1995).

Landsberg, J., and D. S. Gillienson. "Repetitive Sampling of the Canopies of Tall Trees Using a Single Rope Technique." *Australian Forestry* 45 (1982).

Lane, M. A. "Roles of Natural History Collections." *Annals of the Missouri Botanical Garden* 83, no. 4 (1996).

Langhammer, P. F., et al. *Identification and Gap Analysis of Key Biodiversity Areas: Targets for Comprehensive Protected Area Systems*. Gland, Switzerland: IUCN, 2007.

Lapeña, I., I. López, and M. Turdieva. *Guidelines: Access and Benefit-Sharing in Research Projects*. Rome, Italy: Bioversity International, 2012.

Laren, J., and E. Glasener. "The Nuts and Bolts of Labeling." *Public Garden* 2, no. 4 (1987).

Laub, Richard S. "The Natural History Curator: A Personal View." *Curator* 28, no. 1 (1985).

Leadlay, E., ed. *Darwin Manual*. Richmond, Surrey, UK: Botanic Gardens Conservation International, 1998.

Lee, L. "China Valley at the USNA: An Ecologically Designed Depository for Wild-Collected Chinese Germplasm." *Proceedings of the International Symposium on Botanical Gardens*, Kunming, China, 1990.

Le Saout, S., et al. "Protected Areas and Effective Biodiversity Conservation." *Science* 342 (2013).

Lester, Joan. "A Code of Ethics for Curators." *Museum News* 61, no. 5 (1983).

Lewis, C. "A Horticulturist Looks at Management." *Longwood Graduate Program Seminars* 18 (1986).

Li, D. Z., and H. W. Pritchard. "The Science and Economics of Ex Situ Plant Conservation." *Trends in Plant Science* 14, no. 11 (2009).

Liang, M., D. Johnson, D. F. R. P. Burslem, et al. "Soil Fungal Networks Maintain Local Dominance of Ectomycorrhizal Trees." *Nature Communications* 11, no. 2636 (2020). https://doi.org/10.1038/s41467-020-16507-y.

Light, R. B., et al. *Museum Documentation Systems: Developments and Applications.* London: Butterworths, 1986.

Lighty, Richard W. "A History of the North American Public Garden." *Public Garden* 11, no. 1 (1996).

————. "Toward a More Rational Approach to Plant Collections." *Longwood Graduate Program Seminars* 16 (1984).

Lindenmayer, D. B., and G. E. Likens. "Adaptive Monitoring: A New Paradigm for Long-Term Research and Monitoring." *Trends in Ecology and Evolution* 24 (2009).

Linklater, S. "The Care of Archival Photograph Collections." *Dawson & Hind* 14, no. 2/3 (1988).

Littlefield, L. "Woody Ornamental Hardiness Trials at the University of Maine at Orono." Bulletin 599. Orono: Cooperative Extension Service, University of Maine, 1983.

Liu, H., et al. "Translocation of Threatened Plants as a Conservation Measure in China." *Conservation Biology* 29 (2015).

Liu, U., Elinor Breman, Tiziana Antonella Cossu, and Siobhan Kenney. "The Conservation Value of Germplasm Stored at the Millennium Seed Bank, Royal Botanic Gardens, Kew, UK." *Biodiversity Conservation* 27 (2018). https://doi.org/10.1007/s10531-018-1497-y.

Lobdell, Matthew S., and Patrick G. Thompson. "*Ex-situ* Conservation of *Quercus oglethorpensis* in Living Collections of Arboreta and Botanical Gardens." In *Proceedings of Workshop on Gene Conservation of Tree Species—Banking on the Future.* Chicago: U.S. Department of Agriculture, 2016.

Lochnan, K. A. "The Research Function as Philosopher's Stone." *Muse* 1, no. 3 (1983).

Lord, B. "The Purpose of Museum Exhibitions." In *The Manual of Museum Exhibitions,* edited by B. Lord and Gail Lord. Walnut Creek, CA: AltaMira Press, 2002.

Lord, Gail Dexter, and Kate Markert. *The Manual of Strategic Planning for Museums.* Lanham, MD: AltaMira Press, 2007.

Lowe, C. "Managing the Woodland Garden." *Public Garden* 10, no. 3 (1995).

Lundquist, Clive R., R. Sukri, and F. Metali. "How Not to Overwater a Rheophyte: Successful Cultivation of 'Difficult' Tropical Rainforest Plants Using Inorganic Compost Media." *Sibbaldia* 15 (2017).

Lybbert, E. K. "Making Policy Readable." *Administrative Management,* February 1978.

Ma, Y., G. Chen, R. E. Grumbine, A. Dao, W. Sun, and H. Guo. "Conserving Plant Species with Extremely Small Populations (PSESP) in China." *Biodiversity and Conservation* 22 (2013).

Madison Park Conservancy. *Madison Square Park Tree Conservation Plan.* New York: MPC, 2017.

Malaro, M. "Collections Management Policies." In *Collections Management,* edited by A. Fahy. London: Routledge, 1995.

————. "Deaccessioning: The Importance of Procedure." *Museum News* 66, no. 4 (1988).

———. *A Legal Primer on Managing Museum Collections*. Washington, DC: Smithsonian Institution, 1985.

———. "Rationale and Guidelines for Drafting Collections Management Policies." *Longwood Graduate Program Seminars* 16 (1984).

Mallarach, J.-M., ed. *Protected Landscapes and Cultural and Spiritual Values*. Heidelberg, Germany: Kasperek Verlag, IUCN, GTZ, and Ora Social de Caixa Catalunya, 2008.

Maney-O'Leary, Susan. "Preserving and Managing Design Intent in Historic Landscapes." *Public Garden* 7, no. 2 (1992).

Manis, W. "Establishment and Organization of a Plant Research Station." *Longwood Graduate Program Seminars* 3 (1971).

March, S. "New Plants: The Responsibility of Arboreta." *Longwood Graduate Program Seminars* 3 (1971).

Martin, J. *An Information Systems Manifesto*. Englewood Cliffs, NJ: Prentice Hall, 1984.

Martin, T. G., A. E. Camaclang, H. P. Possingham, L. A. Maguire, and I. Chades. "Timing of Protection of Critical Habitat Matters." *Conservation Letters* 10 (2016).

Maschinski, J., and K. E. Haskins, eds. *Plant Reintroduction in a Changing Climate: Promises and Perils*. Washington, DC: Island Press, 2012.

Mathias, M. "Making Use of the Living Collections." *Longwood Graduate Program Seminars* 7 (1975).

Maunder, M., S. Higgens, and A. Culham. "The Effectiveness of Botanic Garden Collections in Supporting Plant Conservation: A European Case Study." *Biodiversity and Conservation* 10, no. 3 (2001).

Maurice, A. C., et al. "Mixing Plants from Different Origins to Restore a Declining Population: Ecological Outcomes and Local Perceptions 10 Years Later." *PLoS One* 8 (2013): e50934.

Maxted, N., and L. Guarino. "Ecogeographic Surveys." In *Plant Genetic Conservation*, edited by N. Maxted, B. V. Ford-Lloyd, and J. G. Hawkes. Dordrecht: Springer, 1997. https://doi.org/10.1007/978-94-009-1437-7_4.

Maxted, Nigel, and Shelagh Kell. "A Role for Botanic Gardens in CWR Conservation for Food Security." *BG Journal* 10, no. 2 (2013).

Mayer, C. "The Contemporary Curator—Endangered Species or Brave New Profession?" *Muse* 9, no. 2 (Summer–Fall 1991).

McCree, M. L. "Good Sense and Good Judgement: Defining Collections and Collecting." *Drexel Library Quarterly* 11 (1975).

McDonald, T., G. D. Gann, J. Jonson, and K. W. Dixon. *International Standards for the Practice of Ecological Restoration—Including Principles and Key Concepts*. Washington, DC: Society for Ecological Restoration, 2016.

McGann, M. "Maintaining the Historic Garden." *Public Garden* 4, no. 3 (1989).

McGuire, D. K. "Garden Planning for Continuity at Dumbarton Oaks." *Landscape Architecture*, January 1981.

McInnes, David. "Commitment to Care: A Basic Conservation Policy for Community Museums." *Dawson & Hind* 14, no. 1 (1987).

McMahan, L. R. "Advice for the Modern Plant Explorer: Pack Your Permits." *Public Garden* 6, no. 4 (1991).

McMahan, L., and E. Guerrant. "Practical Pointers for Conserving Genetic Diversity in Botanic Gardens." *Public Garden* 6, no. 3 (1991).

Medbury, S. "Re-documenting Your Garden's Lost Accessions." *Public Garden* 7, no. 2 (1992).

Medbury, S., and J. R. McBride. "Urban Forestry and Plant Conservation: The Role of Botanical Gardens." *Public Garden* 9, no. 1 (1994).

Meier, L. "The Treatment of Historic Plant Material." *Public Garden* 7, no. 2 (1992).

Meilleur, B., et al. Introduction to "Profiles: Conservation Collections versus Collections with Conservation Values." *Public Garden* 12, no. 2 (April 1997).

Meilleur, Guy. "Retrenching Hollow Trees: An International Practice." http://www.historictreecare.com/wp-content/uploads/2012/05/RETRENCHING-HOLLOW-TREES-FOR-LIFE-131126.pdf/.

Menges, E. S., E. O. Guerrant, and S. Hamzé. "Effects of Seed Collection on the Extinction Risk of Perennial Plants." In *Ex Situ Plant Conservation: Supporting Species Survival in the Wild*, edited by E. O. Guerrant, K. Havens, and M. Maunder. Washington, DC: Island Press, 2004.

Menges, E. S., S. A. Smith, and C. W. Weekley. "Adaptive Introductions: How Multiple Experiments and Comparisons to Wild Populations Provide Insights into Requirements for Long-Term Introduction Success of an Endangered Shrub." *Plant Diversity* 38 (2016).

Menz, M. H., et al. "Reconnecting Plants and Pollinators: Challenges in the Restoration of Pollination Mutualisms." *Trends in Plant Sciences* 16 (2011).

Metcalf, R. L., and W. H. Luckmann, eds. *Introduction to Insect Pest Management.* New York: Wiley, 1975.

Metzger, Laurie. "Historic Tree Collection Management: A New Vision for Old Trees." Master's thesis, University of Delaware, 2014.

Meyer, Abby. "What's Our Backup Plan? A Look at Collections Security." *Public Garden* 33, no. 4 (2018).

Meyer, P. "A Case for Plant Exploration." *Public Garden* 2, no. 1 (1987).

Michener, D. "Collections as a Tool, Not a Purpose." *Public Garden* 11, no. 2 (1996).

———. "Collections Management." In *Public Garden Management*, edited by D. Rakow and S. Lee. Hoboken, NJ: Wiley, 2011.

———. "The Hows and Whys of Verifying a Living Collection." *Public Garden* 6, no. 3 (1991).

Miller, Alden H. "The Curator as a Research Worker." *Curator* 6, no. 4 (1963).

Miller, E. H., ed. *Museum Collections: Their Roles and Future in Biological Research.* Victoria: British Columbia Provincial Museum, 1985.

Milligan, H., S. Deinet, L. McRae, and R. Freeman. *Protecting Species: Status and Trends of the Earth's Protected Areas: Preliminary Report*. London: Zoological Society of London, 2014.

Miner, F. "The Botanic Garden from the Educator's Viewpoint." *Longwood Graduate Program Seminars* 4 (1972).

Mitchell, John F. "The Collections Policy as a Management Tool." *Museum Roundup*, no. 76 (1979).

Mitchell, R. "A Cost Effective Approach to Facilities Maintenance at Walt Disney World." *Longwood Graduate Program Seminars* 18 (1986).

Mori, J. L., and J. I. Mori. "Revising Our Conceptions of Museum Research." *Curator* 15, no. 3 (1972).

Moss, H., and L. Guarino. "Gathering and Recording Data in the Field." In *Collecting Plant Genetic Diversity*, edited by L. Guarino et al. Wallingford, Oxon, UK: CAB International, 1995.

Moss-Warner, C. "Current Design Policies of Botanical Gardens and Arboreta in the United States." *Longwood Graduate Program Seminars* 8 (1976).

Mounce, Ross, Paul Smith, and Samuel Brockington. "Ex-situ Conservation of Plant Diversity in the World's Botanic Gardens." *Nature Plants* 3 (2017).

Munley, M. E. "Asking the Right Questions: Evaluation and the Museum Mission." *Museum News* 64, no. 3 (1986).

Murch, S. J. "In Vitro Conservation of Endangered Plants in Hawaii." *Biodiversity* 5, no. 1 (2004).

Muse. "Being a Curator Is the Most Impossible Profession on Earth!" Winter 1986.

Museum Documentation Association. *Practical Museum Documentation*. 2nd ed. Cambridgeshire, UK: MDA, 1981.

Namoff, S., C. E. Husby, J. Francisco-Ortega, L. R. Noblick, C. E. Lewis, and M. P. Griffith. "How Well Does a Botanical Garden Collection of a Rare Palm Capture the Genetic Variation in a Wild Population?" *Biological Conservation* 143 (2010).

National Germplasm Resources Laboratory. *Plant Exploration Guidelines for FY1997 Proposals*. Beltsville, MD: U.S. Department of Agriculture–Agricultural Research Service, 1995.

National Park Service. *Museum Handbook*. Washington, DC: NPS, 1994.

National Tropical Botanical Garden. "Living Collections." https://ntbg.org/science/collections/living-collections/.

Neal, A., et al. "Evolving a Policy Manual." *Museum News* 56, no. 3 (1978).

Neale, J. R. "Genetic Considerations in Rare Plant Reintroduction: Practical Applications (or How Are We Doing?)." In *Plant Reintroduction in a Changing Climate: Promises and Perils*, edited by J. Maschinski and K. E. Haskins. Washington, DC: Island Press, 2012.

Nevling, Lorin I. "On Public Understanding of Museum Research." *Curator* 27, no. 3 (1984).

New England Wildflower Society. *State of the Plants*. Framingham, MA: NEWS, 2015.

New York Botanical Garden. "*Index Herbariorum*." http://sweetgum.nybg.org/science/ih/.

Nicholson, Thomas D. "NYSAM Policy on the Acquisition and Disposition of Collection Materials." *Curator* 17, no. 1 (1974).

North Carolina Botanical Garden. "Plant Conservation Programs." https://ncbg.unc.edu/research/plant-conservation/.

Noss, R. F. "Protecting Natural Areas in Fragmented Landscapes." *Natural Areas Journal* 7, no. 1 (1987).

Nudds, J. R., and C. W. Pettitt, eds. *The Value and Valuation of Natural Science Collections*. Proceedings of the International Conference, Manchester, 1995. London: Geological Society, 1997.

Nyberg, B. "Eyes in the Sky: Drones Proving Their Value in Plant Conservation." *BG Journal* 16, no. 2 (2019).

Obyrne, Lorraine. "Let's Plan a Collections Policy." *Museum Quarterly* 9, no. 1 (1980).

Ogilvie, F. M. P. *Reference Systems for Living Plant Collections*. Edinburgh: Edinburgh College of Art, 1983.

Oldfield, S., and N. McGough, comps. *A CITES Manual for Botanic Gardens*. 2nd ed. Richmond, Surrey, UK: Botanic Gardens Conservation International, 2007.

Oldfield, S., and A. C. Newton. *Integrated Conservation of Tree Species by Botanic Gardens: A Reference Manual*. Richmond, Surrey, UK: Botanic Gardens Conservation International, 2012.

Ontario Ministry of Citizenship and Culture. "Developing a Research Policy for Museums." In *Museum Notes for Community Museums in Ontario*. Vol. 7. Toronto: OMCC, 1983.

Ontario Ministry of Citizenship and Culture, Heritage Administration Branch. "Writing a Collections Management Policy for the Museum." In *Museum Notes for Community Museums in Ontario*. Vol. 3. Toronto: OMCC, 1983.

Ontario Museum Association, et al. *Museum and Archival Supplies Handbook*. 3rd rev. ed. Toronto: OMA and Toronto Area Archivists Group, 1985.

Ontario Museum Association, Working Committee on Acquisitions and Deaccessions. "Museum Collections: Policy Guidelines for Acquisitions and Deaccessions." *Museum Quarterly* 12, no. 4 (1984).

Orna, E., and C. Pettitt. *Information Handling in Museums*. London: Clive Bingley, 1980.

———. *Information Policies for Museums*. Cambridge: Museum Documentation Association, 1987.

Ornduff, R. "Using Living Collections—the Problems." *AABGA Bulletin* 12, no. 4 (1978).

Øvstebø, Gunnar, Alex Twyford, and Tina Westerlund. "Propagation of Dry Habitat Fern Species Using Spore Collections from Historic Herbarium Specimens." *Sibbaldia* 9 (2011).

Pammenter, N. W., and P. Berjak. "Physiology of Desiccation-Sensitive (Recalcitrant) Seeds and the Implications for Cryopreservation." *International Journal of Plant Sciences* 175 (2014).

Parr, A. E. "Curatorial Functions in Education." *Curator* 6, no. 4 (1963).

———. "Is There a Museum Profession?" *Curator* 3, no. 2 (1960).

———. "Origins, Nature and Purposes of Museum Policy." *Curator* 5, no. 3 (1962).

Parsons, B. "The Role of Woodlands at the Holden Arboretum." *Public Garden* 10, no. 3 (1995).

Pastore, Carla. "Plant Introduction Programs in the United States and Canada." *Public Garden* 2, no. 4 (1987).

———. "Woody Plant Introduction Programs." Master's thesis, University of Delaware, 1988.

Pearce-Moses, R. "Documentation: Compiling a Photographic Thesaurus." *International Journal of Museum Management and Curatorship* 8, no. 4 (1989).

Pearson, R. "Plant Breeding Research." *Public Garden* 2, no. 1 (1987).

Peart, B., and J. G. Woods. "A Communication Model as a Framework for Interpretive Planning." *Interpretation Canada* 3, no. 5 (1976).

Pederson, A., ed. *Keeping Archives.* Sydney: Australian Society of Archivists, 1987.

Peeters, J. P., and J. T. Williams. "Towards Better Use of Genebanks with Special Reference to Information." *Plant Genetic Resources Newsletter* (FAO) 60 (1984).

Pekarik, A. J. "Long-Term Thinking: What about the Stuff?" *Curator* 46, no. 4 (2003).

Pellett, H., and Ken Vogel. "Computerized Method for Collection, Storage and Retrieval of Information on Plant Performance." *AABGA Bulletin*, January 1980.

Pence, V. C. "Cryopreservation of Bryophytes and Ferns." In *Plant Cryopreservation: A Practical Guide*, edited by B. M. Reed. New York: Springer, 2008.

———. "In Vitro Methods and the Challenge of Exceptional Species for Target 8 of the Global Strategy for Plant Conservation." *Annals of the Missouri Botanical Garden* 99, no. 2 (2013).

Pence, Valerie. "From Freezing to the Field: In Vitro Methods Assisting Plant Conservation." *BG Journal* 9, no. 1 (2012).

Pepper, Jane G. "Planning the Development of Living Plant Collections." *Longwood Graduate Program Seminars* 10 (1978).

Pierson, J. C., D. J. Coates, J. G. B. Oostermeijer, S. R. Beissinger, J. G. Bragg, P. Sunnucks, N. H. Schumaker, and A. G. Young. "Genetic Factors in Threatened Species Recovery Plans on Three Continents." *Frontiers in Ecology and the Environment* 14 (2016).

Plant Extinction Prevention Program. "Home." www.pepphi.org/.

The Plant List. "Home." http://www.theplantlist.org.

Porter, D. R. *Current Thoughts on Collections Policy.* Technical Report 1. Nashville, TN: American Association for State and Local History, 1985.

———. *Developing a Collections Management Manual.* Technical Report 7. Nashville, TN: American Association for State and Local History, 1986.

Potter, Kevin M., et al. "Banking on the Future: Progress, Challenges and Opportunities for the Genetic Conservation of Forest Trees." *New Forests* 48 (2017).

Prather, L. A., O. Alvarez-Fuentes, M. H. Mayfield, and C. J. Ferguson. "Implications of the Decline in Plant Collecting for Systematic and Floristic Research." *Systematic Botany* 29, no. 1 (2004).

Price, S. "Home Demonstration Gardens: A Naturalistic Evaluation for Improved Design and Interpretation." *Longwood Graduate Program Seminars* 18 (1986).

Primack, R. B., and A. J. Miller-Rushing. "The Role of Botanical Gardens in Climate Change Research." *New Phytologist* 182 (2009).

Pritchard, Diana J., and Stuart R. Harrop. "A Re-evaluation of the Role of Ex Situ Conservation." *BG Journal* 7, no. 1 (2010).

Pullen, D. R. "Inventorying Historical Collections in the Small Museum." *Curator* 28, no. 4 (1985).

Quek, P., and E. Friis-Hansen. "Chapter 18: Collecting Plant Genetic Resources and Documenting Associated Indigenous Knowledge in the Field: A Participatory Approach." In *Collecting Plant Genetic Diversity: Technical Guidelines—2011 Update*, edited by L. Guarino et al. Rome, Italy: Bioversity International, 2011.

Quigley, J. "Chronicling the Living Collections: The Arboretum's Plant Records." *Arnoldia* 49, no. 1 (1989).

Radford, L., M. Dossman [sic], and D. Rae. "The Management of 'Ad Hoc' Ex Situ Conservation Status Species at the Royal Botanic Garden Edinburgh." *Sibbaldia* 1, no. 1 (2003).

Rae, D. A. H. "Botanic Gardens and Their Live Plant Collections: Present and Future Roles." PhD diss., University of Edinburgh, 1995.

Rae, David. "Fit for Purpose? The Value of Checking Collections Statistics." *Sibbaldia*, no. 2 (2004).

———. "The Value of Living Collection Catalogues and Catalogues Produced from the Royal Botanic Garden Edinburgh." *Sibbaldia* 6 (2008).

———. "What Conservation Role for Botanic Gardens?" *Professional Horticulture* 4 (1990).

Raulston, J. C. "From the Arboretum to the Nursery." *Public Garden* 1, no. 3 (1986).

Raven, P. "A Look at the Big Picture." *Public Garden* 12, no. 2 (1997).

———. "Research in Botanical Gardens." *Public Garden* 21, no. 1 (2006).

———. "Research Programs in Botanic Gardens." *Longwood Graduate Program Seminars* 11 (1979).

Reed, B. M., ed. *Plant Cryopreservation: A Practical Guide.* New York: Springer, 2008.

Reed, B. M., F. Engelmann, M. E. Dulloo, and J. M. M. Engels. *Technical Guidelines for the Management of Field and In Vitro Germplasm Collections.* IPGRI Handbooks for Genebanks 7. Rome, Italy: International Plant Genetic Resources Institute, 2004.

Reibel, D. B. *Registration Methods for the Small Museum.* Nashville, TN: American Association for State and Local History, 1978.

Reichard, S. H. "Learning from the Past." *Public Garden* 12, no. 2 (1997).

Reichard, S., H. Liu, and C. Husby. "Is Managed Relocation of Rare Plants Another Pathway for Biological Invasions?" In *Plant Reintroduction in a Changing Climate: Promises and Perils*, edited by J. Maschinski and K. E. Haskins. Washington, DC: Island Press, 2012.

Ren, Hai, et al. "The Use of Grafted Seedlings Increases the Success of Conservation Translocations of *Manglietia longipedunculata* (*Magnoliaceae*), a Critically Endangered Tree." *Oryx* 50, no. 3 (2016).

Rhoads, A. "Plants vs. Pests: An Integrated Approach for Public Gardens." *Longwood Graduate Program Seminars* 13 (1981).

Richard, G., and P. Wallick. "Computerized Mapping at the Brooklyn Botanic Garden." *Public Garden*, January 1988.

Richardson, S. "Condition Reporting: Charting the Early Warning Signs." *Museum Quarterly* 12, no. 2 (1983).

———. "The Publication of a Statement of Guidelines for the Management of Collections." *Curator* 17, no. 2 (1974).

Richeson, D. R. "An Approach to Historical Research in Museums." *Material History Bulletin* 22 (1985).

Richoux, Jeannette A., et al. "A Policy for Collections Access." *Museum News* 59, no. 7 (1981).

Ritzenthaler, M. L. *Archives and Manuscripts: Conservation*. Chicago: Society of American Archivists, 1983.

Rivers, M. C., N. A. Brummitt, T. R. Meagher, and E. Nic Lughadha. "How Many Herbarium Specimens Are Needed to Detect Threatened Species?" *Biological Conservation* 144, no. 10 (2011).

Roberts, D. A. *Collections Management for Museums*. Cambridge: Museum Documentation Association, 1988.

———. *Planning the Documentation of Museum Collections*. Duxford, Cambridge: Museum Documentation Association, 1985.

Robertson, I. "Botanical Gardens in the Contemporary World." *Public Garden* 11, no. 1 (1996).

Rogers, G. "Plant Nomenclature and Labeling at Botanical Gardens: Some Practical Tips." *Public Garden* 3, no. 2 (1988).

Rogstad, S. H. "Saturated NaCl-CTAB Solution as a Means of Field Preservation of Leaves for DNA Analysis." *Taxon* 41 (1992).

Rose, C. "A Code of Ethics for Registrars." *Museum News* 63, no. 3 (1985).

Royal Botanic Garden Edinburgh. *Collection Policy for the Living Collection*. Edinburgh: RBGE, 2006.

Royal Botanic Gardens, Kew. "A Field Manual for Seed Collectors." https://www.academia.edu/16018755/A_Field_Manual_for_Seed_Collectors_SEED_COLLECTING_FOR_THE_MILLENNIUM_SEED_BANK_PROJECT_ROYAL_BOTANIC_GARDENS_KEW Pl.

———. Seed Information Database. https://data.kew.org/sid/.

Sacchi, C. "The Role and Nature of Research at Botanical Gardens." *Public Garden* 6, no. 3 (1991).

Sakai, A., and F. Engelmann. "Vitrification, Encapsulation-Vitrification and Droplet-Vitrification: A Review." *Cryoletters* 28, no. 3 (2007).

Sandham, J., and B. Morley. "Biological Control Techniques in Large Conservatories." *Public Garden* 10, no. 3 (1995).

Sarasan, L., and J. Sunderland. "Checklist of Automated Collections Management System Features, or How to Go about Selecting a System." In *Collections Management for Museums*, edited by D. Andrew Roberts. Cambridge: Museum Documentation Association, 1988.

———. "Computerized Collections Management." *Longwood Graduate Program Seminars* 16 (1984).

———. *Museum Collections and Computers*. Lawrence, KS: Association of Systematics Collections, 1983.

———. "Why Museum Computer Projects Fail." *Museum News* 3 (1981).

Särkinen, T., M. Staats, J. E. Richardson, R. S. Cowan, and F. T. Bakker. "How to Open the Treasure Chest? Optimising DNA Extraction from Herbarium Specimens." *PLoS ONE* 7, no. 8 (2012).

Saville, D. B. O. *Collection and Care of Botanical Specimens*. Research Branch Publication 1113. Ottawa: Canada Department of Agriculture, 1962.

Sawyers, C. "Where to Start: Plant Records." *Public Garden* 4, no. 1 (1989).

Schmiegel, K. A. "Managing Collections Information." In *Registrars on Record*, edited by Mary Case. Washington, DC: American Association of Museums, 1988.

Schulenberg, R. "Maintaining and Renovating an Outdoor Woody Plant Collection." *Longwood Graduate Program Seminars* 16 (1984).

Schwartz, C., et al. "Keeping Our Own House in Order: The Importance of Museum Records." *Museum News* 61, no. 4 (1983).

Screven, C. "Educational Evaluation and Research in Museums and Public Exhibits: A Bibliography." *Curator* 27, no. 2 (1984).

———. "Exhibit Evaluation: A Goal-Referenced Approach." *Curator* 19, no. 4 (1976).

Seager, P. "Treading Softly." *Public Garden* 8, no. 2 (1993).

Secretariat of the Convention on Biological Diversity. *Bonn Guidelines on Access to Genetic Resources and Fair and Equitable Sharing of the Benefits Arising out of Their Utilization*. Montreal, QC: SCBD, 2002.

———. *Global Biodiversity Outlook 5*. Montreal, QC: SCBD, 2020.

———. *Nagoya Protocol on Access to Genetic Resources and the Fair and Equitable Sharing of Benefits Arising from Their Utilization*. Montreal, QC: SCBD, 2011.

———. *Tkarihwaié:ri Code of Ethical Conduct to Ensure Respect for the Cultural and Intellectual Heritage of Indigenous and Local Communities*. Montreal, QC: SCBD, 2011.

Sellars, W. "A Resource Managers Look at Open Space." *Longwood Graduate Program Seminars* 13 (1981).

Shalkop, Robert. "Research and the Museum." *Museum News* 50, no. 8 (1972).

Shaw, R. L. "The Exhibit Design Process and the Collection: The Forgotten Step." Master's thesis, University of Toronto, 1986.

Shettel, H. "Program Development and Evaluation." *Longwood Graduate Program Seminars* 10 (1978).

Smart, Chris, and Alan Elliott. "Forward Planning for Scottish Gardens in the Face of Climate Change." *Sibbaldia* 13 (2015).

Smith, Adam B., Matthew A. Albrecht, and Abby Hird. "Chaperoned." *BG Journal* 11, no. 2 (2014).

Smith, Paul, and Valerie Pence. "The Role of Botanic Gardens in Ex Situ Conservation." In *Plant Conservation Science and Practice*, edited by S. Blackmore and S. Oldfield. New York: Cambridge University Press, 2017.

Smyth, N., C. Armstrong, M. Jebb, and A. Booth. "Implementing Target 10 of the Global Strategy for Plant Conservation at the National Botanic Gardens of Ireland: Managing Two Invasive Non-native Species for Plant Diversity in Ireland." *Sibbaldia* 11 (2013).

Society for Ecological Restoration. "What Is Ecological Restoration?" https://www.ser-rrc.org/what-is-ecological-restoration/.

Society for Ecological Restoration International Science & Policy Working Group. *The SER International Primer on Ecological Restoration.* Tucson, AZ: SERI, 2004. https://www.ser-rrc.org/resource/the-ser-international-primer-on/.

South African National Biodiversity Institute. *Biodiversity Stewardship: Partnerships for Securing Biodiversity.* Pretoria: SANBI, 2015.

Spongberg, S. A. "The Collections Policy of the Arnold Arboretum: Taxa of Infraspecific Rank, and Cultivars." *Arnoldia* 39 (1979).

———. "The Value of Herbaria for Horticulture and the Curation of Living Collections." *Longwood Graduate Program Seminars* 16 (1984).

Stansfield, G. "Collection Management Plans." *Museum Professional Group News* 20 (1985).

Stansfield, G., J. Mathias, and G. Reid, eds. *Manual of Natural History Curatorship.* London: HMSO, 1994.

Steere, W. "Research as a Function of a Botanical Garden." *Longwood Graduate Program Seminars* 1 (1969).

Steven, R. "Citizen, Where Art Thou? It's Time to Get the Public Directly Engaged with Saving Threatened Species." *Science for Saving Species* 4 (2017).

Stolton, S., K. H. Redford, and N. Dudley. *The Futures of Privately Protected Areas.* Gland, Switzerland, IUCN, 2014.

Strong, Sir Roy. "Scholar or Salesman—the Curator of the Future." *Muse* 6, no. 2 (1988).

Suarez, A. V., and N. D. Tsutsui. "The Value of Museum Collections for Research and Society." *BioScience* 54, no. 1 (2004).

Swarts, N. D., and K. W. Dixon. "Perspectives on Orchid Conservation in Botanic Gardens." *Trends in Plant Science* 14 (2009).

Swinney, H. J., ed. *Professional Standards for Museum Accreditation*. Washington, DC: American Association of Museums, 1978.

Szaro, R. C., and D. Johnston. *Biodiversity of Managed Landscapes*. Washington, DC: U.S. Department of Agriculture, 1996.

Tankersley, Boyce. "Plant Databases Linked for Botanists and Gardeners." *Nature* 441, no. 7093 (2006).

Taylor, Matthew. "Considerations for Commercial Plant Introduction from Public Gardens." *Sibbaldia* 14 (2016).

Taylor, Roy L. "Is a Plant Introduction Program Right for Your Garden?" *Public Garden* 2, no. 4 (1987).

Taylor, S., and G. Dreyer. "Techniques to Control Vegetation." *Public Garden* 3, no. 2 (1988).

Telewski, F. "Options for Display Labels." *Public Garden*, October 1994.

Thibodeau, F. "Endangered Species in Public Gardens." *Longwood Graduate Program Seminars* 18 (1986).

Thomas, P., and K. Tripp. "Ex Situ Conservation of Conifers: A Collaborative Model for Biodiversity Preservation." *Public Garden* 13, no. 3 (1998).

Thompson, J. M. A., ed. *Manual of Curatorship*. London: Butterworths, 1984.

Thompson, J. M. A., et al. *Manual of Curatorship: A Guide to Museum Practice*. Oxford: Heinemann, 1992.

Thomson, Garry. *The Museum Environment*. 2nd ed. London: Butterworths, 1986.

Timbrook, S. "California Island Research." *Public Garden* 2, no. 1 (1987).

Toomer, S. *Planting and Maintaining a Tree Collection*. Portland, OR: Timber Press, 2010.

Torreya Guardians. "About *Torreya taxifolia*." http://www.torreyaguardians.org/torreya.html.

Toth, E. "Managing Urban Woodlands." *Public Garden* 10, no. 3 (1995).

Tuele, N. "Deaccessioning: Proceed with Caution." *Muse* 7, no. 1 (1989).

Turland, N. J., J. H. Wiersema, F. R. Barrie, W. Greuter, D. L. Hawksworth, P. S. Herendeen, S. Knapp, W.-H. Kusber, D.-Z. Li, K. Marhold, T. W. May, J. McNeill, A. M. Monro, J. Prado, M. J. Price, and G. F. Smith, eds. *International Code of Nomenclature for Algae, Fungi, and Plants (Shenzhen Code) Adopted by the Nineteenth International Botanical Congress Shenzhen, China, July 2017*. Regnum Vegetabile 159. Glashütten: Koeltz Botanical Books, 2018.

Turnbull, J. W. "Seed Collection and Certification." In *International Training Course in Forest Tree Breeding*. Canberra: Australian Development Assistance Bureau, 1978.

Turnbull, J. W., and A. R. Griffin. "The Concept of Provenance and Its Relationship to Infraspecific Classification in Forest Trees." In *Infraspecific Classification of Wild Plants*, edited by B. Styles. Oxford: Oxford University Press, 1985.

Ullberg, A. D., and R. C. Lind Jr. "Consider the Potential Liability of Failing to Conserve Collections." *Museum News* 68, no. 1 (1989).

Ullberg, A. D., and Patricia Ullberg. "A Proposed Curatorial Code of Ethics." *Museum News* 52, no. 8 (1974).

United Nations Food and Agriculture Organization. "International Treaty on Plant Genetic Resources for Food and Agriculture." http://www.fao.org/plant-treaty/en/.

University of Maryland–College Park. "Tree Management Plan 2017." October 26, 2017. https://arboretum.umd.edu/sites/default/files/2017TreeManagementPlan .pdf.

U.S. Department of Agriculture–Agricultural Research Service. "Agricultural Genetic Resources Preservation Research: Fort Collins, CO." https://www.ars .usda.gov/plains-area/fort-collins-co/center-for-agricultural-resources-research/ paagrpru/.

———. "Agricultural Genetic Resources Preservation Research: Fort Collins, CO: Plants." https://www.ars.usda.gov/plains-area/fort-collins-co/center-for -agricultural-resources-research/paagrpru/docs/plants/plant-science-at-the-nationa l-laboratory-for-genetic-resources-preservation/.

———. *Gardens and Collections Policies.* Washington, DC: USDA, 1987.

———. *Plant Exploration Guidelines for FY1997 Proposals.* Washington, DC: USDA, 1995.

———. "Plant Health (PPQ)." https://www.aphis.usda.gov/ppq/permits/plantprod-ucts/nursery.html.

———. "U.S. National Plant Germplasm System." https://www.ars-grin.gov/npgs/.

U.S. Fish & Wildlife Service. "Conserving the Nature of America." 2017. https:// www.fws.gov/info/pocketguide/.

van Zonneveld, M., E. Thomas, G. Galluzzi, and X. Scheldeman. "Mapping the Eco-geographic Distribution of Biodiversity and GIS Tools for Plant Germplasm Col-lectors." In *Collecting Plant Genetic Diversity: Technical Guidelines—2011 Update,* edited by L. Guarino et al. Rome, Italy: Bioversity International, 2011.

Vitt, P., P. N. Belmaric, R. Book, and M. Curran. "Assisted Migration as a Climate Change Adaptation Strategy: Lessons from Restoration and Plant Reintroduc-tions." *Israel Journal of Plant Sciences* 63, no. 4 (2016).

Vogler, D., S. Macey, and A. Sigouin. "Stakeholder Analysis in Environmental and Conservation Planning." *Lessons in Conservation* 7 (2007).

Volis, Sergei. "Complementarities of Two Existing Intermediate Conservation Approaches." *Plant Diversity* 39 (2017).

Volk, G. M., D. R. Lockwood, and C. M. Richards. "Wild Plant Sampling Strate-gies: The Roles of Ecology and Evolution." In *Plant Breeding Reviews.* Vol. 29. New York: Wiley, 2007.

Waddington, J., and David Rudkin, eds. *Proceedings of the 1985 Workshop on Care and Maintenance of Natural History Collections.* Toronto: Royal Ontario Museum, 1986.

Wagner, W. H. "Botanical Research at Botanical Gardens." In *Proceedings of the Symposium on a National Botanical Garden System for Canada,* edited by P. F. Rice. Technical Bulletin. Hamilton, ON: Royal Botanical Gardens, 1972.

Walker, B. "The Curator as Custodian of Collections." *Curator* 6, no. 4 (1963).

Walter, K., and M. O'Neal. "BG-BASE: Software for Botanical Gardens and Arboreta." *Public Garden* 8, no. 4 (1993).

Walters, C. "Orthodoxy, Recalcitrance and In-Between: Describing Variation in Seed Storage Characteristics Using Threshold Responses to Water Loss." *Planta* 242 (2015).

Walters, C., P. Berjak, N. Pammenter, K. Kennedy, and P. Raven. "Preservation of Recalcitrant Seeds." *Science* 22 (2013).

Ward, J. "Managing Natural Areas—One Garden's Approach." *Longwood Graduate Program Seminars* 18 (1986).

Ware, Michael E. *Museum Collecting Policies and Loan Agreements.* Association of Independent Museums Guideline 14. Cheshire, UK: Association of Independent Museums, 1988.

Washburn, W. E. "Collecting Information, Not Objects." *Museum News* 62, no. 3 (1984).

———. *Defining the Museum's Purpose.* Cooperstown: New York State Historical Association, 1975.

———. "A Statement of Policy and Procedures Regulating the Acquisition and Disposition of Natural History Specimens." *Curator* 17, no. 2 (1974).

Watson, G., and G. Ware. "What Botanical Gardens Can Contribute to Urban Forestry Research." *Public Garden* 9, no. 1 (1994).

Waylen, K. *Botanic Gardens: Using Biodiversity to Improve Human Well-Being.* Richmond, Surrey, UK: Botanic Gardens Conservation International, 2006.

Weil, S. E. "Deaccession Practices in American Museums." *Museum News* 65, no. 3 (1987).

Weinstein, G. "Wildflower Research." *Public Garden* 4, no. 2 (1989).

Welton, L., et al. *Guidelines for Acquisition and Management of Biological Specimens.* Lawrence, KS: Association of Systematics Collections, 1982.

Wentz, P. "Museum Information Systems: The Case for Computerization." *International Journal of Museum Management and Curatorship* 8 (1989).

Werner, O., and G. M. Schoepfle. *Systematic Fieldwork.* 2 vols. Newbury Park, CA: Sage, 1987.

Westwood, A., E. Reuchlin Hugenholtz, and D. M. Keith. "Perspective: Re-Defining Recovery: A Generalized Framework for Assessing Species Recovery." *Biological Conservation* 172 (2014).

Wheatcroft, Penelope. "Merely Rubbish: Disposal of Natural History Collections." *Museums Journal* 87, no. 2 (1987).

Wheeler, S. "Information Management in a Small Collection." *Curator* 30, no. 2 (1987).

White, Derryll. "Collections Policies: The Basics of Collections Management." *Museum Roundup,* no. 76 (1979).

White, P. S. "A Bill Falls Due: Botanical Gardens and the Exotic Species Problem." *Public Garden* 12, no. 2 (1997).

———. "In Search of the Conservation Garden." *Public Garden* 11, no. 2 (1996).

Whiteley, Andrew R., et al. "Genetic Rescue to the Rescue." *Trends in Ecology & Evolution* 30, no. 1 (2015).

Wikipedia. "DMAIC." Last updated June 28, 2021. https://en.wikipedia.org/wiki/DMAIC.

———. "Plant Tissue Culture." Last updated May 17, 2021. https://en.wikipedia.org/wiki/Plant_tissue_culture.

———. "*Torreya taxifolia*." Last updated April 30, 2021. https://en.wikipedia.org/wiki/Torreya_taxifolia.

Wilcove, D. S. "Endangered Species Management: The US Experience." In *Conservation Biology for All*, edited by N. S. Sodhi and P. R. Ehrlich. New York: Oxford University Press, 2010.

Wilkinson, N., and J. B. Akerman. "Garden Archeology: The Restoration of E. I. du Pont's Garden at Eleutherian Mills." *Longwood Graduate Program Seminars* 6 (1974).

Willan, R. L. *A Guide to Forest Seed Handling (with Special Reference to the Tropics)*. FAO Forestry Paper 20/2. Rome, Italy: FAO, 1985.

Williams, C., K. Davis, P. Cheyne, and N. Ali. *The CBD for Botanists: An Introduction to the Convention on Biological Diversity for People Working with Botanical Collections*. Version 4. Richmond, Surrey, UK: Royal Botanic Gardens, Kew, 2012.

Williams, D. W. *A Guide to Museum Computing*. Nashville, TN: American Association for State and Local History, 1987.

Wilson, E. O. *The Future of Life*. New York: Knopf, 2002.

Wise, G. "Do Plants Speak for Themselves? An Interpretive Plant for the Garden." *Longwood Graduate Program Seminars* 9 (1977).

Woods, J., et al. *Joy of Planning: The Approach to Interpretive Planning*. Ottawa: Parks Canada, 1976.

World Flora Online. "Home." http://www.worldfloraonline.org.

World Intellectual Property Organization. *Documenting Traditional Knowledge: A Toolkit*. Geneva: WIPO, 2017.

Wright, T. *Large Gardens and Parks: Maintenance, Management and Design*. New York: HarperCollins, 1982.

Yang, Meipu. "Manuals for Museum Policy and Procedures." *Curator* 32, no. 4 (1989).

Yinger, B. "Objectives and Funding of Ornamental Plant Explorations." *Longwood Graduate Program Seminars* 16 (1984).

Young, J. A., and C. G. Young. *Collecting, Processing and Germinating Seeds of Wildland Plants*. Portland, OR: Timber Press, 1986.

Zimmerman, C. "Developing a Deaccessions Policy." *Dawson & Hind* 12, no. 4 (1986).

———. "Historical Documentation: Going beyond the Obvious." *Dawson & Hind* 10, no. 4 (1981).

———. "Preparing a Collections Management Procedures Manual." *Dawson & Hind* 15, no. 1 (1988/1989).

———. "Preparing an Education and Interpretation Policy." *Dawson & Hind* 14, no. 1 (1987/1988).

Zobel, B. "Gene Conservation—as Viewed by a Forest Tree Breeder." *Forest Ecology and Management* 1 (1978).

Zuk, J. D. "Confessions of a Collections Advocate." *Longwood Program Seminars* 16 (1984).

Zuk, J., et al. "Displays: Some Successes, Some Failures." *Public Garden* 2, no. 3 (1987).

Index

accessions, 106, 107, 109, 118–122; bar codes for, 119–120; and labels, 118–119; and numbers, 107, 117, 118, 119, 123; RFID transponders, 121

access, 23–24, 64, 90

accessions policy, 17

acquisitions, 47, 48, 51; basic tenets of collecting, 47; bequests, 52; Convention on Biological Diversity and, 48, 52; criteria and organizational themes, 17, 18, 19, 21, 28; desiderata, 49; exchanges, 55–61; field collecting, 62–93; gap analysis, 49–50; general recommendations, 52; gifts, 6, 20, 36, 52–53; import permits, 56; *Index Seminum*, 55, 56, 58, 59; invasive plants, 24, 25, 26, 51, 56, *61*, 97; limits, 48; loans, 20, 21, 54; personal collections, 30, 94; phytosanitary certificates, 55, 56, 60, 78, 210; plant propagation, 39, 43, 93–94; purchases, 54–55; seed storage, 57, 58–59; shares and subscriptions, 54; standards, 17, 21; surplus plants, 59–

61; transfers, 57; verification, 57, 92, 123. *See also* ethics; field collecting; invasive plants; propagation; recommendations

American Public Gardens Association, 10, 60,137, 150, 225, 259, 309, 311, 328, 354

Ancient Tree Forum, 179

arboreta: ArbNet, 176, 242; Arnold, operations manual, 37; Arnold cultivars policy, 42; Arnold desiderata, 49; Arnold expedition toolkit, 70–71, 80; Arnold deaccessioning in, 134; Arnold inventory operations, 144–146; Arnold landscape plan, 166–169, 176; Arnold, plant introduction in, 351; Donald E. Davis, oak metacollection, 311; Holden, reserve in, 267; Hoyt, 350; Morton, gap analysis, 242, *in situ* conservation, 274–275, oak metacollection, 311; North Carolina State University, plant introduction in, 182; plant introduction, 177; Scott, cartography

Botanic Gardens Conservation International (BGCI), 50, 57, 123, *124*, 150, 210, 225, 242, 243, 252, 259, 264, 266, 270, 285, 311, 319, 320, 321, 326, 329, 354

Brahms software, 140, 200

breeding: biology, 234, 280, 282, 286, 312, 324; conservation, 261, 269, 275, 310, 312; field collecting, 66, 67, 111; gene banks, 182, 188, 284, 306, 307; genetic diversity, 171; herbaria, 247; invasive character, 95; plant improvement, 159, 160; plant introduction, 348, 349, 350, 352, 353; restoration, 330

CAD. *See* computer-aided design

cartographic file/record, 9, 115, 122, 126–128, 131, 139, 140, 141–143, 144, 173, 222

catalog, 9, 17, 34, 36, 103, 105, 106, 107, 109, 111, 117, 122, 129–132; activity files, 130–131; collections evaluation, 250, 251, 253, 254; collections interpretation, 222; collections research, 237; conservation, 261; creating, 129; data capture, 132; data files, 117; emergency preparedness, 209; exhibits, 227; inventory, 143–146; output, 133–134; pest management, 163; preservation interface, 173; procedures manual, 136; and propagation, 130; recommendation, 153; research for, 245, 246; seed bank, 200

CBD. *See* Convention on Biological Diversity

Center for Plant Conservation (CPC): citizen science, 225; collecting diversity, 73, 286; conservation collections, 258–259, 282; evaluate restoration sites, 324; field gene bank, 183, 189; gene bank flowchart, 283; germplasm biosecurity, 211; national collection, 211, 311; networking, 354; rare plant finder, 285; resource, 185, 288, 289, 330; seed storage, 195–198, 287, 293; species recovery, 321–323

CITES. *See* Convention on International Trade in Endangered Species of Wild Fauna and Flora

citizen science, 223–225, 261, 354, 355

climate change, 11, 26, 66, 161, 164, 174, 177, 178, 217, 223, 243, 259, 261, 270, 272, 318, 322, 323, 324, 326, 329, 334, 335, 340, 341

climate modeling, 177, 178, 267, 341–343

clones: accessions, 51, 118; conservation, 260; cultivars 42; gene bank, 305, preservation, 171, 206; historic, 192; plant introduction, 351, 353; research, 236, 302; restoration, 325, 330,

collections: and museums, 4; concepts, 5; curatorial elements of, 6, 7; defined, 5; living, 5, 6; management, 5; manager, 9; synoptic status, 6

collections committee, 30, 31

collections management manual, 35–38; areas of concern, 35, 36; definition, 35; example, 37–38; format, 35; writing, 36

collections management plan, 6, 14, 31–35; components, 31; consultant, 33; example, 33, 34; IMLS support, 32; preparation, 32; preparation team, 32, 33; site work, 32; strategic plan, 14. *See also* Institute of Museum and Library Services; recommendations

collections management policy, 6, 13, 14, 15–31; access, 21, 22, 23, 24, 29; accessioning, 17; accountabilities,

About the Author

Timothy C. Hohn is the retired chair of the horticulture department at Edmonds Community College, where he taught for 21 years. Previously, he was the curator of the University of Washington's Washington Park Arboretum and Center for Urban Horticulture and the first curator of plants for the Wildlife Conservation Society at the Bronx Zoo. He received a master's degree in public garden management from the University of Delaware's Longwood Graduate Program in 1986. He currently lives in the Puget Sound area of Washington.